ENVIRONMENTAL CONDITION OF THE MEDITERRANEAN SEA:
EUROPEAN COMMUNITY COUNTRIES

ENVIRONMENT & ASSESSMENT

VOLUME 5

The titles published in this series are listed at the end of this volume.

Environmental Condition of the Mediterranean Sea:

European Community Countries

Edited by

F. B. de Walle
M. Nikolopoulou-Tamvakli
W. J. Heinen

Netherlands Organization for
Applied Scientific Research (TNO),
Centre for Environmental Research,
Delft, The Netherlands

With major contributions from:
V. Silano
Ministry of Health, Rome, Italy
M. Vassilopoulos
Greek Delegation to the EC, Brussels, Belgium
L. A. Romaña
Ifremer, Toulon-La Seyne-sur-Mer, France
A. Estevan
Consultant, Madrid, Spain

With the support of:
European Investment Bank
Luxembourg, Luxembourg

KLUWER ACADEMIC PUBLISHERS
DORDRECHT / BOSTON / LONDON

A C.I.P. Catalogue record for this book is available from the Library of Congress.

ISBN 0-7923-2468-4

Published by Kluwer Academic Publishers,
P.O. Box 17, 3300 AA Dordrecht, The Netherlands.

Kluwer Academic Publishers incorporates
the publishing programmes of
D. Reidel, Martinus Nijhoff, Dr W. Junk and MTP Press.

Sold and distributed in the U.S.A. and Canada
by Kluwer Academic Publishers,
101 Philip Drive, Norwell, MA 02061, U.S.A.

In all other countries, sold and distributed
by Kluwer Academic Publishers Group,
P.O. Box 322, 3300 AH Dordrecht, The Netherlands.

Cover photo: ©ANP/ANSA

Printed on acid-free paper

All Rights Reserved
© 1993 Kluwer Academic Publishers
No part of the material protected by this copyright notice may be reproduced or
utilized in any form or by any means, electronic or mechanical,
including photocopying, recording or by any information storage and
retrieval system, without written permission from the copyright owner.

Printed in the Netherlands

FOREWORD

This study investigates the extent of the pollution of the Mediterranean Sea with respect to the four bordering EC countries - Spain, France, Italy, and Greece. The environmental pressures and economic impacts are examined and the institutional/legal framework is described together with all the necessary environmental expenditures. The book is written in such a way that separate chapters are devoted to each of the four countries, after an initial section summarizing the major commonalities. These chapters are organized in parallel formats so that it is possible to examine the same topic country by country.
Good references are provided for the reader who is not familiar with the subject of Mediterranean pollution. For specialists, the book provides a useful overview of adjacent fields other than their own speciality; for policymakers, the chapters provide sufficient foundations for decision-making; for the investment planner and banks, it provides budget and investment needs; and for the private sector, it gives an insight into the possibilities for corporate applications of environmental technologies.

Four specialists - Prof. V. Silano (Italy), Dr. M. Vassilopoulos (Greece), Dr. L.A. Romaña (France), and Mr. A. Estevan (Spain) - contributed sustantially by evaluating the necessary information from national documents on environmental policies and strategies.
The book was developed on the basis of an integrated environmental study financed by the European Investment Bank, to which we would like to express our sincere thanks.

The Editors
Delft, The Netherlands
June, 1992

TABLE OF CONTENTS

FOREWORD v

EXECUTIVE SUMMARY 1

A. General overview of the environmental
   quality of the Mediterranean Sea 1

B. Spanish Mediterranean environment 11

C. French Mediterranean environment 15

D. Italian Mediterranean environment 21

E. Greek Mediterranean environment 26

1. GENERAL OVERVIEW OF THE ENVIRONMENTAL
   QUALITY OF THE MEDITERRANEAN SEA 33

   F.B. de Walle
   J.J. Lomme
   M. Nikolopoulou-Tamvakli

1.1. The state of the environment 34
1.1.1. Physical characteristics of the
       Mediterranean Sea 34
1.1.2. Chemical and biological characteristics
       of the Mediterranean Sea 36
1.1.2.1. Nutrients and eutrophication 36
1.1.2.2. Heavy metals in water and sediments 39
1.1.2.3. Petroleum hydrocarbons 41
1.1.2.4. Chlorinated hydrocarbons 45
1.1.3. The seafood chain 46
1.1.3.1. Benthic communities 46
1.1.3.2. Shellfish 47
1.1.3.3. Heavy metals in benthic and pelagic
         organisms 47
1.1.3.4. Petroleum hydrocarbons in biota 49
1.1.3.5. Organochlorines in biota 50
1.1.3.6. Reptiles and mammals 50
1.1.4. The coastline 52
1.1.5. Inland water 53
1.1.6. The atmosphere 54
1.1.7. Land use/forest 55
1.1.8. Nature reserves 57
1.1.9. Historic sites 57

1.2. Pressures on the environment 59
1.2.1. Land-based pollutants 59
1.2.2. Regional contributions to pollution loads 59
1.2.3. Organic matter 60
1.2.4. Nutrients 61
1.2.5. Specific organics 61
1.2.6. Metals 63
1.2.7. Suspended matter 63
1.2.8. Pesticides 64

1.2.9.	Radioactive discharges	64
1.2.10.	Microbial pollution	64
1.2.11.	Pressures from economic development	65
1.2.12.	Other pressures	66
1.3.	Economic impacts	67
1.3.1.	Health impacts	67
1.3.2.	Tourism	70
1.3.3.	Marine production	71
1.3.4.	Agricultural and forestry impacts	72
1.4.	The institutional/legal framework	73
1.4.1.	Existing action	73
1.4.1.1.	Mediterranean Action Plan	73
1.4.1.2.	EC Policy on environmental protection	74
1.4.1.3.	Fourth Environmental Action Program	75
1.4.1.4.	CORINE Program	75
1.4.1.5.	European Regional Development Fund Environmental Program	76
1.4.1.6.	MEDSAP	76
1.4.2.	Environmental protection standards	77
1.4.3.	Discrepancies between law and reality	79
1.5.	Past trends	80
1.6.	The state of the environment: projections	81
1.6.1.	The Blue Plan scenarios	81
1.6.2.	Predictions with the scenarios	83
1.6.2.1.	Population growth	84
1.6.2.2.	Industrial development	84
1.6.2.3.	Oil pollution	85
1.7.	Measures and investment requirements	86
1.7.1.	Waste	87
1.7.2.	Sewage treatment	88
1.7.3.	Oil pollution of the Mediterranean Sea	89
1.7.4.	Industrial pollution policy	91
1.7.4.1.	Criteria for heavy metals	91
1.7.4.2.	Cadmium	92
1.7.4.3.	Mercury	93
1.7.5.	Management of nature sites	94
1.7.6.	Management of marine living resources	95
1.7.7.	Policy recommendations	96
1.7.8.	Overall investment requirements	97
1.8.	Conclusions	97
	References	99
	Figures	104
	Tables	129

2.	SPANISH MEDITERRANEAN ENVIRONMENT	180
	A. Estevan M. Nikolopoulou-Tamvakli A. Cofiño-Molina	
2.1.	State of the environment	180
2.1.1.	Water quality	180
2.1.1.1.	Heavy metals in marine waters and sediments	180
2.1.1.2.	Heavy metals in marine organisms	181
2.1.1.3.	Organics in marine sediments and seawater	182
2.1.1.4.	Organics in marine organisms	183
2.1.1.5.	Bacteria and viruses in the marine environment	184
2.1.1.6.	Freshwater environment	187
2.1.2.	State of the coastal region	188
2.1.2.1.	Erosion	188
2.1.2.2.	Forest fires	189
2.1.2.3.	Natural interest areas	190
2.1.3.	Air pollution	190
2.1.4.	Urban and industrial solid waste	191
2.2.	Pressures on the environment	192
2.2.1.	Municipal discharges	192
2.2.2.	Industrial discharges	194
2.2.3.	River discharges	194
2.2.4.	Maritime traffic	195
2.3.	Economic impacts	196
2.3.1.	Economic impacts of water pollution on tourism	196
2.3.2.	Other economic impacts	197
2.4.	The institutional/legal framework	198
2.5.	Existing/planned action and investments	200
2.6.	Summary and conclusions	203
	References	204
	Figures	206
	Tables	230
3.	FRENCH MEDITERRANEAN ENVIRONMENT	256
	L.A. Romaña M. Nikolopoulou-Tamvakli	
3.1.	State of the environment	256
3.1.1.	Water quality	257
3.1.2.	Sediments	260
3.1.3.	Living matter	264
3.1.4.	The quality of the bathing waters	265
3.1.5.	Air pollution	266

3.2.	Pressures on the environment	267
3.2.1.	Municipal discharges	268
3.2.2.	Industrial discharges	270
3.2.3.	River discharges	272
3.2.4.	Maritime discharges	273
3.3.	Economic impacts of low environmental quality	273
3.4.	The institutional/legal framework for environmental protection	275
3.5.	Past and future trends in environmental expenditures	277
3.6.	Conclusion	278
	References	278
	Figures	280
	Tables	318
4.	**ITALIAN MEDITERRANEAN ENVIRONMENT**	329

V. Silano
J.J. Lomme

4.1.	State of the environment	329
4.1.1.	Heavy metals	329
4.1.1.1.	Heavy metals in coastal seawaters	329
4.1.1.2.	Heavy metals in marine sediments	331
4.1.1.3.	Heavy metals in marine organisms	332
4.1.2.	Organic contaminants	335
4.1.2.1.	Organics in marine waters and sediments	335
4.1.2.2.	Organics in marine organisms	336
4.1.3.	Bathing waters	337
4.1.4.	Eutrophication	338
4.1.5.	The coastal strip	340
4.1.5.1.	Erosion	340
4.1.5.2.	Destruction and deterioration of forests	341
4.1.6.	Environmental quality of selected areas	343
4.1.6.1.	The Po basin	343
4.1.6.2.	Lambro, Olona and Seveso basin	345
4.1.6.3.	Adige, Arno and Tiber basins	345
4.1.6.4.	Province of Naples	347
4.1.7.	Damage to fauna and flora	347
4.2.	Pressure on the environment	348
4.2.1.	Municipal discharges	348
4.2.2.	Industrial discharges	348
4.2.3.	Maritime discharges	349
4.2.4.	River discharges	349
4.2.5.	Pressures in selected areas	350
4.2.5.1.	The Po basin	350
4.2.5.2.	Basin of Lambro, Olona and Seveso rivers	351
4.2.5.3.	Province of Naples	352
4.2.5.4.	Lagoon of Venice	353

4.3.	Economic impacts	354
4.3.1.	Health impacts	354
4.3.2.	Tourism	355
4.3.3.	Historic sites and natural landscapes	356
4.3.4.	Marine production	356
4.3.5.	Agriculture and zootechnics	357
4.4.	The institutional and legal framework	358
4.4.1.	Environmental authorities	358
4.4.2.	The adoption of EC directives	360
4.4.3.	Marine waters	361
4.5.	Past and present trends in environmental expenditures	362
4.5.1.	Environmental expenditures by the central government	362
4.5.2.	Expenditures by regional governments	363
4.6.	State of the environment: projections	364
4.6.1.	Demography	364
4.6.2.	Agro-food and manufacturing industry	365
4.6.3.	Tourism and transportation	365
4.6.4.	Liquid and solid pollution loads	366
4.7.	Estimate of investment requirements and further operational costs	366
4.7.1.	The Lambro, Olona and Seveso basin	366
4.7.2.	The Po basin	367
4.7.3.	Province of Naples	367
4.7.4.	The Lagoon of Venice	368
4.7.5.	Water treatment in the rest of Italy	368
4.7.6.	Solid-waste disposal	368
4.7.7.	Soil conservation	369
4.7.8.	Conclusion	369
	References	369
	Figures	372
	Tables	399
5.	**GREEK MEDITERRANEAN ENVIRONMENT**	425

M. Vassilopoulos
M. Nikolopoulou-Tamvakli

5.1.	State of the environment	426
5.1.1.	State of the sea	426
5.1.1.1.	Saronikos Gulf/Elefsis Bay	427
5.1.1.2.	Gulfs of the northern Aegean Sea	434
5.1.1.3.	Pagassitikos Gulf	439
5.1.1.4.	Euboikos Gulf	439
5.1.1.5.	Patraikos Gulf	439
5.1.1.6.	Aegean Sea	440
5.1.1.7.	Ionian Lagoons	441
5.1.2.	State of the freshwater	441
5.1.3.	State of the coastal strip	443
5.1.4.	Air pollution	446

5.2.	Pressures on the environment	448
5.2.1.	Human settlements	448
5.2.2.	Industrial activities	449
5.2.3.	Agricultural activities	450
5.2.4.	Fisheries and aquaculture	451
5.3.	Institutional/legal framework	452
5.4.	Past and future trends	456
5.4.1.	Sewage treatment plants	456
5.4.2.	Port reception facilities	457
5.5.	Conclusions	459
	References	460
	Figures	464
	Tables	479

ACKNOWLEDGEMENTS 500

ABBREVIATIONS 502

CURRICULA VITAE 503

APPENDIX: Investment needs 505

SUBJECT INDEX 512

EXECUTIVE SUMMARY

The present study presents an overview of the most critical issues concerning the pollution of the Mediterranean as influenced by the 4 coastal European Community countries: Spain, France, Italy and Greece. The information of this overview is obtained from literature sources and interviews with many experts and officials in each of the Mediterranean countries. First, a general overview is presented, followed by detailed descriptions for Spain, France, Italy and Greece, respectively.

A. GENERAL OVERVIEW OF THE ENVIRONMENTAL QUALITY OF THE MEDITERRANEAN SEA

More than 350 million people in 18 countries live along the 2.96 million km^2 Mediterranean Sea. Approximately 3.3 million tons of organic matter as biochemical oxygen demand (BOD), 350 million tons of suspended solids together with heavy metals and organics (i.e. 0.8 million tons of petroleum hydrocarbons) are annually discharged into the Mediterranean. The semi-enclosed Mediterranean Sea has a volume of 3.75 million km^3 and an average 80-year renewal period of its water. The lack of significant tides and strong coastal currents results in reduced dispersion of pollutants in coastal waters.
The Mediterranean marine environment is directly linked to the meteorological conditions of the area. It is subjected to the action of winds, often strong, which affect the level of surface currents, the vertical mixing of the water masses, the upwelling and the resuspension of contaminated sediments. The outflow of rivers into the sea fluctuates very much during the annual cycle and between years as a result of meteorological conditions.
The sea currents are weak near the surface and decrease

considerably with depth, bringing a limited dilution of the discharged pollutants. This is the opposite of the Atlantic Ocean where the tides displace large water masses which allow the dispersion of contaminants.

Many environmental indicators have shown a declining quality of the waters and the sediments. Eutrophication has greatly increased and toxic algae and red tides occur more frequently. Substantial investments have already been made for the construction of wastewater treatment plants and port reception facilities. Quality improvements in certain regions have been observed, but further effort and investments are necessary to halt the general trend of environmental degradation.

A.1. State of the environment

While the Mediterranean is oligotrophic, large amounts of nutrients from wastewater discharges result in severe eutrophication along the coast of the Emilia-Romagna region (where the Po River flows into the Adriatic Sea), in the Gulf of Trieste and in the Bay of Naples; also in the Berre Lagoon, in the northern and eastern coast of the Saronikos Gulf and in the Bay of Thessaloniki (Thermaikos Gulf). Eutrophication has serious consequences for the marine ecosystem and affects the tourist industry (aesthetic problems, loss of water clarity, dead animals on the shore, foul smells, toxic shellfish, etc.).

Heavy metals pose another threat to the marine environment. Typical levels of mercury (Hg) in the open water of the Mediterranean Sea range from 7 to 25 ng Hg/l. High concentrations, however, are observed in Roquetas de Mar (Gulf of Almería): 88 ng/l; Mojácar (eastern coast of Almería): 84 ng/l; Sète (Gulf of Lions): 65 ng/l. These values are near or above the 70 ng/l limit defined by the European Community (EC) for heavily polluted seawater.
In the sediments typical values for open water are 0.01 to

0.03 mg Hg/kg (dry weight) while near Barcelona and in the Gulf of Fos concentrations up to 4.9 mg/kg and 4.4 mg/kg, respectively, have been measured.

Comparisons of mercury levels in sardines, mackerel, tuna and anchovy both inside and outside the Mediterranean have led to the conclusion that Mediterranean pelagic fish contain roughly twice the mercury concentration as Atlantic specimens of similar size. The striped mullet <u>Mullus barbatus</u> contained an average of 2.81 mg Hg/kg wet weight near the Argentario promontory (Tyrrhenian Sea), up to 1,100 µg Hg/kg wet weight near Palma de Mallorca, an average of 540 µg Hg/kg wet weight along the Slavic coasts of the Adriatic Sea, and 193 µg Hg/kg wet weight in the inner part of the Thermaikos Gulf (close or above the 200 µg/kg wet weight limit of the EC Directive).

High mercury concentrations, close or above the 1 mg/kg dry weight limit of the Oslo-Paris Convention for strongly polluted shellfish, have been found in mussels (<u>Mytilus</u> sp.) from the Ligurian Sea (0.18-0.96 mg/kg dry weight), the Gulf of Trieste (0.28-1.3 mg/kg dry weight) and near Marseille (0.87 mg/kg dry weight).

For open waters the typical cadmium concentrations range from 0.005 to 0.150 µg/l. Elevated cadmium concentrations (> 1 µg/l limit defined by the EC for very heavily polluted waters) have been measured in El Grau de Castellón (1µg/l), near the Rhône Delta (1.1 µg/l) and in the area of Ancona (1.41 µg/l).

The naturally occurring level of cadmium in the sediments ranges from 0.1 to 2.0 mg/kg dry weight, while in polluted areas (Portman, Gulf of Fos) values of 3.2 and 5.8 mg/kg dry weight, respectively, have been recorded.

Very high concentrations of cadmium in mussels (<u>Mytilus</u> sp.), higher than the Oslo-Paris Convention limit of 5 mg/kg dry weight for strongly polluted mussels, have been measured in the Turkish Aegean (6.6-12 mg/kg dry weight) and at the mouths of the French rivers Orb and Hérault (12.5 mg/kg dry weight).

For the entire Mediterranean the average hydrocarbon concentration is 2.0 µg/l. The average Mediterranean concentration is 100 times the average North Sea concentration. Near the coast of Spain and France, concentrations of petrogenic hydrocarbons are in the range of 1.1-4.5 µg/l for the aliphatic fraction and 0.1-0.8 µg/l for the aromatic one.

In the Ebro Delta, mussels (Mytilus galloprovincialis), oysters (Ostrea edulis) and clams (Venus gallina) had an unresolved complex mixture (UCM) of the saturated and aromatic fractions of 100-300 µg/g dry weight. These concentrations were equivalent to those in mussels in the most polluted harbors and bays in California.

The concentrations of pelagic tar in the Mediterranean vary from 0.5 to 130 mg/m^2. Normal values for offshore areas are below 5 mg/m^2, while in nearshore waters concentrations can be much higher (10-100 mg/m^2). The mean concentrations of pelagic tar in the western and central Mediterranean showed a sharp decline with time from 37 mg/m^2 in 1969 to 9.7 mg/m^2 in 1974 and 1.75 mg/m^2 in 1987 as a result of the adoption of international conventions, technical improvements in the oil shipping industry and the installation of coastal port reception facilities for oily waste waters.

Contamination by chlorinated hydrocarbons is mainly found near the EC countries in the northwestern part of the Mediterranean, the Tyrrhenian, and the Aegean Sea. Polychlorinated biphenyl (PCB) concentrations for open waters of the Mediterranean average about 2 ng/l. At Cortiou near Marseille the average concentration measured was 57 ng/l while in the Po and Adige rivers the maximum concentration found was 42 ng/l.

High concentrations of PCB in sediments have been recorded at Cortiou (785-21,615 µg/kg dry weight), around Barcelona between the Llobregat and Besós rivers (maximum: 524 µg/kg dry weight), and near the Athens outfall in the Saronikos Gulf (maximum: 775 µg/kg dry weight).

Elevated concentrations of PCB in mussels (Mytilus galloprovincialis) were measured in the northern Aegean Sea (580 µg/kg wet weight), while in the striped mullet (Mullus barbatus) from the Tyrrhenian Sea concentrations up to 3,950 µg/kg wet weight were recorded. It should be noted that the 1979 U.S. Food and Drug Administration (FDA) maximum permissible concentration is 2 mg/kg wet weight for PCBs in the edible portion of fish and shellfish.

High insecticide dichlorodiphenyl trichlorethane (DDT) concentrations have been measured in sediments near the Athens outfall in the Saronikos Gulf and around Barcelona between the Llobregat and Besós rivers (1,900 and 117 µg/kg dry weight respectively).

In the Ebro Delta a concentration of 11,696 µg/kg dry weight of DDT has been found in the liver of the hake (Merluccius merluccius).

The Mediterranean with 25,000 species is a rich biotope, one of the richest in the world. The geographical isolation of certain areas (particularly islands) has brought about the appearance of many endemic species, unknown elsewhere. The Mediterranean is also a key area in the migration path of many birds. The survival of these species depends on the conservation of their host biotope. With increasing human activity along the coasts of the Mediterranean, animals lose part of their habitat and breeding areas. Some large animals, e.g. dolphins, are viewed as competitors of fishermen.

Accidental or deliberate killing by fishermen and human disturbance of the breeding areas of the Mediterranean monk seal (Monachus monachus) have been the most important reasons for its disappearance in several regions. The monk seal was once common along the entire coast but its population is now reduced to between 400 and 600 animals, most of which live off the coasts of Greece, Morroco and Turkey.

Pollution or destruction of marine plants (Posidonia spp.) has reduced the spawning grounds of many fish species.

A.2.   Pressures on the environment

The pollution that enters the Mediterranean originates from sources located on land (domestic and industrial, via wastewater discharges), from rivers, agricultural run-off, maritime discharges and atmospheric deposition. The coast of the western and northern Mediterranean is potentially the most affected.

Approximately 23% of the total pollution load is discharged into the northwestern Mediterranean from the bordering countries Spain, France and Italy. Up to 35% is discharged into the Adriatic primarily from Italy.

The northwestern basin also receives major river inputs from the Ebro (with a river basin area of 86,800 km^2) and the Rhône (with a river basin area of 90,000 km^2). The Tiber (with a river basin area of 17,170 km^2) flows into the Tyrrhenian Sea and the Po (with a river basin area of 75,000 km^2) into the Adriatic Sea. Phosphorus and nitrogen loads largely derive from river inputs (75 to 80%).

The estimated value of 820,000 tons per year of oil spilled in the Mediterranean (1.2% of which via atmospheric input) is 17% of the total oil pollution in the oceans (4.5-5 million tons per year) while the area represents only 0.7% of the global water surface.

The discharge of mercury is largely due to river inputs with only 8% from coastal sources. Approximately half of the chromium and two-thirds of the zinc load is carried by rivers. Most of the metal loads originating in the coastal zone derive from industrial sources and to a lesser extent from domestic sewage.

Approximately 15% of the suspended solids stems from surface run-off within the coastal area while the rest is carried by the major rivers.

Of the total radioactive tritium input, 44% is discharged into the northwestern Mediterranean and 51% into the Adriatic. Up to 85% of the tritium and 40% of the other radionuclides stem from power plants located on major rivers in the northern part of the Mediterranean basin.

A.3.   Plans and measures

The following plans or measures have been taken or are planned to improve the Mediterranean environment.

a) Wastewater Treatment Plants
   Further construction of wastewater treatment plants is required to treat up to 65-90% of the discharges using as a minimum primary and preferably secondary treatment together with extended outfalls of sufficient dilution ability. Improved operation, supervision and enforcement of regulations in existing plants will increase their efficiency.

b) Port Reception Facilities
   Additional port reception facilities are required to receive and treat contaminated bilge waters and tank cleanings. Advanced treatment units can greatly reduce hydrocarbon effluent concentrations.

c) Coastal Zone Planning
   Conflicting land uses exist in the coastal zone. Planning is required for siting of urban areas, industrial complexes, tourist facilities and nature reserves.
   Although there are plans to direct or curb urbanization, lack of enforcement or supervision results in numerous dispensations for housing or tourist establishments. In 20 years nearly 2,000 kilometers of coastline have been sacrificed or reallocated in this way. Future scenarios show greatly increased siting competition between coastal power plants and tourist facilities. Better information on the extent of coastal zone degradation by improper disposal of hazardous waste is required.

d) River Basin Management

Most of the contamination and pollution of the EC countries enters the Mediterranean through the large rivers. Not all rivers have an effective management structure to disburse investment funds for treatment plants, monitor the river quality and set enforceable discharge goals.

e) Biotope Protection

The rich genetic heritage of the Mediterranean regions, of both wild species and cultivated or domestic varieties, is seriously threatened. The application of the Barcelona Convention protocol on "specially protected areas" will help to protect coastal and marine biotopes, particularly through the expansion and improvement of the biosphere reserve network, the creation of biotope reserves and the adoption of a regional conservation strategy.

f) Resource Management

Overfishing is threatening the Mediterranean ecosystem. Optimization of fisheries requires effective international cooperation and information about existing fish stocks (demersal and pelagic species) and their migration and reproduction cycle. Priority should be given to concerted action between countries exploiting the same resource, and the formulation of measures to limit fishing activity and ensure distribution of this resource, together with supervision of their implementation.

Experimental closing and opening of different areas and seasons and license restriction may be used beneficially for the protection and management of the overexploited trawl and coastal resources in the Mediterranean, together with habitat enhancement (e.g. artificial reefs).

g) Better Coordination of National, Regional and Local Authorities

Fragmentation of state, regional and local authorities often hampers projects in the environmental field. Necessary mechanisms and institutions for financing and competence to develop and apply regulations are required. Transfrontier issues in the environmental field should be addressed systematically. Environmental impacts of infrastructure projects in the coastal zone should be required during their planning phase. Better enforcement by local authorities is necessary for effective implementation of national and international policies.

h) Environmental Awareness Enhancement

Effective environmental policies are only sustained by public awareness of their importance. Educational materials should disseminate objective and relevant information to the public about the possibilities and constraints of the local and regional environment in which they live, directed at various age groups and stressing the fact that one generation inherits from another.

The EC has several environmental protection activities that need expansion:

a) Fourth Environmental Action Program: it integrates the environmental dimension into the Community's economic, industrial, agricultural and social policies;
b) CORINE Program: it gathers information on the state of the environment, it coordinates activities on the collection of data, it ensures the consistency of information;
c) European Regional Development Fund Environmental Program: it is intended to assist in the development and structural adjustment of lagging regions and in the conversion of declining regions by means of promoting

the implementation of Community environmental policy;
d) Mediterranean Strategy and Action Plan (MEDSAP or MEDSPA): it integrates projects designed to protect the Mediterranean environment. Furthermore, it increases public awareness through demonstration projects, information campaigns and training courses.

A.4.   Further studies

a) Biological-Effect Data
Limited data exist on the chronic effects of low levels of heavy metals and persistent organics on individual species, food chains and biotopes. A literature study should systematically tabulate the effects of relevant studies and compare them with ambient levels in the Mediterranean. Limits or guidance values established by other countries for edible species or ambient levels should be systematically compared with Mediterranean levels. Physical dimensions and requirements of biotopes and their effect on species propagation should be more clearly established.

b) Input Data
Most of the surveys on marine discharge of domestic, industrial and agricultural origin into the Mediterranean use conventional parameters such as BOD, chemical oxygen demand (COD), suspended solids, heavy metals, organochlorines and hydrocarbons. More up-to-date data and better estimates should be used to update the existing surveys and to expand it with the inclusion of blacklist substances together with the different organic nitrogen and phosphorus species. Data collected from additional marine sites should be used to measure and calculate the atmospheric deposition in the different Mediterranean regions more quantitatively. Calculations and measurements should further establish the impact of improper disposal of hazardous waste on the water quality. Little information exists on the inputs through

urban storm-water run-off.

c) Quality Improvement or Trends due to Environmental Projects

A systematic tabulation of water quality improvements after the realization of environmental projects will allow the estimation of the effectiveness of similar additional projects. Environmental projects can also be compared among themselves to determine their relative cost-effectiveness. The financial benefits in terms of human (bather) health, protection of fishing grounds and land resources should be calculated more quantitatively.

B. SPANISH MEDITERRANEAN ENVIRONMENT

The main environmental problems of Spain are land management (erosion, desertification) and water management (irrigation, flood control, water quality, saltwater intrusion). In the Mediterranean region (5 autonomous regions, 12 provinces) land-use pressures by human habitation, tourist facilities, industry, agriculture and waste disposal result in habitat destruction and local water pollution problems. Divisions of governmental responsibility at the coastal, regional and provincial level hamper implementation of effective environmental programs.

B.1. State of the environment

Heavy-metal contamination especially mercury (Hg), lead (Pb), zinc (Zn) and cadmium (Cd) of shore waters was severe at Roquetas de Mar (88 ng Hg/l, 12 µg Pb/l), Mojácar (84 ng Hg/l, 25 µg Pb/l), Alicante (70 µg Zn/l) and El Grau de Castellón (1.0 µg Cd/l). The limits defined by the EC for heavily polluted waters are 70 ng Hg/l, 10 µg Zn/l, 5 µg Pb/l and 0.5 µg Cd/l. Metal concentrations in sediments are high near Portman (1,160 mg/kg dry weight for Pb; 3.2 mg/kg for Cd) due to past mining activities (which were stopped on March 30, 1990) and near Barcelona (230 mg/kg for Pb; 4.9 mg/kg for Hg) due to industrial discharges. Food-chain magnification occurs for several heavy metals. The highest values are recorded for mercury (1,230 µg/kg wet weight) in shrimps (<u>Aristeus antennatus</u>) and lead (125,000 µg/kg wet weight) in mussels (<u>Mytilus galloprovincialis</u>). The most affected organisms are found near Guardamar, Algeciras and Palma de Mallorca (mercury) and Portman and Cartagena (lead).

Elevated bottom-sediment concentrations for DDT (maximum: 117 µg/kg dry weight) and PCBs (maximum: 524 µg/kg) are detected around Barcelona between the Llobregat and Besós rivers. The highest DDT concentration (48 µg/kg wet weight) in <u>Mytilus galloprovincialis</u> is found at Cullera near the mouth of the Júcar River. In the Ebro Delta the liver of the hake (<u>Merluccius merluccius</u>) has the highest DDT (11,696 µg/kg dry weight) and PCB values (26.5 µg/g dry weight).

The greatest petroleum hydrocarbons in seawater were observed in Alicante (maximum: 8.26 µg/l) and Portman (maximum: 6.5 µg/l). Significant marine sediment hydrocarbon (C_{14} to C_{24}) concentrations are found in the Tordera Delta at a maximum of 565 µg/kg. <u>Mytilus</u> accumulates the highest hydrocarbon concentrations (500-3,200 µg/g dry weight as UCM saturate fraction) near Barcelona.

Bacterial and viral contamination has occurred at beaches near municipal discharges. A general overview conducted by the EC in 1987 showed high coliform densities exceeding the EC limits near Peñíscola and Puerto de Alcudia. In 1979 enteropathogenic viruses were isolated from seawater in the urban area of Barcelona, ranging from 0.12 to 1.60 MPN CU (expressed as the most probable number of cytopathogenic units) per liter.

Many rivers have been contaminated by municipal, industrial

and agricultural discharges. The Ebro shows a steady decline in the General Quality Index from 84 at the uppermost station at Palazuelos to 55 at Pina after which some restoration occurs. The mean water flow (1.6×10^{10} m^3 per year) has decreased by approximately 14% since the early years of this century as a result of the construction of dams for irrigation projects. Retention of sediments behind dams has halted the delta formation.

The Mediterranean coastal area and interior regions are vulnerable to soil erosion due to climatological conditions, type of soil, insufficient drainage and land-use practices. In Almería and Granada desertification and severe soil erosion affect 72% and 51% of the territory, respectively. Erosion was enhanced by 12,837 forest fires in 1985 which affected an area of 469,426 hectares (in the whole of Spain). Biotope preservation is established in 93,968 hectares of specially protected natural interest areas and in 123,445 hectares of partially protected natural interest areas.

Coastal land degradation occurs as a result of waste disposal. The coastal region generates 3.9 million tons of urban solid waste per year of which 36.7% is dumped uncontrolled, 26% dumped controlled, 28% composted and 9% incinerated. The coastal region of Spain has 29 of the 40 compost plants, which produce 473,800 tons of compost every year accounting for 70% of the Spanish total. Little information is available on hazardous waste disposal.

B.2. <u>Pressures on the environment</u>

As a result of tourism the population of the coastal municipalities increases from 6.75 million in the winter to 12.67 million in the summer. The coastal municipalities discharge 500×10^6 m^3 of waste water per year of which two-thirds are being treated (70%). During the summer the maximum waste flow increases to 2.5×10^6 m^3 per day of which 49% can be treated. In the winter the minimum flow is

1.35×10^6 m^3 per day of which 91% is treated. By the latest counts there were 147 sewage treatment plants and 236 submarine outfalls. The median percentage of the sewage treated was 67% and ranged from 1 to 80%. The municipal load discharged into the Mediterranean is 166×10^3 tons of BOD per year and 249×10^3 tons of suspended solids per year.

Most of the industry is located in Cataluña. In 1987 MOPT estimated an industrial discharge of 634×10^6 m^3 of waste water per year, 90.5×10^3 tons of BOD per year and 140×10^3 tons of suspended solids per year. An earlier survey by the Ministry of Industry and Energy in 1981 estimated a load of 70×10^3 tons of BOD per year and 96×10^3 tons of suspended solids per year.

The largest river discharging into the Mediterranean is the Ebro. The highest BOD discharge is through the Ebro and Besós rivers, while the highest suspended solids load is carried by the Vinalopó and Ebro rivers. The major rivers discharge a load of 53×10^3 tons of BOD per year and 157×10^3 tons of suspended solids per year.

Estimates for aggregate discharges (municipalities, industries, rivers) are 310×10^3 tons of BOD per year and 546×10^3 tons of suspended solids per year.

B.3.   *Plans and measures*

Marine pollution can affect tourism income as bathing in polluted seawater poses a health risk (eye infection, skin infection). Economic losses also occur as a result of erosion (crop reduction) and forest fires.

The environmental protection in Spain is governed by an administration division of power at the functional and territorial level. At the central governmental level, there are 13 institutions regulated by 5 different ministries. In the 5 Autonomous Regions with a Mediterranean coastline there are more than 20 departments with environmental jurisdiction. At the central administration level, MOPT (Ministry of Public Works and Urban Planning) spent 5.4 billion ptas., MINER (Ministry of Industry and Energy) 0.6 billion ptas., MISACO (Ministry of Health and Consumer Affairs) 0.08 billion ptas. in 1986 on environmental issues.

In Cataluña 10.1 billion ptas. were invested in 18 sewage treatment plants between 1975 and 1985. The total forecasted need between 1983 and 1996 is 145 billion ptas. The autonomous region of Valencia invested 7.3 billion ptas. between 1975-85. A total of 20 billion ptas. is required between 1986 and 1995. In Murcia 5.5 billion ptas. are required for investments between 1986-91. Andalucía has an investment plan for 3.3 billion ptas.

Planned port reception facilities for ship wastes require 6 million ECU and their absence is considered an increasing problem.
Other environmental issues to which attention needs to be given are: toxic waste disposal facilities; river basin management; coastal zone management; and wetland conservation projects.

C. FRENCH MEDITERRANEAN ENVIRONMENT

C.1. State of the environment

The French coastal zone occupies 2.5% of the national territory while 11% of the total French population lives in that zone. The demographic concentration is acccompanied by an industrial concentration. A major input of pollutants results from the Rhône River discharge, the largest river in the western Mediterranean. Consequently, the problems of preserving the marine coastal environment are numerous.

The maximum nitrate concentrations in coastal water are observed in the Fos-Berre zone (18 µmoles/l), influenced by the discharge of the Rhône and Durance rivers. The lowest phosphate concentrations occur in the Bays of Cannes and Villefranche (0.04 µmoles/l). In the lagoons of Languedoc-Roussillon region the nitrate, ammonia and silicate concentrations are at an intermediate level, but phosphate has the highest levels of all sites (2.62 µmoles/l). Cortiou has high phosphate values (0.83 µmoles/l) near the outfall of the Marseille municipal sewage treatment plant.

During monitoring by the national Network of Observation of the marine environment in 1979 the highest recorded mercury concentration in water was 0.44 µg/l (in the Thau Lagoon) while the highest cadmium concentration was 0.86 µg/l (near Cortiou, Marseille). In a survey by the EC in 1987 the maximum mercury concentration was 65 ng/l (at Sète) – a significant decrease compared to 1979 values. The most elevated concentration of cadmium was 1.1 µg/l and that of lead as high as 10.2 µg/l (both near the Petit Rhône Delta). The most significant concentrations of copper and zinc were 15.5 µg/l and 23 µg/l, respectively (both in Ajaccio).

The values of organic pollutants in coastal water have been decreasing significantly over the last 10 years and certain parameters (pesticides and organochlorines) are reaching detection limits. The organochlorine biocide lindane decreased from the maximum value of 115 ng/l in 1977 to 13 ng/l in 1985. The highest hydrocarbon concentration was detected at Cortiou at the end of the Marseille sewage outfall (104 µg/l), followed by the mouth of the Rhône River (18-23 µg/l). The most elevated PCB concentrations have also been recorded at the end of the Marseille sewage outfall at Cortiou, reaching 57 ng/l, and at the mouth of the Rhône with values of 11 ng/l. Elsewhere the concentrations of PCB in water are generally below the detection limit (2 ng/l).

Significant values of mercury in sediments have been measured in the Berre Lagoon (maximum: 3.16 mg/kg in the Saint-Chamas Bay) and in the Gulf of Fos (maximum: 4.4 mg/kg in the zone of Port-de-Bouc-Lavéra). Around the exit of the sewage outfall of Cortiou in a radius of up to 1.5 kilometers a zone with toxic concentrations of mercury, copper, lead and zinc has been defined.

Considerable accumulation of hydrocarbons of around 10,000 mg/kg has been observed in the sediments of the Berre Lagoon (Vaine Lagoon) and around the exit of the sewage outfall of Cortiou (in a radius of 1 kilometer). Excessive pollution by PCBs has been found in the sediments of Cortiou Cove (up to 21,615 µg/kg).
A regulation to reduce mercury wastes from the electrolytic industry, established in 1974, has resulted in a tenfold decrease in the concentration of mercury in living matter (from 0.3 to 0.03 mg/kg dry weight). Mussels exposed to urban waste waters in the Marseille area have 0.87 mg Hg/kg dry weight, close to the upper limit of 1 mg/kg dry weight as established by the Oslo-Paris Convention for strongly polluted shellfish.

Concentrations of organic pollutants (hydrocarbons, PCB and DDT) in mussels and oysters have been decreasing but are still high in the more polluted zones. The maximum hydrocarbon concentration of 15 mg/kg dry weight found in the Rhône Delta is three times greater than that found in the Bay of the Seine (French Atlantic coast). The PCB maximum of 1,440 µg/kg dry weight measured in Toulon is twice as high as the limit of 700 µg/kg dry weight established by the Oslo-Paris Convention for strongly polluted shellfish.

Since 1975, water samples from all French coastal bathing areas were analysed annually for various bacteriological parameters. The percentage of French Mediterranean beaches complying with the European Directive has increased from 83.6% in 1976 to 95.7% in 1986.

C.2.   Pressures on the environment

The French Mediterranean coastal regions experience nearly a doubling of their population during the summer. In Corsica the population increases by a factor of 2.3. Consequently, the measures against domestic pollution must take into consideration both the native and tourist population. The treatment percentage of the domestic discharges was 28% in 1985, releasing 3.55 million population equivalents into the sea. Most of the domestic discharges are concentrated in the department of Bouches-du-Rhône; however the load has decreased due to the installation of sewage treatment plants.

The industrial pollution in the Mediterranean coastal waters comes predominantly from the chemical industries, the refineries, the paper industries, the alcohol distilleries and the wine-liquor industries. Recent tabulated data from the Agence de Bassin Rhône-Méditerranée-Corse show a discharge of 22.5×10^3 tons of BOD per year and 22.7×10^3 tons of suspended solids per year. The suspended solids decline between 1982 and 1986 was 31%, while the BOD was reduced 12% during the same period of time, due to increased wastewater treatment.
In the coastal region Provence-Alpes-Côte d'Azur the industrial discharges are mainly concentrated in Marseille, Toulon and Nice and in particular among the Gulf of Fos-Berre Lagoon (one of the most important coastal industrial zones of France).

The rivers also contribute considerable pollutant quantities to the Mediterranean. The 16 major rivers discharge a load of 212.4×10^4 tons of suspended solids per year and 290×10^3 tons of BOD per year, whereby the Rhône accounts for 94% of the organic load. There is an important decrease in discharge of suspended solids and no change in BOD discharge between 1981 and 1986. Part of the decrease in discharge of suspended solids can be attributed to the 19%

decrease in flow rate during that period and the sedimentation of the particles in the riverbed. As a result of the discharges into the Rhône, both ammonia and phosphate concentrations increase steeply downstream from Lyon. The CNR (Compagnie Nationale du Rhône) data collected between 1979 and 1985 show that the average ammonia concentration exceeds the limits for class 1B (the moderate quality class) and phosphate exceeds that for class 2 (the poor quality class). The load from the Rhône River entering the Mediterranean varies between 100-400 tons of BOD per day, 80-200 tons of total nitrogen per day and 10-25 tons of phosphate (as phosphorus) per day.

France's coastal lagoons are of major economic importance. In 1984 3,000 tons of fish were caught in the Berre Lagoon. The enrichment of the lagoon with organic matter and nutrient discharge of urban, industrial and agricultural origin resulted in increased eutrophication. In 1966 the diversion of a great part of the outflow of the Durance River towards the Berre Lagoon completely changed the ecosystem. Since 1972, major efforts have been made to reduce the industrial and urban discharges. As a result levels of trace pollutants have decreased in fish. Eutrophication, however, is still important and causes anoxic benthic conditions and fish kills.

The significance of the Thau Lagoon for the shellfish farming is very large. The official production figures of 7,000 tons of oysters per year and 5,000 tons of mussels per year may greatly underestimate the actual catch. Annual production is likely to be around 17,000 to 20,000 tons of oysters and 8,000 tons of mussels.

Massive shellfish mortalities occurred three times during the past 15 years due to anoxic conditions in the bottom layer.

C.3.   *Plans and measures*

The Ministry of the Environment is the leading administrative body, coordinating all actions (legal, technical, financial) concerning the environment. An active role in planning, project implementation and monitoring is played by three other administrative bodies (the Regional Branch of the Ministry of Industry, the Regional Water Agency, the municipalities).

Law 64-1245/1964 contains the measures that must be taken to reduce water pollution.

Laws 76-599 and 76-600 provide for the prevention of sea pollution by discharges from ships and air craft and the burning of refuse at sea.

Law 85-661/1985 regulates the installations for the protection of the environment.

Law 86-2/1986 covers the proper development, protection and usage of the coastal regions.

The Ministry of the Environment has formulated two major environmental goals:
- to increase the current efficiency of pollution removal of 40% in 1985 to 60% in 1995, using more advanced treatment processes;
- to protect coastal areas and lake shores; and additionally protect unique forests and agricultural and urban landscapes, and develop national parks.

As of 1982, the installed capacity of the treatment plants reached 3.05 million population equivalents in the departments of the French Mediterranean region. The treated domestic pollution load in population equivalents from 1975 to 1984 was about half the installed total treatment capacity, indicating poor operating conditions. The percentage of treated waste water is expected to increase due to the construction and completion of new treatment plants (in Antibes, la Ciotat, Saint-Cyr-sur-Mer, Bormes-les-Mimosas, Sainte-Maxime, Saint-Tropez, Cavalaire and Bastia) and the operation of the constructed plants in Perpignan, Marseille (1.5 million population equivalents) and Nice.

A total of 62.3 million francs was given by the "Commission

des Aides de l'Agence de Bassin Rhône-Méditerranée-Corse" as grants, 45 million in advance payments, 4.6 million in loans for a total of 112 million francs in 1986 for the construction of treatment plants.

Grants for industrial treatment facilities were 30 million francs, advanced payments 17.8 million, loans 2.7 million for a total of 50.5 million francs. A special program was established to combat water pollution in the heavily industrialized zone of Fos-Berre Lagoon, which resulted in a 94% reduction in COD discharges between 1972 and 1983.

France has gradually provided the Mediterranean ports with hydrocarbon reception facilities. The recently constructed Fos-sur-Mer station operates according to the most strict conditions concerning the discharge of hydrocarbons into the sea. The reception stations of Marseille-Marignane, Lavéra and Sète have only gravity type separators. The estimated cost of additional reception facilities is $1,550,000 (U.S.).

D.   **ITALIAN MEDITERRANEAN ENVIRONMENT**

D.1.   State of the environment

Accumulation of mercury in marine sediments (above 1 mg/kg) has been measured at the mouths of the rivers that flow into the upper Tyrrhenian, at the upper part of the Adriatic Sea and around Naples Harbor. Peak concentration levels for zinc (greater than 200 mg/kg) are observed in front of the Lagoon of Venice, while the highest content in lead (greater than 60 mg/kg) is found in front of the Po Delta and in the Gulf of Trieste.
At the island of Ischia (province of Naples) coastal water was heavily polluted with mercury. At Lido di Ostia (near Rome), Pesaro and along the coast of the Ancona Province elevated values of zinc and copper were measured. Along the

Ancona Province, coastal water was heavily or very heavily polluted with cadmium.

The upper Tyrrhenian Sea, besides receiving many anthropogenic pollutants, is associated with a large area of cinnabar-rich ores which were mined intensely for years. The weathering of discharged sediments has evidently increased the concentration of mercury in the terrestrial and marine environment. Near the Argentario promontory elevated concentrations of mercury have been recorded in the mussel Mytilus galloprovincialis (average 1.83 mg/kg wet weight), the lobster Nephrops norvegicus (average 1.92 mg/kg wet weight) and in the striped mullet Mullus barbatus (2.81 mg/kg wet weight). In the Bay of Follonica elevated concentrations of lead, iron and nickel have been measured in M. galloprovincialis (up to 16.4, 270 and 6.9 mg/kg dry weight, respectively). The maximum mean concentration of zinc in fish is found in the Tyrrhenian Sea near the Sicilian coasts (7.0 mg/kg wet weight).

Organic contaminants are present in significant values along the coast. In the southern part of the Gulf of Venice and in the area influenced by the Po River tributaries considerable concentrations of PCB in sediments (20-80 or > 80 µg/kg) have been measured. In the Gulf of Venice elevated concentrations of DDT (> 30 µg/kg) have been found. In the Gulf of Pozzuoli (west of Naples) sediment concentrations of polycyclic aromatic hydrocarbons exceeding 40 mg/kg were measured (maximum: 170 mg/kg near Bacoli).

PCB residues predominated in all samples of mussel (Mytilus galloprovincialis), lobster (Nephrops norvegicus), anchovy (Engraulis encrasicolus), striped mullet (Mullus barbatus) and tuna (Thunnus thynnus) collected in fishing areas between Trieste and Ancona. In all samples measurable amounts of DDT, DDE, benzene hexachloride and dieldrin were detected.

Marine bathing waters are in compliance with acceptability

limits prescribed by the Presidential Decree 470 of 1982 for 86% of the tested samples, with 37% and 25% of the polluted samples exceeding fecal coliform and fecal streptococci limits, respectively.

Marine eutrophication and occurrence of "red tides" is a major problem in the Adriatic Sea which is classified as hypertrophic, especially along the 200 kilometers of coast south of the Po Delta. Dead fish (due to anoxic conditions) on the beaches and high algal densities have a negative impact on tourism.

Studies on coastal erosion reveal that 48% of the coasts have regressing and only 6% advancing shorelines. Intensive erosion has been observed near the mouths of the rivers due to a diminished supply of fluvial sediments to the sea (retained behind river dams). The loss of woodland (mainly due to forest fires) on sloping coastlines and land subsidence (geological or induced by excessive groundwater pumping) are also damaging the coastal strip.

The marine water quality is greatly affected by pollutants from the Po. Significant concentrations of heavy metals near the Po mouth were found each year (between 1975 and 1980) due to industrial discharges. The average concentration of BOD of 7.3 mg/l measured near the mouth indicates extensive organic pollution. Considerable levels of nutrients were also present (1.35 mg/l nitrate and 0.23 mg/l total phosphorus). Accumulation of pesticides and organochlorines in fatty tissues of living organisms was found all along the Po River, exceeding the U.S. FDA standard limit of 2 mg/kg wet weight below the confluence with the Adda and Lambro River. The freshwater quality for fish life is low throughout the entire Po basin. The quality in only 18% of the tested municipalities could support life and reproduction of salmon and carp. The quality in only 43% of the tested municipalities was adequate for irrigating the more susceptible crops. Bathing-water quality was satisfactory in only 13% of the cases. Surface waters for production of drinking water were below the minimal quality in 73% of the tested municipalities.

D.2. Environmental pressures

The main marine environmental pressures are discharges from municipalities, industry and polluted rivers.
The domestic discharge equals 816×10^3 tons of BOD per year, with Campania, Latium and Sicily the most contributing regions. The maximum potential industrial load released into the Mediterranean was calculated at 158.5×10^4 tons of BOD in 1981, mostly from the Tuscany, Latium, Emilia-Romagna and Veneto regions. Every year approximately 50,000 tons of phosphorus and 660,000 tons of nitrogen are discharged into the Italian coastal waters. The North Adriatic basin receives about 48% of the total phosphorus load.
Among the 16 major Italian rivers discharging organics into the Mediterranean, the Po (258×10^3 tons of BOD per year), the Tiber (109×10^3 tons of BOD per year) and the Adige (38×10^3 tons of BOD per year) are the major ones. The nutrient load discharged into the Mediterranean by the Po, Tiber, Adige and Arno rivers is estimated at 114×10^3 tons of inorganic nitrogen per year and 17×10^3 tons of total phosphorus per year. The Po River accounts for 71% of the inorganic nitrogen and 63% of the total phosphorus load of all Italian rivers. Metals discharged by the Po River into the Adriatic Sea have been estimated at: 243 tons of arsenic per year; 1,554 tons of copper per year; 944 tons of chromium per year; 1,312 tons of lead per year; 2,646 tons of zinc per year; 89 tons of nickel per year; and 65 tons of mercury per year.

The Po Valley accounts for more than 50% of all industrial waste produced in Italy, or about 25 million tons per year. The aggregate waste processing capacity only suffices to cope with some 25% of the waste in the Po Valley. Each year the province of Naples produces more than 1,700,000 tons of degradable urban solid waste and sludge, and more than 910,000 tons of special waste, including around 440,000 tons which can be classified as toxic or hazardous; at the same time, there is a serious lack of facilities to treat and dispose of the waste. The total amount of solid wastes produced in the Lagoon of Venice is estimated at 1,957,000 tons per year of which 1,614,000 is of industrial origin, 129,000 is biological sludge, 2,700 is hospital waste and 214,000 is ash from power stations.

D.3.   Plans and measures

A low environmental quality can have an adverse effect on the health of the population, as 334 cases of typhoid fever were reported in Naples in 1979. Income from tourism decreased by 20 billion lire after one nearby industrial explosion. The damage caused by air pollution to historic sites has direct economic consequences.

The legal aspects of the environment are highly complex and fragmented. Over 600 national laws and decrees and 300 regional ones regulate environmental matters. The Ministry of the Environment exists only since July 1986. The Government has been adopting EC directives dealing with the quality of bathing, drinking and fishing waters.

Between 1982 and 1987 national environmental expenditures have been stable at around 1% of the Gross National Product. Approximately half of these expenditures comes from the central (State) and regional governments and the other half from the local governments and private sector.
Environmental expenditures by the central government reached 2,500 billion lire in 1985. Approximately half has been devoted to natural disasters. The other half is used for water protection and wastewater treatment, solid-waste disposal, forestation, hydraulic works, soil protection and environmental protection.

Projections were made for 3 possible scenarios concerning demography, agro-food industry, manufacturing, energy, tourism, transportation and their pollution loads for the years 2000 and 2025. As a consequence of the moderate continuing scenario, BOD generation will increase from 2,549 kilotons in 1985 to 3,246 kilotons in 2000 and 6,731 kilotons in 2025. The treatment efficiency of domestic waste waters is thereby expected to increase from the present 31% to 76% in 2000 and 100% in 2025. The aggravated continuing scenario has a lower BOD production but a degree of sewage treatment that will only satisfy 62% of the need in 2000.

The costs for a number of measures in the basin of the Lambro, Olona and Seveso rivers amount to 4,800 billion lire, to be spread over a 5-year period. The proposals concerning the Po basin comprise 2,100 billion lire for wastewater treatment, 1,869 billion lire for aqueduct works, over 300 billion lire for projects on cleanup of aquifers, cleanup of soils, studies on environmental damage, monitoring, and sewer extensions. A plan for the province of Naples requires 1,027 billion lire. The overall 10-year reclamation cost of the Lagoon of Venice is 3,043 billion lire. Approximately 15,000 billion lire over a 5-year period will be needed to improve sewerage and purification systems in the rest of Italy. Improved solid-waste disposal requires 7,250 billion lire for a 5-year period and soil conservation 74,000 billion for the next 30 years.

E.   **GREEK MEDITERRANEAN ENVIRONMENT**

E.1.   State of the environment

Due to the mountainous nature of Greece there is a strong competition between industry, agriculture and tourism for land in coastal areas. These pressures have resulted in environmental degradation and water and air pollution.

In general, Greek waters are oligotrophic. However, in polluted gulfs such as Saronikos the water is enriched in nutrients and organic matter. In the northern Saronikos Gulf (Elefsis Bay) the ratios of phosphate, silicate, ammonia and nitrate to background values (Aegean Sea) are 5.11, 4.15, 15.80 and 7.00, respectively. In the inner Saronikos Gulf the ratios are 2.50, 1.39, 4.10 and 2.60.

In a EC survey on bacteriological bathing-water quality of 58 Greek beaches in 1987, 18 beaches exceeded the EC guide number (recommended criteria limit) but were in compliance with the required standard, while 1 (Corfu) exceeded the required standard.

Tar pollution is a major problem for the Greek coasts, especially in the Ionian Sea which has average pelagic concentration of tar balls of 150 $\mu g/m^2$.

Parts of the Saronikos Gulf (Elefsis, Keratsini and Drapetsona Bays and Piraeus Harbor) show high heavy-metal concentrations in the sediments. Maximum heavy-metal concentrations in the sediments of Drapetsona Bay reach 1,000 times background levels for copper, 3,500 times background for iron and as high as 700 times background for zinc. In Keratsini Bay maximum organic carbon levels in the sediments reached 30 times background.

Concentrations of PCB and the common chlorinated pesticides investigated in specimens of the striped mullet (<u>Mullus barbatus</u>) from the Saronikos Gulf showed the most elevated values in the vicinity of Piraeus Harbor. They contained on the average 15 times as much PCBs and DDT as those from the cleanest area.

Elefsis Bay receives industrial effluents (mainly from oil refineries, steel works, etc.) along its northern shore and domestic wastes through the east channel from the Keratsini Athens outfall. These inputs generate severe ecological stresses in the bay, the most acute being the summer anoxic conditions.

Organic matter concentrations measured in the municipal

sewage outfalls in the Thermaikos Gulf (Thessaloniki) have a COD between 412 and 720 mg/l, exceeding the allowed maximum value of 250 mg/l. COD values range from 68 to 255 mg/l in the industrial wastewater discharges. In the mouths of the Axios, Loudias and Aliakmon rivers, which flow into the Thermaikos Gulf, COD values are 25, 100 and 50 mg/l respectively. For the entire Thermaikos Gulf a discharge of 45,000 kg of BOD per day and 59,500 kg of COD per day was measured.

High heavy-metal concentrations have been detected in the muscle of fish and mussels in the Thermaikos Gulf. For the entire bay the average concentrations in fish were 390 µg/kg wet weight for lead, 124 µg/kg wet weight for cadmium and 193 µg/kg wet weight for mercury whereas the average concentrations in mussels from coastal waters near the industrial zone were 460, 103 and 86, respectively. The bioconcentration factor found in mussels was 1.5 for lead, 3.4 for cadmium and 2.97 for mercury.

The water quality parameters of the six main rivers (the Nestos, the Strymon, the Axios, the Aliakmon, the Pinios and the Acheloos) are within the EC directives with the exception of the bacteriological quality of the Axios and Strymon rivers due to domestic sewage discharges. The average concentration of fecal coliforms was 4,288/100 ml and that of fecal streptococci as high as 2,179/100 ml in the Axios River in 1985, while the values in the Strymon River were 6,518/100 ml and 3,580/100 ml respectively.

The main pollution problems in the coastal zone are the municipal refuse and hazardous wastes, most of which are disposed of in an uncontrolled way using nonsanitary landfills. Problems are the production of leachates with BODs above 20,000 mg/l, the production of flammable and explosive gases from anaerobic decomposition, and foul smells.

The variable rain distribution in time and space, the unequal distribution of activities and, consequently, of consumption patterns have caused water deficits in some regions. Coastal groundwater overpumping has caused saltwater intrusion in aquifers in many coastal regions.

Air pollution problems, especially the smog formation, are severe in Athens. Air pollution has increased during the past 10 years in Thessaloniki and is becoming noticeable in a number of other big cities. Traffic contributes to about 75% to Athens air pollution. Extensive investments to reduce traffic congestion and car emissions are required.

E.2.   Environmental pressures

The rapid increase in human settlements (some of them without license by the authorities) in the coastal areas created a number of environmental problems because the necessary infrastructure in the coastal cities was not developed in parallel.
Tourism has become a major economic activity. Between 1962 and 1982 the number of foreign tourist arrivals in Greece increased by 890% to more than 5 million per year. Internal tourism has also increased significantly. Approximately 90% of all tourism activities occur in the coastal areas, resulting in additional environmental degradation.
A domestic load of 5 million population equivalents is being discharged to the Aegean and Ionian Seas.
Solid waste, especially in tourist areas, is another problem.
Industrial activities have rapidly increased. Almost 80% of the Greek industry is located in the coastal zone. All four large existing refineries, three of the four fertilizer plants and all the metallurgical industries are located there. Most can be found in the Greater Athens and Thessaloniki areas. The general practice has been to dispose of the untreated liquid wastes in the sea, hence creating significant water pollution. The estimated industrial and agricultural load is 18 million population equivalents. Industrial estates were established as a means of decon-

gesting the capital and major cities and developing the semi-urban areas. Wastewater treatment units are currently in operation in 6 of the industrial estates. They are situated in the prefectures of Achaia, Heraklion, Thessaloniki, Kilkis, Xanthi and Komotini. The wastewater load of the above six industrial estates is 300,000 population equivalents. The designed BOD treatment efficiency of the above major six industrial estates is 90-95%.
Agriculture also contributes to the pollution of the coastal seas through run-off of fertilizers and pesticides. In 1985 2.2 million tons of fertilizers and 2,800 tons of pesticides were added to agricultural soil.

The fisheries production in Greece increased to 138×10^3 tons in 1986. Some of the spawning and nursery grounds are partially affected by domestic or agricultural pollution.
Aquaculture has become an important economic activity. The total area covered by aquaculture is estimated at 131,700 hectares, with 70,000 hectares planned for further development. The most productive lagoon is threatened by municipal and agricultural discharges.

E.3.   Plans and measures

Environmental protection is included in the Greek Constitution. According to the Law 360/1976 the policy for physical planning and environment is shaped and monitored by a National Council (NCPPE) chaired by the Prime Minister and composed of by the Ministers of National Economy, Finance, Agriculture, Sciences and Culture, Industry and Energy, Social Affairs, Public Works, Regional Planning and Mercantile Marine.
In order to strengthen the implementation and control procedures a new Ministry of the Environment, Regional Planning and Public Works was created by Law 1032 in 1980. This Ministry is the main executive body. In addition there are administrative units in almost every Ministry which have responsibility for environmental protection.

Law 1069 of 1980 has compelled each municipality of more than 10,000 inhabitants to create a municipal enterprise for water supply, wastewater collection and treatment. They have the ability for project financing and fee collection.

Law 1298/90 supports industrial antipollution measures by subsidizing 40% in Athens to 50% near the borders of the cost of treatment plants. The differentiation on a regional basis intends to decentralize production capacity.
Relevant incentives are also planned for the introduction of clean technologies.

Law 1650 of 1986 determines the overall framework for environmental protection. The Law distinguishes 3 industrial categories, in relation to their level of pollution. The standards are defined at a regional level (prefectures). Most prefectoral decrees conform to EC directives on effluent quality.

Treatment plants for 14 coastal cities have been (or are being) constructed. The World Bank and the European Investment Bank (EIB) have given several loans i.e. $15.1 million for the Thessaloniki and Volos sewerage projects. The World Bank has funded 6% of the cost of Greek town (I) projects, while $62 million has been provided by the EIB for the Athens sewerage project. Another source of finance has been the European Fund for Regional Development which until the end of 1984 provided up to 11.5 billion drachmae (in grants) covering 40% of the cost of many environmental projects.
In 1985 the Public Investment Fund gave grants for water supply, sewer networks and treatment plants of 28.9 billion drachmae (which was 10% of the whole investment program). As of 1986 the State increased to 37% its contribution in the total cost of water supply and sanitation, as estimated in the 1983 five-year plan. Up to 1.5 billion ECU of future investments are required for plants under construction or in the planning stage.

Port reception facilities will be constructed in 8 cities. The total cost is estimated to be over 2 billion drachmae. Until the construction of the above facilities has taken place, alternative temporary solutions have been developed. To abate accidental oil release, contingency plans have been adopted. Several coastal antipollution centres have the necessary means and equipment.

Environmental quality can be improved by regular monitoring of coastal waters to guide further action. Integrated river basin management is required to reduce the pollution discharge and to reduce loads of nitrogen, phosphorus and pesticides.

1. GENERAL OVERVIEW OF THE ENVIRONMENTAL QUALITY OF THE MEDITERRANEAN SEA

More than 350 million people in 18 countries live along the 2.96 million km^2 Mediterranean Sea. Approximately 3.3 million tons of organic matter as biochemical oxygen demand (BOD), 350 million tons of suspended solids together with heavy metals and organics (i.e. 0.6 million tons of petroleum hydrocarbons) are annually discharged into the Mediterranean.

The Mediterranean is a rich biotope, with 25,000 species one of the richest in the world. The geographical partitioning or isolation of certain areas (particularly islands) has brought about the appearance of many endemic species, unknown elsewhere.
The Mediterranean is also a key area in the migration path of many birds. The survival of these species depends on the conservation of their necessary biotope.

However, many indicators have shown a declining quality of the waters and the sediments. Eutrophication has greatly increased and toxic algae and red tides occur more frequently. Many investments have already been made for the construction of sewage treatment plants and port reception facilities. Quality improvements in certain regions have been observed (i.e. declining hydrocarbon concentrations). But further effort and investments are necessary to halt the general trend of environmental degradation.

The present study gives an overview of the most critical issues concerning the pollution of the Mediterranean. The information of this overview is obtained from literature sources and interviews with many experts and officials from the Mediterranean countries.
In this first chapter a general overview is presented of

the state of the Mediterranean environment, the environmental pressures, the economic impacts of the pollution, the institutional/legal framework that exists to handle the pollution, the past trends with regard to environmental quality, future projections of the state of the environment, and measures and investment requirements to prevent a further degradation of the Mediterranean environment. The chapters 2, 3, 4 and 5 discuss in detail all these environmental issues for the four Mediterranean European Community (EC) countries: Spain, France, Italy and Greece, respectively.

It must be noted that most studies of the coastal Mediterranean environment focus on polluted sites. This deforms the picture of the situation in the entire Mediterranean by giving too much importance to these sites as compared to the whole area.

1.1. The state of the enviromment

The present section gives an overview of the following components: the sea; the seafood chain; the coastline; inland water; the atmosphere; land use/forest; nature reserves; and historic sites.

1.1.1. Physical characteristics of the Mediterranean Sea

Figure 1.1 shows the Mediterranean area and the regional entities of the Mediterranean, adjacent sea areas and the bordering countries. The numbers given to the regional seas are important, because they are used in almost all of the literature stemming from the United Nations Environment Programme (UNEP) and in much literature based upon UNEP sources.

The average depth of the Mediterranean is 1,500 meters with

a maximum of 5,000 meters (Table 1.1). The sea has a volume of 3.75 million km^3 and an average renewal period of 80 years for its water in the direction of the Atlantic. There are three important sills in the Mediterranean: the Strait of Gibraltar (365 meters depth); the Strait of Sicily (350 meters); and the Dardanelles (100 meters). The first one, with a width of 15 kilometers, separates the Mediterranean from the Atlantic Ocean and makes it an almost enclosed sea; the second divides it into a western and an eastern basin; the third separates it from the Sea of Marmara and the Black Sea. The Adriatic Sea is the shallowest with an average depth of 237 meters (Table 1.2). The summer surface currents show counterclockwise patterns with the major cyclonic gyres, Cyrenaican, Tyrrhenian and Algerian-Provencial, due to the prevailing southeasterly winds. The major effect of these gyres is that the pollution of one country is carried to the bordering country (Figure 1.2).

The input of pollutants into Mediterranean waters has a major impact because this sea is semi-enclosed. Due to lack of significant tides and, as a rule, also lacking strong coastal currents, pollutants discharged in coastal waters are not sufficiently dispersed or diluted. In the open sea, on the other hand, pollutants are generally more dispersed and diluted as a result of stronger currents. Therefore pollutants from ports, industrial outfalls, ships, rivers, land run-off and atmospheric fallout can accumulate in coastal areas.

The water input into the Mediterranean from run-off is 455 x 10^9 m^3 per year while the net input through the Strait of Gibraltar amounts to 2,500 x 10^9 m^3 per year (Figure 1.3). Of the 154 x 10^9 m^3 drawn off and used per year, 72% is used for irrigation, 10% for drinking water and 16% for industries (including thermal power stations) not linked to the domestic supply. The 4 EC countries account for 61.6% of the freshwater supply and 57.5% of the freshwater demand.

On account of the climate and virtual enclosure of the

Mediterranean, the temperature of Mediterranean water, except for the surface layer, does not fall below 13° C, even in the winter (Figure 1.4).

The water of the sea is therefore subject to a very high evaporation (140 cm per year), compensated only for 75% by rainfall and river inflows (UNEP/WG.171/3, 1987). The evaporation rate varies over the entire area. In the Aegean Sea, the Adriatic Sea and the Ligurian Sea, the evaporation balance is zero whereas a high evaporation rate is found in the eastern Mediterranean, the Gulf of Sirte and in the central western Mediterranean.

Consequently, the salinity of the Mediterranean is one of the highest that exists. The average salinity of the Mediterranean is $38°/_{oo}$ (against $35°/_{oo}$ in the Atlantic Ocean) (UNEP/WG.160/11, 1987) and increasing towards the east, where the rainfall is lowest. The overall salinity remains, however, constant due to the inflow of less saline water from the Black Sea and the Atlantic Ocean (without which the level of the Mediterranean would drop by about 1 meter per year). In addition, very saline water flows along the bottom towards the Atlantic.

The drainage area around the Mediterranean is shown in Figure 1.5, with the most important rivers the Ebro, the Rhône and the Po located in the EC countries.

Construction of riverworks and sewage treatment plants can be damaged by earthquakes especially in Greece and southern Italy (Figure 1.6).

1.1.2. Chemical and biological characteristics of the Mediterranean Sea

1.1.2.1. Nutrients and eutrophication

The Mediterranean Sea is characterized by relatively low mean concentrations of nutrients in the surface waters (less than 0.03 µmoles/l PO_4^{3-}-P and less than 0.1 µmoles/l NO_3^--N, as shown in Table 1.3), rather weak vertical circulation and, in general, narrow shelves, which all combine to make this sea one of the most nutrient-depleted basins of the world. This is manifested by an oligotrophic pelagic environment and its extremely transparent water of a bright azure color, which gives the famous beauty of "mare nostrum". The highest algal growth rates have been observed in the Adriatic Sea (alcohol biomass method) due to Po inflow, the Strait of Gibraltar because of Atlantic inflow (^{14}C method), and the Aegean Sea (chlorophyll method) due to Black Sea inputs (Table 1.4).

Eutrophication can generally be understood as an overenrichment of a water body by nutrients, resulting in excessive growth of algae and in depletion of oxygen concentrations after algal die-offs, resulting in extensive fish and benthic organism kills.

In the Emilia-Romagna coastal waters of the Adriatic Sea phosphorus was found to be the limiting factor in 87% of the 6,038 samples collected during 1978-1984. In many Mediterranean waters the ratio of nitrogen to phosphorus is usually above 19 : 1 while the assimilatory optimum is supposed to be 15 : 1, also indicating that phosphorus is usually limiting. Bimodal phosphorus or nitrogen limitations were established by Salamun-Vrabec and Stirn (1982) who noted that nitrogen in the Adriatic Sea is often limiting in the winter. The availability of sufficient dissolved iron may trigger the development of dinoflagellate red tides.

Due to the bimodal nature, Stirn (1988) states that it is not correct to define phosphorus as the "limiting nutrient" and thus the cause of eutrophication. The focus on phosphorus is probably stemming from the fact that phosphorus can be removed more easily by precipitation from waste waters than nitrogen compounds.

Cultural eutrophication usually produces a highly unsteady state, involving severe oscillations (i.e. heavy blooms of algae followed by die-offs, which in turn trigger another bloom on release of nutrients). Thus phosphorus and nitrogen may rapidly replace one another as limiting factors during the course of the transitory oscillations.

Severe eutrophication problems are occurring in 3 major parts of the Mediterranean (North Adriatic, Gulf of Lions, Saronikos Gulf). More localized eutrophication occurs in many coastal lagoons or small bays near sewage discharges. Serious consequences for the marine ecosystem are mass mortalities of benthic organisms, including fish and shellfish. This can seriously affect the tourist industry due to aesthetic problems, loss of water clarity, dead animals on the shore, foul smells, probably toxic shellfish, etc.
Eutrophication can also cause health effects from algal toxins (intra- or extrametabolites), from direct exposure or by accumulation in seafood organisms.

The most eutrophic area is the northwest Adriatic. Red tides and algal blooms occurred in the Adriatic Sea more frequently after the first major red tide of 1969. A major bloom in September 1978 resulted in a huge fish kill in which thousands of tons of dead molluscs and fish were collected along the coastal area of Ravenna.
An extensive bloom in June 1977 of chloroflagellate Chattonella sp. of 3.10^7 cells/l resulted from a high Po flow (up to 8,000 m^3/sec) and overenrichment. During the decomposition of the dead algae, anoxic conditions caused mass mortality of benthic macrofauna in a large area of the central and western North Adriatic. The reoxygenation was restored back to normal in December.
Giant algal mucus, extended up to several meters in length, covered macro-epifauna, killed sponges and entangled their associate fauna during a bloom in the Gulf of Trieste in 1983 (Stachowitsch, 1984). A great number of dead gobiids (Gobius jozo) as well as juvenile Trachinidae and small flatfish were found dead in the sediment surface. A major red tide with Gymnodinium corii in October 1984 extended from the Po River to the Conero promontory for 200 kilome-

ters in length and several kilometers wide, with a maximum of 850 mg/m^3 of total chlorophyll near the Po mouth (Marchetti et al., 1988).

The second major area is the Gulf of Lions along the French coast. In the summer of 1983, Aubert and Aubert (1988) observed a "red tide" dinoflagellate bloom (Gonyaulax polyedra) along the coast of the Alpes-Maritimes that affected the health of frequent bathers.
Nutrient inputs from the Rhône River were 8 to 10 times higher for nitrates and 3 to 10 times higher for phosphates, compared to upwelling deep waters. Nutrients brought by coastal effluents constitute only 10% of the river inputs (Bellan-Santini and Leveau, 1988). The maximum productivity in the Gulf of Lions is estimated at 460 mg C/m^3.day.

The third largest area affected by eutrophication is the Saronikos Gulf. Algal die-offs and oxygen depletions and benthic die-offs have been noted in Elefsis Bay (Zarkanellas, 1979).

1.1.2.2.   Heavy metals in water and sediments

The elements copper, nickel and cadmium are, according to Spivack et al. (1983), substantially more concentrated in the surface waters of the Mediterranean than in similar nutrient-depleted waters in the open ocean. The distribution of these elements shows the same general trend as the salinity (Figure 1.7). Outside the Strait of Gibraltar the concentrations are all lower than in the Mediterranean.
The concentrations of the metals increase towards the east. The distribution of silicate differs from that of the metals. Immediately inside the Strait of Gibraltar the concentration drops by a factor of 3. However, as with the metals there is a general rise in concentration towards the east.
There are a number of possible sources of the fluxes of

these elements, both natural and anthropogenic: river transport of weathering products; mobilization from shelf sediments; both rainfall and fallout from the atmosphere; and municipal as well as industrial waste discharges.

Emelyanov and Shimkus (1986) investigated the distribution of trace elements in the upper bottom-sediment layer of the Mediterranean Sea. The results for chromium (Cr) are given in Figure 1.8. It shows major concentrations from 150 to more than 200 mg/kg in the Adriatic, the Aegean, the Tyrrhenian, the Alboran, and large areas in the northwestern Mediterranean. Figure 1.9 shows that copper (Cu) concentrations in the sediments of more than 100 mg/kg can be found in the central part of the Mediterranean and in the South Levantin. Large areas of these concentrations are also present in the northwestern part of the Mediterranean and in the Tyrrhenian Sea. Nickel (Ni) concentrations of more than 150 mg/kg are found in the Aegean Sea, the North Levantin and in the South Adriatic (Figure 1.10). High zinc (Zn) concentrations of more than 200 mg/kg in the sediments have been detected in the western part of the central Mediterranean and along the coast of Egypt (Figure 1.11).

Cadmium (Cd) concentrations in Mediterranean sediments have average values ranging from 0.1 to 2.3 µg/g dry weight (Table 1.5). Coastal lagoons have often higher concentrations with values from 10 to 32 µg/g dry weight. In the western harbor of Alexandria values range from 7 to 64 µg/g dry weight.

Heavy metals often accumulate in the sediments of the sea by precipitation, sedimentation, ingestion and deposition. The concentrations found in the sediments are for that reason often much higher than those found in the overlying water. In open waters cadmium values range from 5 to 150 ng/l and for coastal waters reach up to 2,000 ng/l (Table 1.6). The background cadmium concentration in the sediments is 0.15 µg/kg dry weight, while in polluted lagoons common values are up to 50 µg/kg dry weight.

Figure 1.12a shows cadmium concentrations of more than 0.3 µg/l in the coastal waters near Málaga, Barcelona, Marseille, Genoa, Gulf of La Spezia, South Adriatic and Venice. Figure 1.12b shows mercury (Hg) concentrations of more than 50 ng/l for the same regions.

In the sediments, the mercury concentration is 1,000 times the concentration in open waters and 500,000 times the concentration in coastal waters (on a weight basis) (Table 1.7).

Typical levels of total mercury (Hg-T) in the open water of the Mediterranean Sea range from 7 to 25 ng Hg-T/l and in the coastal waters up to 100 ng Hg-T/l. In the sediments these values are 0.01 to 0.03 mg Hg/kg dry weight for open, and up to 45 mg Hg/kg dry weight for coastal waters (Table 1.7).

Some polluting substances such as mercury tend to be strongly associated with suspended particulate matter. The values of these substances found in the sediments reflect the accumulated discharge in that specific area. Maserti et al. (1988) found in the neighborhood of a chlor-alkali plant (Rosignano Solvay, Livorno, Italy) a much higher concentration of mercury in the sediment (0.51 mg/kg) than in an unaffected area near Corsica (0.02 mg/kg). At both areas they found comparable mercury concentrations in the seawater (7-9 ng/l and 6-8 ng/l, respectively).

1.1.2.3. Petroleum hydrocarbons

Petroleum hydrocarbons continue to be introduced into the sea, as a consequence of transporting, refining, producing, using of oil and other common activities in the Mediterranean area (Figure 1.13). Especially the coast of the northwestern Mediterranean is potentially the most affected by pollution, due to the presence of refineries, ports, industrial activities and river discharges.

Oil pollution also occurs as the result of infrequent tanker accidents or offshore platform blowouts. However, the persistent fouling of the seas by systematic pollution with oil residues, discharged from the land (industrial and urban wastes) and from ships of all types, is by no means less serious.

If the yearly input of 0.6 million tons of petroleum hydrocarbons was equally distributed in the top 100 meters of the open Mediterranean, an average hydrocarbon concentration of 0.2 µg/l would result (UNEP/IOC, 1988).

Dynamic processes such as current, wave and tidal movements together with wind have a pronounced effect on the behavior of oil in the sea. An average loss to the atmosphere by evaporation of 20-30% of spilled oil inputs and 10% of land-based residues was estimated.

The hydrocarbon concentrations in the surface microlayer are in the range of 0.57 to 13.48 µg/l (n-alkanes by GC), 6.8 to 61 µg/l (UCM: unresolved complex mixture by GC) and 0.25 to 5.15 µg/l (aromatics by UV-fluorescence) as shown in Table 1.8.
In the microlayer the dissolved polycyclic aromatic hydrocarbons (PAH) range from 0.074 to 0.226 µg/l; this represents on the average 55% of the total concentration (Saliot and Marty, 1986). In 1974 Benzhitsky and Polikarpov (1976) found the highest average concentrations of 9.98 µg/m^2 in the Ionian Sea and 7.06 µg/l in the Tyrrhenian Sea. Relevant information was obtained on the accumulation of hydrocarbons in the air-seawater interface. Enrichment factors up to 50 were observed in the surface microlayer (Sicre et al., 1984), although the effect was more evident for the biogenic hydrocarbons.

A tabulation of 465 measurements in the entire Mediterranean gave an average of 2.0 µg/l (Table 1.9). More than 53% of the samples were above 0.4 µg/l with an average of 3.7

µg/l. The average Mediterranean concentration is 100 times the average North Sea concentration. The geographical distribution of the hydrocarbon concentrations in 5°x5° quadrants (Figure 1.14) shows the maximum average value of 3.75 µg/l in the eastern Mediterranean.

In the western Mediterranean, samples collected nearshore frequently show concentrations above 10 µg/l, particularly if they are taken close to industrialized areas or river mouths. Near the coast of Spain and France, concentrations of petrogenic hydrocarbons are in the range of 1.1-4.5 µg/l for the aliphatic fraction and 0.1-0.8 µg/l for the aromatic one.

In the Adriatic, concentrations range from 0.1 µg/l and below in unpolluted zones, to 50 µg/l in polluted areas. In the central Mediterranean the highest values (10-28 µg/l) are reported for areas where bilge discharge may occur.

Values ranging from 0.1 to 2.6 µg/l are reported for Greek coastal waters, while concentrations in the range of 1 to 2.6 mg/l were reported for harbor areas (measured by infrared spectroscopy). However, some studies in areas quite far from major land-based industrial activities, such as offshore in the Aegean Sea, show concentrations exceeding 10 µg/l. High concentrations of dissolved hydrocarbons were found south of Cyprus (25-40 µg/l) and southeast of Crete (10 to above 40 µg/l), probably the result of offshore contamination by ship traffic.

Surface oil slicks were present in more than 10% of the observations throughout the Mediterranean in 1981. These data, compared with MAPMOPP (Marine Pollution Monitoring Pilot Project) results from other seas, provide evidence of a relatively extensive surface pollution in the Mediterranean.

Tar contamination is usually found near areas in the Mediterranean where deballasting of oily waters and release of oily compounds into the sea was permitted until 1978.

Tar formation has been estimated to occur for approximately 30% of the spilled oil in the Mediterranean. Tar residues are often stranded on the shores. In an enclosed sea there is an increased probability of suspended tar contacting shores as compared to open ocean.
The concentrations of pelagic tar in the Mediterranean vary from 0.5 to 130 mg/m^2. Normal values for offshore areas are below 5 mg/m^2, while in nearshore waters concentrations can be much higher (10-100 mg/m^2). Mean quantities of tar on beaches are in the range of 0.2 to 4,388 grams per frontal meter of beach (UNEP/IOC, 1988).

The most tar-contaminated areas in the Mediterranean are in the northeast between Cyprus and Turkey and in the Gulf of Sirte off the coast of Libya, where the mean tar content was 1,847 and 6,859 µg/m^2, respectively. The least polluted areas were the southwestern Mediterranean and the northern Ionian Sea as far east as halfway between Crete and Cyprus, with mean tar concentrations of 236 and 154 µg/m^2, respectively. During the last decade a significant reduction in tar pollution occurred at some of the Mediterranean coastlines such as that of Paphos, Cyprus, where beach-stranded tar quantity declined from a mean of 268 g/m^2 in 1976-1978 to 67 g/m^2 in 1983. The (mean) concentrations of pelagic tar in the western and central Mediterranean showed a sharp decline with time from 37 mg/m^2 in 1969 to 9.7 mg/m^2 in 1974 and 1.2 mg/m^2 in 1987. The reasons for the tar reduction are the adoption of international conventions, technical improvements in the oil shipping industry, and the installation of coastal port reception facilities for oily waste waters (Golik et al., 1988).

Petroleum hydrocarbons in sediments have been studied by several investigators (Table 1.10). The highest concentrations were found in the Côte Bleue near Fos-sur-Mer (952 µg/g).

Hydrocarbon concentrations in sediments along the Spanish

coast outside harbors, oil terminals and river mouths are in the range of 1 to 62 µg/g for aliphatics and 2 to 66 µg/g for aromatics. Levels of pyrolytic-like PAH range from 0.3 to 2.3 µg/g dry weight. Background levels of petrogenic hydrocarbons are 1.2 µg/g for aliphatics and 0.6 µg/g for aromatics.

The sediments in the eastern Mediterranean are less polluted by hydrocarbons. For instance, concentrations of 0.114 to 1.35 µg/g are reported in sediment samples collected off Cyprus at 90 meters depth.
The hydrocarbon concentrations encountered in this area indicate a moderate contamination compared with other sites in which petroleum contamination has been assessed such as 2-1,200 µg/g for the New York Bight and 45-730 µg/g for the California Bight (UNEP/IOC, 1988).

Burns et al. (1985) determined residence times of hydrocarbon residues in the upper 100 meters of the Mediterranean water column and calculated 0.6 years for surface particles, 1.0 year for sediment trap material and 10.5 years for seawater and sediment concentrations. The estimated total hydrocarbon input is 716,000 tons per year of which 49% is transported to the sediments (Table 1.11).

1.1.2.4. Chlorinated hydrocarbons

Sediments are a depository of chlorinated hydrocarbons in the aquatic ecosystem. The levels of these pollutants in the water phase are determined by adsorption and desorption processes, sedimentation and exchange at the sediment-water interface. Figure 1.15 presents summarized data of pollution of the sediments in the Mediterranean Sea caused by insecticide dichlorodiphenyl trichlorethane (DDT), benzene hexachloride (BHC) and polychlorinated biphenyl (PCB). It clearly shows that the pollution by these chemicals is mainly found near the EC countries in the northwestern part of the Mediterranean and in the Tyrrhenian and Aegean Sea

(Picer, 1988).

1.1.3. The seafood chain

1.1.3.1. Benthic communities

The Mediterranean benthic communities are quite variable, due to the diverse environmental conditions found in the Mediterranean (climate, bathymetry, hydrology, substrate type, nutrient level, local contributions of terrestrial matter, etc.).
Different benthic communities are established on soft-bottom or on rocky substrates. The communities established on soft-bottom substrates are determined by the nature of the sediments and the process of sedimentation, which is determined by the quality and quantity of sediment influx, and hydrodynamic conditions. Both of these can be modified directly or indirectly by human activity. The communities established on hard-bottom substrates are conditioned by the nature of the substrate and the hydrodynamic conditions. This has resulted in a tremendous diversity in these communities. The conditioning factors do not evolve independently of each other. Pollution, sedimentary influx and direct human activity all interfere, and could modify the community in a profound manner (Bellan, 1983).

The enormous filtering capacity of the benthic fauna (e.g. certain ascidians filter 10-12 l/h) makes some of them vulnerable. Destruction can occur by excessive eutrophication. Such damage constitutes the removal of an invaluable marine self-purification system. Recovery from an episodic benthic mortality is only gradual. When these events occur with increasing frequency, ecosystem diversity is eventually lost, resulting in simple impoverished communities.

The pollution source may be suppressed by halting the dumping of dredged materials and installing sewage treatment prior to outfall discharge. Recovery of polluted areas occurs after reducing the pollutant inputs, due to dispersion, dilution and cleaning by hydrodynamic action (Bellan, 1983).

1.1.3.2. Shellfish

Shellfish water can be contaminated by domestic sewage from outfalls, discharge from polluted rivers or drainage from polluted areas. Sewage can cause pollution of the shellfish-growing waters by contributing pathogenic microorganisms.
The sanitary quality of shellfish is affected by physical and biological environmental factors.
The physical factors are:
- the location of the sewage discharge relative to the shellfish-growing areas;
- the quantity and concentration of the sewage effluent;
- the dilution and dispersion characteristics of the receiving waters.

The biological factors are:
- the microbiological content of the sewage discharge;
- the viability of the different microorganisms in the sea;
- the biological processes of the shellfish;
- the development of algal toxins (UNEP/WG.160/10, 1987).

A monitoring program exists within the framework of the Mediterranean Pollution Monitoring and Research Program VII, or MED POL VII, measuring the microbiological quality of shellfish and shellfish-growing areas. Table 1.12 shows a summary assessment according to the EC criteria. Between 1977 and 1981 approximately 40% of the stations surveyed had shellfish of satisfactory quality.

1.1.3.3. Heavy metals in benthic and pelagic organisms

Information about concentrations of heavy metals in Mediterranean marine organisms is available, often stemming

from MED POL programs. Table 1.6 and 1.7 give the average values of cadmium and mercury concentrations in plankton in the open sea: 2,000 µg/kg dry weight for cadmium and 15-560 µg/kg dry weight for total mercury (Hg-T). Crustaceans have 50 µg/kg wet weight for cadmium and 20-300 µg/kg wet weight for total mercury (in the northwestern Mediterranean and the Tyrrhenian Sea values are four times higher). Molluscs have 40-2,000 µg/kg wet weight for cadmium and 70-200 µg/kg wet weight for total mercury (in the Tyrrhenian and Adriatic Seas values are four times higher).
Several heavy-metal concentrations are detected in the flesh of mussels (Mytilus sp.) from different regions of the Mediterranean Sea (Table 1.13). The samples from the Turkish Aegean contained the highest concentrations of cadmium (Cd) (6.6-12 µg/g dry weight) whereas those from the Ligurian Sea contained the highest concentrations of copper (Cu) (2.4-154 µg/g dry weight), zinc (Zn) (97-644 µg/g dry weight) and lead (Pb) (2.4-117 µg/g dry weight). The samples from the Gulf of Trieste contained the highest concentrations of mercury (Hg) (0.28-1.3 µg/g dry weight).

The highest cadmium concentration in different organisms was found in the red mullet Mullus surmuletus. The mean value in 218 samples was 140 µg Cd/kg wet weight with a standard deviation of 83 µg/kg (Table 1.14). Cadmium concentrations in selected seafood and other marine species from the Mediterranean showed values as high as 280 µg Cd/kg wet weight in the tuna Thunnus thynnus from Valencia/Castellón (Table 1.15). In the Netherlands a concentration of 50 µg Cd/kg wet weight in fish is the maximum allowable in seafood. High concentrations were also found in the sardines Sardina pilchardus (120 µg Cd/kg wet weight) and the mackerel Scomber scombrus (130 µg Cd/kg wet weight) from the Aegean Sea.

Mercury concentrations in tuna (Thunnus thynnus) increase with body weight and age (Figure 1.16). The mercury concentration per kg of fish from the Mediterranean is much higher than from Atlantic fish. The Mediterranean tuna starts to accumulate mercury much faster and continues to maintain a higher accumulation rate.

High to very high concentrations of mercury in composite samples of molluscs (Table 1.16) were found in the mussel Mytilus galloprovincialis from the northwestern Mediterranean (up to 1,260 µg Hg-T/kg wet weight), and especially from the Adriatic Sea (mean value 870 µg Hg-T/kg with a maximum of 7,000 µg Hg-T/kg wet weight).

Crustaceans with high mercury levels were found in the northwestern Mediterranean (mean value of 1,080 µg Hg-T/kg wet weight) and in the Tyrrhenian Sea (mean value of 1,110 µg Hg-T/kg wet weight) as shown in Table 1.17. Fishes with high mercury levels were found in the same areas (Table 1.18). In the northwestern Mediterranean, striped mullet (Mullus barbatus) reached values up to 5,600 µg Hg-T/kg wet weight, bonito (Sarda sarda) had mean values of 1,000 µg Hg-T/kg and tuna (Thunnus thynnus) had mean values of 1,100 µg Hg-T/kg. In the Tyrrhenian Sea Mullus barbatus had a mean concentration of 1,440 µg Hg-T/kg wet weight and a maximum concentration of 7,050 µg Hg-T/kg.

Table 1.19 gives mercury concentrations in eggs and liver of Mediterranean birds. The higher a bird is placed in the food chain the more mercury accumulation takes place. The mean concentration found in the liver of birds (tertiary consumers) feeding in the highly polluted Lagoon of Santa Gilla near Cagliari was 39,420 µg Hg-T/kg wet weight. The highest concentration in eggs of birds was also found in the Santa Gilla Lagoon (7,760 µg Hg-T/kg wet weight).

1.1.3.4. Petroleum hydrocarbons in biota

Petroleum hydrocarbon levels found in biota are often an order of magnitude greater than those in sediment samples. The mussel (Mytilus) generally has the highest accumulations (Table 1.20).

In a study of petroleum hydrocarbons in molluscs from the

Ebro Delta, mussels (Mytilus galloprovincialis), oysters (Ostrea edulis) and clams (Venus gallina) had an unresolved complex mixture (UCM) of the saturated and aromatic fractions of 100-300 µg/g dry weight. These concentrations were equivalent to those in mussels in the most polluted harbors and bays in California (UNEP/WG.160/11, 1987).

Relatively high concentrations of petroleum hydrocarbons were found in bivalves from the same area (190-215 µg/g dry weight). Pelagic fish showed lower concentrations in tissue samples (less than 10 µg/g dry weight). Mussels, collected from a drilling platform in the Ebro Delta, showed concentrations of up to 20-30 µg/g of n-alkanes.

The average PAH concentrations in muscles and livers of fish collected from Iskenderun Bay (Turkey) were 0.13 and 0.79 µg/g dry weight, respectively.

Hydrocarbons in mussels from Mar piccolo di Taranto (Italy) ranged from 0.5 to 10.1 µg/g with an average of 2.7 µg/g wet weight. If these concentrations were to be transformed into dry weight figures, they become approximately five times higher.

1.1.3.5. Organochlorines in biota

Pesticides can also be found in marine organisms as shown in Table 1.21. The concentrations are very high in the area around Barcelona with values of 244 µg/kg wet weight for DDTs and 1,292 µg/kg wet weight for PCB. The lowest concentrations are found in the waters of Turkey with values of 5.4 µg/kg wet weight for DDTs and 0.4 µg/kg wet weight for PCB.

1.1.3.6. Reptiles and mammals

Mediterranean animals lose part of their habitat with increasing human activity along the coasts. Some large animals are viewed to compete with commercial fisheries. For example, fishermen often try to reduce dolphin stocks. The loggerhead turtle and the monk seal are prime examples of threatened reptiles and mammals, respectively.

The loggerhead turtle

Caretta caretta has lived on this planet for 90 million years but faces extinction during the last few decades. The loggerhead turtle's numbers have reduced dramatically. A small number remains near the Daylan Delta in Turkey and many get caught every year in the nets of Spanish fishermen near the Balearic Islands.

Only 800 remain, as compared to 1,400 twelve to fifteen years ago (MEDWAVES No 8, 1987). They reproduce on the Greek island of Zakynthos, in the Ionian Sea. On Cyprus' northwest beaches about 300 loggerheads breed. Historically their numbers were much larger there.

Urbanization and tourist development of many of the islands' beaches have decreased the population of these turtles, depriving them of their traditional breeding beaches. The Kalamaki area has plans for 30,000 beds at a distance of only 150-200 meters from the nesting beaches. Several years ago turtles laid eggs throughout the length of the large Bay of Laganas, in Zakynthos (14 kilometers). Today they have restricted themselves to Sekania and Daphni (two small and isolated beaches 800 meters long) where 64% of all the nests can be found.

In January 1987 the Greek Ministry of the Environment established a buffer zone and a zone of restricted building activity in the area of Laganas, but this legal protection is valid for only 3 years. The local population does not support the idea of strict regulation for fear of losing their major source of income from tourism (MEDWAVES No 8, 1987).

The monk seal

Monachus monachus is one of the most endangered species in the Mediterranean Sea. The monk seal had a rapid decline in numbers and has already become extinct in several areas where, until some decades ago, it was thriving (southern France, mainland of Italy, Sicily, Cyprus, Israel and

Egypt). Their numbers are decreasing in most other areas. Sizable populations can still be found in Greece, Turkey and Morocco. The species is restricted to very remote areas where disturbance by humans is low. They are mostly found in waters around very inaccessible caves, probably the only territory left where they can rest, breed and suckle their pups.

Local pollution can be an important reason for the disappearance of the monk seal. Hunting is the most important reason. It is easy to approach the seals while they are sleeping in the caves.

Seals in search of food often damage nets and catch. They may also become entangled in fishing gear and drown.

Even the larger colonies (Greece, Turkey and Mauritania) have a high adult mortality and reduced breeding success. None can be considered securely protected.

An effective conservation strategy should include the following elements: protection of the caves used by the monk seal; regular census of population; creation of buffer zones in feeding areas; bans for fishing in monk seal areas; creation of a network of monk seal reserves; adoption of legal measures and coordination of all relevant activities (MEDWAVES No 8, 1987).

1.1.4. The coastline

The coastlines are valuable environments, with great biologic and geographic diversity, due to their morphologic partitioning. The total length of the Mediterranean coastline is estimated at about 46,000 kilometers. The coastline of Greece is the longest with 15,021 kilometers, due to the indented coast and many islands. The flat, usable area accounts for only about 40% of the total coast. All the demographic and economic development is concentrated on these 40%.

The total population in coastal regions in 1985 was 133 million, including an urban population of 82 million. Additional pressures in the coastal area are:

- tourism, with strong increases in the summer along beaches;
- industry and energy, with among other things
  . 58 main loading and unloading oil ports,
  . 50 refineries,
  . 62 thermal power plants;
- fisheries and aquaculture (around 1 million hectares).

The natural areas have decreased in size. The 70 "specially protected areas" currently registered (natural or regional parks, nature reserves, etc.) are not everywhere as protected as they should be and require more vigilance of authorities and favorable public opinion (UNEP/MAP, 1988).

1.1.5.   Inland water

Water is a scarce resource in the Mediterranean basin. Water appears as one of the main factors limiting agricultural development in the south and east of the basin. Its many functions together with limited supply have created conflicts.
The water supply and demand in the Mediterranean drainage basin is shown in Table 1.22. The supply in the four EC countries is large enough to satisfy the demand. In Spain, France and Greece the water drawoffs are 53%, 33% and less than 70% of the stable or stabilized resources, repectively. Only in Italy is this index more than hundred percent (112%). Figure 1.17 shows the relative positive balance of the countries north of the basin, and the insufficient water supply in the southern and eastern countries (UNEP/MAP, 1988).

Irrigated areas in the Mediterranean countries currently cover more than 16 million hectares and have increased over the past 15 years at an average rate of 200,000 hectares per year, entailing additional water requirements in the order to 2,000 million m^3 per year. In 1985 irrigated land in Spain covered 3.217, in France 1.17, in Italy 3 and in

Greece 1.08 million hectares (UNEP/MAP, 1988). Irrigation return flows contain nutrients (eutrophication) and pesticides.

1.1.6. The atmosphere

A major part of the land-based pollution enters the Mediterranean through atmospheric deposition. Pollutants of major concern are heavy metals and metalloids (such as lead, cadmium, mercury, arsenic and tin); petroleum hydrocarbons; chlorinated hydrocarbons; and pathogenic microorganisms. Such airborne elements or substances are present in the particulate (in the aerosol-particle size range) and/or in the gaseous form.

The highest atmospheric concentrations of lead (Pb) and cadmium (Cd) (Table 1.23) were found in Monaco (171 ng Pb/m^3, 4.5 ng Cd/m^3) and Marseille (305 ng Pb/m^3, 5.9 ng Cd/m^3). The mercury concentration in aerosols in the western Mediterranean (0.24 ng/m^3) was 2.4 times higher than the concentration in Sicily and 3.7 times higher than the concentration in the North Atlantic (Table 1.24).

The mercury levels over the Mediterranean open sea, and consequently the air-sea exchanges, are probably only slightly affected by the presence of industrial and human activity. The lowest concentrations were observed above the open sea (2 ng/m^3). Values of 4 ng/m^3 were measured in the atmosphere of rural areas, far from possible anthropogenic and anomalous natural sources. Very high values (200 ng/m^3) were found above deposits of weathered cinnabar resulting from mining activity. The concentration of mercury around the chlor-alkali complex of Rosignano Solvay (Livorno, Italy) was 20 ng/m^3. Values in various urban areas range from 5 to 15 ng/m^3, with lower values for areas along the coast (Ferrara and Maserti, 1988).

The atmospheric lifetime of such materials is generally

long enough (more than 1 day) to allow them to be transported far from their sources (more than 1,000 kilometers). On the other hand these lifetimes are often too short (less than 1 month) to allow uniform horizontal and vertical mixing of pollutant material in the global atmosphere.
Although the importance of this transport path can nowadays be inferred, the data base available is too limited to allow quantitative estimates of atmospheric fluxes to the sea (Buat-Ménard, 1986).

High concentrations of n-alkanes are present in the atmosphere in particulate form above the Ligurian Sea (Table 1.25).

1.1.7.   Land use/forest

Urbanization and infrastructures (roads, railways, etc.) are expected to cover 70,000 km^2 in the year 2025 for the basin as a whole. They reduce the amount of agricultural land and the limited and desirable areas in coastal regions. Table 1.26 shows the utilization of the land in the Mediterranean countries. Arable land in Spain covers 31% of the total surface while an equal amount of land is covered with forest. For France these values are 32% and 27%, respectively, for Italy 31% and 22%. Greece has 23% arable land and 20% forest.

The Mediterranean soil is vulnerable to erosion. Erosion of farmed or grazed slopes is often the result of overgrazing and multiple land use. Overgrazing occurs both on grassland and in forests. The breakup, degradation or disappearance of the protective plant cover enhances erosion. The Figures 1.18 and 1.19 show the distribution of erosion potentials within the Mediterranean watershed. Greece suffers the most from erosion among the EC countries. The southern countries have 10 times more erosion than the northern countries.
Erosion classified as "serious" (from 5 t/ha/year to more than 50 t/ha/year) in agricultural and other land covered

35% of the farmland in the watershed of the Mediterranean in 1980. The estimated amount of lost sediments was 300 million tons per year.

The losses are mainly concentrated in the few months of heavy rainfall. Appropriate control methods include steps and terraces. Soil erosion causes the silting up of downstream reservoirs and dams. The life span of these structures (usually between 50 and 100 years) is shortened in regions experiencing heavy soil erosion (Italy, Turkey, Greece, Spain, the Magreb countries), thus reducing the volume of the stored water reserves. Silt retention behind dams deprives some coastal zones (deltas, beaches) of the sediment inputs, needed to maintain shorelines and prevent regression.

Conversely, sediments reaching the sea worsen coastal pollution, as they provide an adsorption medium for many organic and inorganic pollutants and accumulate in the coastal sediments.

Another threat to the soil is salinization, waterlogging or alkalization of irrigated land, leading to productivity losses and gradual soil sterilization in a span of 5 to 15 years. The cause of salinization is poor management of irrigation and drainage networks and high evaporation. This reduces the capacity of the soil required for agricultural production. Currently the salinization of irrigated Mediterranean land exceeds 30% in some countries (such as Greece and the Nile Valley in Egypt) and reaches 50% in the Euphrates in Syria.

The increase in population, concentrated especially in the coastal area, will lead to an intensification of soil exploitation resulting in soil degradation (UNEP/MAP, 1988).

The ecological function of the forest is to protect land on slopes, regulate water patterns and preserve plant and

animal genetic resources. The production of (fire)wood and the use as grazing land is important for the economy.
The forests are also used as pleasure and leisure areas, and enjoyed for its beauty in the countryside.
In the northern region of the basin the forests are subject to forest fires to such an extent that replanting is already insufficient to compensate for losses.
To the south and east of the Mediterranean, on the other hand, demographic pressures together with overgrazing by domestic animals are the main pressures that threaten the the forest.
To combat these trends, soil management or rehabilitation (replanting) is required (UNEP/MAP, 1988).

1.1.8.   Nature reserves

The increasing human activities in the coastal zone endanger many Mediterranean plants and animals as well as whole areas considered vital for their survival. In 1982 the governments of the Mediterranean countries decided to protect the breeding grounds (Table 1.27) of some 500 species, of which almost 100 are unique to the Mediterranean. Some protected marine and coastal zones are wintering homes for migratory birds; others are sanctuaries to preserve genetic diversity; still others are to safeguard rich but fragile ecosystems. The Mediterranean monk seal, the marine turtle, the Dalmatian pelican and the peregrine falcon, together with some other species, all desperately need the protection of their natural habitat (Medit. Coordinating Unit, 1985).

1.1.9.   Historic sites

The link between the sea and the economic and social activities existed already in ancient times. The sea has been used extensively for fisheries and freight transport. A number of ships never reached their destination. The exact amount of shipwrecks is not known but it must be several hundredthousands. Many shipwrecks have already been discovered in the Mediterranean Sea. Figure 1.20 shows that most wrecks are found in the north of the basin along the busy trade routes.

A ship is a world on its own. So, a shipwreck is an unique opportunity to learn about the everyday life in ancient times (Parker, 1988).

Historic settlements

The preservation and rehabilitation of historic settlements in the coastal area has been the focus of a Mediterranean Action Plan Priority Actions Program project since 1984 which has produced case studies on six types of settlements (The Siren, 1986):

- towns of European, Islamic or mixed cultures, such as Genoa, Fez and Nicosia, that illustrate the problems facing their historic centers;
- historically ethnic town quarters such as le Panier in Marseille, Hafsia in Tunis, Lalrhun in Algiers and Trinidad-Perchel in Málaga;
- small towns of historic and environmental value, such as Safranbolu, Turkey;
- historic traditional villages like those on the island of Santorini, Greece;
- architectural sites such as Basra in Syria; and
- historic settlements where there is a conflict between historic, traditional and architectural values and modern development, such as Yaffa in Israel.

Common problems shared by these settlements are conflicts in institutional authority and lack of coordination between national, regional and local authorities, insufficient information and lack of financial resources. The project stressed the need to solve these common problems and to strengthen public participation through publicity campaigns. Comprehensive planning is required to preserve this heritage and retain its attractiveness to tourism.

More than 100 coastal historic sites are mentioned in Table 1.28 (UNEP/WG.171/3, 1987).

1.2. Pressures on the environment

Many human activities result in the discharge of waste or degradation of the environment. In this section we subdivide these activities in: land-based pollutants; and other pressures.

1.2.1. Land-based pollutants

Table 1.29 gives an estimation of land-based pollutant loads from different sources. The main pollutants are: organic matter; nutrients; specific organics such as pesticides; metals; suspended matter; radioactive substances; and microbial agents.
The broad categories of pollution sources are: domestic sewage; industrial wastes; agricultural run-off; river discharges; radioactive discharges; and airborne emissions.
Table 1.30 gives an overview of the pollutants originating from major industrial sectors considered for the pollution load assessment of Table 1.29.
A large part of the data used is from 1977. Since that year a number of sewage treatment plants and other facilities have been built, but also the population and the industrial production increased. For that reason the used data, although old, are still valid.

1.2.2. Regional contributions to pollution loads

There is a strong imbalance in the value added by manufacturing industries in the Mediterranean basin. Of the more than $200,000 million in 1983, about 80% was generated in the northern countries and only 20% in the southern and eastern countries. Approximately 23% of the total pollution load (99 tons per year) is discharged into the northwestern Mediterranean, from the bordering countries Spain, France and Italy. Up to 35% is discharged into the Adriatic (151 tons per year) with Italy, the former Yugoslavian states and Albania as bordering countries (Table 1.31). The regional entities of the Mediterranean Sea are given in Figure 1.1.

Of the total radioactive tritium, 44% is discharged into the northwestern Mediterranean and 51% into the Adriatic. With regard to other radionuclides a large part (37%) is discharged into the Tyrrhenian Sea, with France and Italy as bordering countries.

The northwestern basin is not only bordered by three industrialized countries but also receives major river inputs from the Ebro (with a river basin area of 86,800 km^2 and a mean flow rate of 615 m^3/sec) and the Rhône (with a river basin area of 90,000 km^2 and a mean flow rate of 1,700 m^3/sec). The Tiber (with a river basin area of 17,170 km^2 and a mean flow rate of 210 m^3/sec) flows into the Tyrrhenian Sea and the Po (with a river basin area of 75,000 km^2 and a mean flow rate of 1,500 m^3/sec) into the Adriatic Sea (Figures 1.21 to 1.23).

The Mediterranean pollution problems can largely be attributed to a limited number of significant point sources along coastlines. Industrial centers, municipalities and several rivers are the major sources in this respect (UNEP/ECE/ UNIDO/FAO/UNESCO/WHO/IAEA, 1984).

1.2.3.   Organic matter

Approximately 60 to 65% of the total organic matter load stems from sources in coastal regions while the remainder is carried by rivers. Organics in domestic sewage are more degradable than industrial and agricultural organics, as reflected in different ratios of biochemical to chemical oxygen demand (BOD/COD). Figure 1.24 shows the heavy BOD load of the countries in the north of the Mediterranean. Italy discharges the highest amount of BOD from among the northern countries (in 1977: 400,000 tons per year) as shown in Table 1.32. This Table also shows that the northwestern Mediterranean receives the highest amount of BOD per kilometer of coastline (336 tons per year). Table 1.33 shows that France discharges twice as much BOD load as Spain and Italy (UNEP/ECE/UNIDO/FAO/UNESCO/WHO/IAEA, 1984).

1.2.4. Nutrients

Phosphorus and nitrogen loads largely derive from river inputs (75 to 80%). The riverborne inputs to the Adriatic are estimated at 79,000 tons of phosphorus per year (8% from natural origin) and 250,000 tons of nitrogen per year (30% from natural origin). Each year 7,000 tons of phosphorus and 150,000 tons of nitrogen are carried by the Rhône into the Mediterranean.

The major contributors in the coastal area are domestic sewage and agricultural run-off. "Point" sources along the coastal zone contribute less than 5% of the total nitrogen/ phosphorus loads (in less developed areas this is 12% for phosphorus and 25% for nitrogen).

Atmospheric inputs contribute for 1-2% of the total phosphorus and 8-18% of the nitrogen load. These values are significantly higher in coastal areas close to large industrial agglomerations. High levels of ammonia evaporate from highly fertilized agricultural land (UNEP/UNESCO/FAO, 1988).

The total phosphorus load of the regions, stemming from several sources, is shown in Figure 1.25. This Figure shows the heavy loads discharged in the northern part of the Mediterranean. The northwestern Mediterranean has a phosphorus discharge per littoral kilometer of 15 tons per year, much higher compared to 2.7 tons per year in Italy and Spain (Table 1.32).

1.2.5. Specific organics

Detergents in wastewater discharges are largely due to household use. One-third of the total load comes from coastal municipalities while the other two-third is con-

tributed by the population living within the river catchment areas. Phenols as well as mineral oil discharges result from industrial activities, with contributions from coastal refineries and oil terminals predominating (UNEP/ECE/UNIDO/FAO/UNESCO/WHO/IAEA, 1984).
The Figures 1.26 and 1.27 show the distribution of the oil terminals and the refineries. In section 1.2.2.3 it was shown that high concentrations of hydrocarbons can be found in the neighborhood of these facilities. However, with technological developments new installations are less polluting than old ones as shown in Table 1.34. The discharge of oil is reduced by a factor of 44 for coastal refineries and by a factor of 60 for inland refineries when the values from before 1960 are compared with the values since 1969.

According to the most probable estimates, the total input of petroleum hydrocarbons in the Mediterranean Sea is about 820,000 tons per year (for 1980 and 1981), as detailed in Table 1.35 (EEC, 1982). Table 1.36 gives estimates of inputs of petroleum hydrocarbons from different sources into the Mediterranean. The total amount of 635,000 tons per year is 185,000 tons per year lower than in Table 1.35. It is estimated that 20% of the spilled oil evaporates, 16% forms tar and strands, 36% sinks to the sediments and 28% disappears through biodegradation and biological uptake (Table 1.37)
The estimated value of 820,000 tons per year of oil spilled in the Mediterranean is 17% of the total oil pollution in the oceans (4.5-5 million tons per year) while the area represents only 0.7% of the total surface of the world's oceans. Its particular characteristics (semi-enclosed sea, with slow surface currents, etc.) makes it more vulnerable to oil pollution damages. The pelagic tar level, for example, is much higher than levels detected in other seas (EEC, 1982).
The inputs of hydrocarbons from the atmosphere range from 1.3 to 26.1 mg/m^2.year (Table 1.38). The wet and dry

deposition of petroleum hydrocarbons (wet: 1.67-16.7 mg/m^2.year, dry: 0.94-9.4 mg/m^2.year) in the Ligurian Sea is twice as high as near the Balearic Islands (wet: 0.84-8.4 mg/m^2.year, dry: 0.47-4.7 mg/m^2.year), calculated for total hydrocarbons.

1.2.6. Metals

The discharge of mercury is largely due to river inputs with only 8% coming from coastal sources. Approximately half of the chromium and two-third of the zinc load are carried by rivers. As shown in Table 1.39 the atmospheric input into the Mediterranean Sea is also very important. The atmospheric input of lead (Pb) (5,000-30,000 tons per year) and zinc (Zn) (4,000-25,000 tons per year) into the Mediterranean is more important than the dissolved-phase input.

Most of the metal loads originating in the coastal zone derive from industrial waste waters and, to a lesser extent, from domestic sewage (UNEP/ECE/UNIDO/FAO/UNESCO/WHO/IAEA, 1984). Figure 1.28 shows the regional inputs of chromium, stressing that the major part of the input is from the northern Mediterranean countries. The Figures 1.29 and 1.30 show the distribution of mineral mines in the Mediterranean. The highest heavy-metal pressures on the environment are situated in the northern countries.

Table 1.40 gives estimates on regional inputs of mercury in the Mediterranean. High amounts of mercury are discharged into the northwestern Mediterranean (33 tons per year) and especially into the smaller Adriatic (41 tons per year). The Aegean Sea also receives a high load (14.3 tons per year).

1.2.7. Suspended matter

Large amounts of suspended solids are carried from the watersheds into the Mediterranean Sea. Approximately 15% stems from surface run-off within the coastal area, while the rest is transported by the major rivers. (The Nile near Aswan: 62-88 million m^3 per year; the Rhône: up to 31.5 million m^3 per year; and the Po: 18-800 million m^3 per year (Emelyanov and Shimkus, 1986)). Comparatively, minor contributions originate from domestic and industrial sources; however they are generally more toxic (UNEP/ECE/ UNIDO/FAO/UNESCO/WHO/IAEA, 1984).

1.2.8. Pesticides

Every year about 90 tons of persistent organochlorine compounds are carried by surface run-off, directly or through rivers, into the Mediterranean Sea. Approximately one-third stems from DDT compounds, BHC compounds and from other organochlorines. Cyclodienes account for only about 5% of the total.
Table 1.41 gives amounts of pesticide consumption by the agricultural sector in 11 countries in the Mediterranean. It shows a very high consumption of pesticides in Italy (UNEP/ECE/UNIDO/FAO/UNESCO/WHO/IAEA, 1984).

1.2.9. Radioactive discharges

Up to 85% of the tritium and 40% of the other radionuclides stem from power plants located on major rivers in the northern part of the Mediterranean basin (Figure 1.31 and Table 1.42) (UNEP/ECE/UNIDO/FAO/UNESCO/WHO/IAEA, 1984).
The Ebro and the Po carry a large amount of tritium to the Mediterranean (1,620 and 1,650 curies per year, respectively, in 1987).

1.2.10. Microbial pollution

Domestic waste waters represent by far the largest microbial sources, resulting in the contamination of many beaches and requiring bathing restrictions. As discussed in section 1.3.3, the load of domestic sewage is highest in the northern Mediterranean countries. For that reason these countries suffer most from microbial pollution (UNEP/ECE/UNIDO/FAO/UNESCO/WHO/IAEA, 1984).

1.2.11. Pressures from economic development

People require space and the necessary infrastructure to live. Governments and supra-national organizations like the EC do much to help the economic development of the Mediterranean countries and regions. But economic development and environmental protection often clash. Long-range economic development should take environmental issues into consideration. Unfortunately this is often not the case, as shown by the "Integrated Mediterranean Programs" (IMPs) of the EC in collaboration with the governments of France, Italy and Greece.

At this moment there are already some 29 IMPs approved by the European Commission. The term "integrated" stands mostly for the integration of money flows. The programs are financed by several funds from the EC, loans of the European Investment Bank and contributions of national, regional and local authorities. The programs contain economically viable projects, but there are some projects that are harmful to the environment.

For instance, the Mikri Prespa National Park in the northwest of Greece is endangered by fish farms located near colonies of internationally recognized species of rare birds. Drained wet grasslands around the lake, removal of 15,000 trees, road construction and intensified agriculture resulted in lower water quality from salinization and from nutrient run-off. The above project was financed with more than $4.8 million from the EC and $1.7 million from the Greek government. The colonies of black ibis and spoonbill are almost completely destroyed and a colony of the bank swallows is irreversibly damaged. By the RAMSAR Convention, the National Park has the status of a protected wetland.

Some IMPs include projects to improve the environment, for instance, the cleanup of polluting sources in the Saronikos Gulf in the IMP-Attica. Some environmental protection measures for the famous Gorge Samaria are included in the IMP-Crete.

There should be a much stronger formal link between the IMPs and MEDSAP (Strategy and Action Plan for the Protection of the Environment in the Mediterranean Region) to conform to environmental quality goals. The European Commission set up MEDSAP to demonstrate that environmental policies are an integral part of the Community policy. Lack of regulatory manpower at the EC level, country level and regionally can endanger incorporation of environmental aspects in financing decisions. The environmental legislation in most Mediterranean countries is only recent and there exists little prior experience with the enforcement of the environmental laws (Logemann, 1988).

1.2.12. Other pressures

In addition to the discharge of wastes, there are other pressures on the Mediterranean Sea environment. One of them is the introduction of exotic species suitable for marine aquaculture. Aquaculture is a significant space occupier (in the nearby future up to 1 million hectares) along the already crowded Mediterranean coastline. In addition, the accidental escape of these non-native species can have a very significant effect on the biota of the Mediterranean. These introductions combined with the migration of Red Sea species through the Suez Canal has already had an impact on commercial catches.

Environmental degradation is also caused by the exploitation of the living resources of the sea. This is most apparent in the case of the destruction of the _Posidonia_ beds by trawling. The damage caused by scratching apparatus to reefs during fishing for red coral is of major concern. Turtle stocks are damaged by floating long lines. Total

destruction is caused by the use of explosives to stun fish (IMO/FAO/UNESCO/WMO/WHO/IAEA/UN/UNEP, 1987).
The resource itself can be damaged by overexploitation. For some species fishing has now reached the point of overfishing. Stergiou (1988) shows that in the Greek coastal waters the effort to catch cephalopods in 1985 was 300% higher than in 1975. However, it only resulted in 200% higher catches. For trawler fishing these values are a 126% higher effort for 10% lower output. The same author comes also to the conclusion that gadoid resources (hake especially) are highly overexploited in Greek waters. This conclusion is confirmed by models based on biological and catch-per-unit-effort data (Stergiou and Panos, 1988).

1.3.   Economic impacts

In this section we will look at the environmental impacts on health; tourism; marine production; and agriculture/forestry and their economic repercussions. Specific information about the impacts for each of the EC countries is found in chapters 2, 3, 4 and 5.

1.3.1.   Health impacts

Marine pollution poses a risk to public health during recreational bathing in microbiologically polluted coastal waters. An increasing awareness about public health risks has led to stepped-up control measures by the relevant country authorities. Countries have agreed on a common minimum standard. Each country shows an increase in the number of acceptable beaches (IMO/FAO/UNESCO/WMO/WHO/IAEA/UN/UNEP, 1987).
Mujeriego et al. (1982) found skin infections in 1979 at 24 beaches in Málaga and Tarragona with a morbidity rate of 2 percent followed by ear and eye infections with a morbidity rate of 1.5 percent.
Fattal and Shuval (1986) found a swimming-associated

morbidity on Israeli beaches in 1983. The morbidity was associated with seawater bacterial densities. There was a significant relationship between the presence of staphylococci, enterococci and E. coli and symptoms of enteric (gastrointestinal) morbidity, particularly among the 0- to 4-year-old children.
Foreign tourists are particularly susceptible to local endemic diseases, since they normally have lower levels of acquired immunity compared to local populations (ICP/CEH 001 m06, 1986).

There exists a strong relationship between the distance of the sample point away from the source of pollution and the number of fecal coliforms and coliphages. The bacteria of fecal origin are accepted as indicators of pollution. The dispersion is by physical phenomena such as superficial currents and winds. The Figures 1.32 and 1.33 relate the concentrations of bacteria and phages to the distance to the pollution source. The bacterial densities decrease much faster with increasing distance than the concentration of phages (Borrego et al., 1982). Thus fecal coliforms may not always be suitable as indicators.

Eutrophication can lead to the development of toxic algae (chloromonads and chrysomonads, species of the dinoflagellates, present in some "red tides"). This can cause human illness following the consumption of molluscs, exposure to aerosols from such waters or direct ingestion. Symptoms range from respiratory problems, high temperature, pains in joints, skin rash, to pruritus and gastroenteritis. Often brief hospitalization is required, occasionally resulting in death (UNEP/UNESCO/FAO, 1988).

Public health problems in connection with the consumption of raw shellfish still arise regularly in the Mediterranean region, though no major epidemic has been reported for a number of years. Table 1.43 gives the characteristics of principal bacterial and viral fish- and shellfishborne diseases in man.

The main shellfish-producing countries in the region currently have strict legislation controlling the quality of both shellfish areas and the shellfish themselves at the postharvest marketing stage, either specifically or as part of general food quality and safety regulations.

Humans are also exposed to cadmium (Cd) from ambient air, drinking water, tobacco and food. On the average, approximately 5% of ingested cadmium is absorbed, but some people absorb cadmium at a much higher rate. The critical concentration in the kidney cortex of 200 µg Cd/g wet weight is reached by intake of about 200 to 400 µg Cd per day, as estimated by a WHO Task Group (1980). Applying a safety factor of only 5, this would result in a tolerable intake of 40 to 80 µg Cd per day (UNEP/WG.160/9, 1987).

Mercury (Hg) poisoning is still the major public health hazard almost exclusively due to ingestion of contaminated seafood (IMO/FAO/UNESCO/WMO/WHO/IAEA/UN/UNEP, 1987). One of the most serious mass poisoning incidents involving congenital disease and death due to exposure to methyl mercury occurred at Minamata and Niigata in Japan (Kuwabara, 1984). Methyl mercury damages the sensory part of the nervous system. According to WHO (1976) it is expected that 5% of an adult population will have overt symptoms when the blood level of total mercury is between 0.2 and 0.5 mg/l. This corresponds to a hair concentration of 50-125 mg Hg/kg or to a long-term daily intake of 3-7 µg Hg/kg body weight in the form of methyl mercury. Both foetus and infants are more sensitive than adults to the toxic effects of methyl mercury (UNEP/WG.160/8, 1987).

Methyl mercury in the hair of Italian fishermen from polluted areas (3.45-25.31 mg/kg) is much higher than in unpolluted areas (0.17-2.1 mg/kg). From the 7 fishermen of the polluted area, 6 belong to the "risk group" (Table 1.44).

1.3.2 Tourism

The Mediterranean is the world's most popular tourist area: out of 250 million international tourists, around 80 to 100 million come to the Mediterranean. Table 1.45 shows that in 1983 Italy earned the most from tourism ($9,034 million), followed by France ($7,226 million), Spain ($6,836 million) and Greece ($1,176 million) (UNEP/MAP, 1988).

The "mare nostrum", with its oligotrophic pelagic environment and its extremely transparent water of a bright azure color, is the major attraction.

Tourism has become an important component in the economic structure of many Mediterranean countries. Tourism has generated substantial social and economic benefits in terms of significant contributions to the national economies of these countries. In many countries tourism is a relatively substantial source of foreign exchange and a considerable source of employment particularly of unskilled labor.

A high environmental quality (e.g. keeping the Mediterranean Sea oligotropic, i.e. preventing eutrophication) is essential for tourism. On the other hand, the quality of the environment is threatened by tourism itself.

Coastal development, tourism and the relevant economic activities which put pressure on the environment should therefore be carefully evaluated and balanced (Gerelli et al., 1987).

For example, the Emilia-Romagna region has the highest density of tourists of Italy, with as many as 40 million of "attendances" in the summer along the 100 kilometers of coast. Starting with 1975, algal blooms became more and more frequent in the waters along the coast. These blooms, mainly consisting of diatoms and dinoflagellates, are extensive enough to have adverse effects on the marine ecosystems and cause serious problems for tourism and considerable damage to fisheries (UNEP/UNESCO/FAO, 1988).

The opinion of recreationists about the aesthetic quality of coastal waters is determined mostly by the seawater transparency and the presence of floatable materials.

Pollution which has an observable effect in reducing the attractiveness of the beaches and the inshore waters is likely to have an adverse effect upon the popularity of the particular locality. Any outbreak of illness attributable to contaminated seawater, either through bathing or the consumption of seafood, tends to receive widespread international publicity with adverse consequences for the area. These are matters about which international tourists are highly sensitive (WHO/UNEP, 1982).

1.3.3. Marine production

In 1980 the Mediterranean yielded 859,000 tons of fish without counting the undeclared, locally sold or consumed catches. Since 1980 the catches have remained quite constant. Table 1.46 shows that Italy is by far the most important fishing country of the Mediterranean. With 352,000 tons per year, it yields more than twice as much as the second most important Spanish fishery. Greece comes as third with 76,000 tons.
The catches in the Mediterranean represent approximately 1% of the world total. In terms of commercial value the percentage is higher, due to the higher quality of the types of fish.
Enrichment of the waters with nutrients from the land leads to higher productivity of the sea. However, there are limits on this increased productivity. Eutrophication and the resulting massive algal die-off and oxygen deficiency often lead to a great loss of fish resources on fishing grounds.

Overfishing also leads to an economic loss. As soon as the fishing effort is higher than the optimum, the profitability decreases at constant prices.

Coastal pollution contributes to reduced stocks; pollution or destruction of marine plants (_Posidonia_ spp.) has reduced the spawning grounds of many fish species (MAP

Brochure, 1986).

Trace amounts of oil components in seawater interfere with the sexual behavior of marine animals. Salmon fry avoid oil concentrations as low as 1.6 mg/l, often occurring around river mouths, and this therefore disrupts their migration patterns. Synergistic effects particularly between aromatic hydrocarbons and trace metals may also occur (UNEP/WG.160/11, 1987).

Table 1.47 shows the long-term effects of cadmium on the marine biota. Even low concentrations of cadmium can reduce the productivity of the sea, and thus reduce the catches.

1.3.4. Agricultural and forestry impacts

Forests have limited economic value in terms of timber production and as requirement for coastal tourism. However, they are important components of the natural environment. They protect the soil from erosion and help to stabilize slopes. They also play a vital role in nutrient cycling and in maintaining soil fertility. In addition, trees utilize carbon dioxide (CO_2) from the atmosphere, thus reducing the greenhouse effect.

In the Mediterranean EC countries the main forest damage results from forest fires. Each year fires damage large areas of land. In the EC 648,648 hectares of land were affected by fires in 1985, including 282,311 hectares of forest (Table 1.48). Spain suffers more than twice as much from fires than the other EC countries. It affects an area of 147,000 hectares of forest and 209,000 hectares of farmland or natural land. The absolute number of fires in Italy (16,903) was much higher than in the next highest country (Spain with 9,770 fires).

The incidence of these fires varies greatly from year to year, as a result of climate differences (see Figure 1.34). The economic loss due to forest fires in Spain is large (Table 1.49). For many regions it varies between 0.5 and 0.9% of the regional agricultural income. The Gallicia region lost 1.4% of its agricultural income as a result of forest fires (CEC, 1987).

Soil erosion also leads to economic losses. It is estimated that 300 million tons of fertile soil per year disappears from agricultural land in the Mediterranean watershed. This can lead to reduced productivity of the land. The remaining soil is often less fertile and needs a higher input of fertilizers. The economic losses in Spain due to soil erosion are large in Andalucía and Castilla-León (Figure 1.35).
As already discussed in section 1.2.7, soil erosion can lead to the silting up of reservoirs and dams, resulting in a shorter life span of these works.
The biggest threat for agricultural land is salinization, resulting from an inadequate drainage of irrigated land. This leads to productivity losses and soil sterilization which may occur in a span of 5 to 15 years.

1.4. <u>The institutional/legal framework</u>

1.4.1. Existing action

1.4.1.1. Mediterranean Action Plan

The Mediterranean is bordered by 18 countries. To protect it, a Mediterranean Action Plan (MAP) was adopted in 1975, in Barcelona, under the auspices of the United Nations Environment Programme (UNEP) (see Figure 1.36).
MAP has been coordinated by a central unit located in Athens since 1982, and is financed by 17 coastal states (only Albania is not involved so far), in accordance with their Gross National Product (GNP), and also by the EC.
The plan consists of three components:
- the scientific component (study of the environment) with the MED POL scientific research and monitoring program;
- the socioeconomic component (integrated planning) including the Blue Plan and the Priority Actions Program (PAP);

- the institutional component, which covers all legal activities of the application of the Convention for the Protection of the Mediterranean Sea against Pollution, signed in Barcelona in 1976, and its protocols, four of which already have been signed:
    - Protocol for the Prevention of Pollution of the Mediterranean Sea by Dumping from Ships and Aircraft which came into effect in 1978,
    - Protocol concerning Cooperation in Combating Pollution of the Mediterranean Sea by Oil and other Harmful Substances in Cases of Emergency which also came into effect in 1978,
    - Protocol for the Protection of the Mediterranean Sea against Pollution from Land-based Sources which came into effect in 1983 (see Tables 1.50 and 1.51) The total cost of applying the protocol is estimated at up to $15 billion within the first 10 to 15 years (Medit. Coordinating Unit, 1985),
    - Protocol concerning Mediterranean specially protected areas which was signed in 1982 (see Table 1.52) (MAP Brochure, 1986).

1.4.1.2. EC Policy on environmental protection

Protection of the marine environment forms an essential part of the EC policy on environmental protection, since the sea is in fact the ultimate receptor of a large number of toxic substances. This policy has a large number of different instruments due to the complexity and the number of problems.

The measures taken can be grouped under four main headings:
- regulatory measures for reducing pollution by requiring limit values for discharge or quality objectives for certain dangerous substances, as:
    - Council Directive 76/160/EEC of December 8, 1975, concerning the quality of bathing water, and
    - Council Directive 79/23/EEC of October 30, 1979, on the quality required of shellfish waters;

- action programs for dealing with large-scale pollution (for example, the action program on the control and reduction in pollution caused by hydrocarbons discharged at sea);
- specific action for the prevention of sea pollution; and
- multicountry cooperation and assistance.

All the measures mentioned above have already and will continue to have beneficial effects on the marine environment. In addition, supplemental actions are anticipated at the EC and international level. These policy measures should be accompanied by effective enforcement actions to ensure that the results are in accordance with the plans. The enforcement should also be strengthened by a public awareness campaign in order to achieve a voluntary and complementary effort to protect the sea. This means that everyone should become more and more aware of the fact that the protection of the sea is very much in their own interest, for themselves, as well as for their children and later generations (Barisich, 1987).

1.4.1.3. Fourth Environmental Action Program

The Community's Fourth Environmental Action Program started in 1987 and it will continue until 1992. This phase of the Community environmental policy integrates the environmental dimension into the Community's economic, industrial, agricultural and social policies. It also formulates preventative policies which strike a balance between economic and social developments on the one hand and the environmental policies on the other. This, in turn, implies a clear commitment to formulating and enforcing the standards needed for the protection of the environment (CEC, 1986).

1.4.1.4. CORINE Program

The CORINE program of the European Economic Community has

three explicit objectives:
- gathering information on the state of the environment under a number of priority themes of communitywide scope;
- coordinating activities in member states and at the international level for the collection of data and the coordination of information;
- ensuring the consistency of information and improving data comparability.

The acronym "CORINE" comes from the following words: COoRdination - INformation - Environment (CORINE Program, 1986).

1.4.1.5. European Regional Development Fund Environmental Program

The EC adopted an European Regional Development Fund Environmental Program. This program is intended to assist in the development and structural adjustment of lagging regions and in the conversion of declining regions by means of promoting the implementation of Community environmental policy.

To this end, the program does not deal with environmental problems and actions as such, but rather with those environmental issues which have an important effect on the economic development of assisted areas in the Community.

In principle, the Community program will address two types of issues: environmental problems which destroy or weaken the basis of economic development; and environmental potentials and opportunities which can be used to foster economic development and job creation (Wettmann et al., 1987). In 1990 the EC adopted the integrated ENVIREG program for that purpose.

1.4.1.6. MEDSAP

The main objective of MEDSAP or MEDSPA (Mediterranean Strategy and Action Plan) is to coordinate and link indi-

vidual projects designed to protect the Mediterranean environment. The measures taken must form a unified and programmatic entity, channel efforts and increase effectiveness.
The program attaches particular importance to increasing public awareness through demonstration projects, information campaigns and training courses.
Actions to protect the Mediterranean should cover the whole region and deploy all instruments established by the international organizations active in this field as well as those established by the Community through cooperative agreements with the other Mediterranean countries.

In 1986 and 1987 MEDSPA participated in 20 projects in the 4 Mediterranean countries of the EC, with a total amount of 1,562,607 ECU.
The priorities of MEDSPA for 1990 in the EC countries are:
- collection, treatment, storage and disposal of effluents and solid wastes from coastal cities with less than 100,000 inhabitants and small islands;
- collection, treatment, storage, recycling and safe disposal of sewage sludge and toxic and hazardous waste;
- treatment of bilge waters (ballasting/deballasting and tank washing) of hydrocarbon residues and other chemical substances which originate from activities at sea;
- protection of sensitive ecosystems in coastal areas.

Incorporating the "environmental" dimension into the Integrated Mediterranean Programs (IMPs) is one of the concerns of MEDSAP: firstly, by checking that development projects are in harmony with the environment and, secondly, by giving more importance to specific measures to protect and improve the environment (MEDSAP, 1987).

1.4.2. Environmental protection standards

Standards are established in order to limit or prevent the exposure to pollutants. They are also a means of achieving certain quality objectives. The EC standards establish limits for pollution that must not be exceeded in the member countries. These are established by means of laws, regulations or administrative procedures, or by mutual agreement or voluntary acceptance.

The country chapters provide information about the directives that are adopted in the Mediterranean EC countries.

The "EC environmental quality standards", with legally binding force, prescribe the levels of pollution not to be exceeded in a given environment or part thereof (see Tables 1.53 to 1.55).

A rational basis is available for establishing marine recreational water criteria as well as guidelines and standards using sound epidemiological data.

Shuval (1986) concluded that bathing in sewage-polluted seawater or freshwater can cause a significant excess of gastrointestinal disease. The disease rates show a high degree of correlation with enterococci and $E.\ coli$ concentrations in the seawater.

The U.S. Environmental Protection Agency recommended the following criteria (Shuval, 1986):
 marine water   :   3 enterococci / 100 ml
 freshwater   :   20 enterococci / 100 ml or
                    77 $E.\ coli$ / 100 ml

For the Mediterranean Sea the UNEP interim environmental quality criteria for bathing waters use fecal coliform bacteria. It states that 50% of the samples must contain less than 100 fecal coliforms per 100 ml, and 90% of the samples must contain less than 1,000 fecal coliforms per 100 ml (Table 1.56) (UNEP/IG.56/5, 1985).

The principal investigators of MED POL VII recommend the following interim criteria (UNEP/WG.160/10, 1987):

- for shellfish, in terms of fecal coliforms (FC) per gram

of shellfish flesh:
- 0 - 2 FC/g   sale permitted
- 3 - 10 FC/g   temporary prohibition of sale
- above 10 FC/g   sale prohibited

- for satisfactory water of shellfish-growing areas, in terms of fecal coliforms (FC) per 100 ml of water:
  - less than 10 FC/100 ml in 80% of the samples
  - less than 100 FC/100 ml in the remaining 20% of the samples (Table 1.57) (or in 99% of all samples).

1.4.3. Discrepancies between law and reality

The effective application of government decisions sometimes deviates from the intentions expressed by adopted laws.

Control of urbanization

There are intentions to direct or curb urbanization. However, lack of enforcement or supervision and numerous dispensations have occurred for housing or tourist establishments. In 20 years nearly 2,000 kilometers of coastline have been sacrificed in this way.

Supervision of manufacturing or transport activities

Regulations for industrial plants and measures regarding maritime transport have been established. However, their implementation can be hampered by inadequate enforcement. Plans and their implementation concerning industrial wastes are clearly inadequate. Their destruction, storage and transport often create an environmental hazard. There is also a discrepancy between regulations and deballasting practices for ships in transit.

Wastewater treatment plants

Land-based pollution requires adequate measures. Wastewater treatment-efficiency data are often not available. Many treatment plants are not adequately operated.

Management
Fragmentation of jurisdiction of state, regional and local authorities hampers projects in the environmental field. Necessary mechanisms and institutions for financing and fee collection are required. Transfrontier issues in the environmental field should be addressed systematically (UNEP/MAP, 1988).

1.5.   Past trends

The first generation environmental policies had the following characteristics:
- pollution abatement is concentrated on the worst pollution;
- each problem is dealt with separately, often under the pressure of important environmental damages (react-and-cure policy);
- a "catching-up" policy is developed aimed at dealing with all pollution sources with "add-on" and often imported technology;
- macroeconomic costs of environmental protection are rather low.

Southern European governments have spent 0.5% of their GNP on environmental protection in recent years. According to Table 1.58, France spent 0.8% of its Gross Domestic Product (GDP) (2,970 million ECU) on environmental protection and Spain 0.2% (175 million ECU). The expenditure of private sector, municipalities and provinces double this percentage. However, a large part of these expenditures is devoted to hydraulic structures. Furthermore, various inefficiencies exist. A recent survey, for example, reveals that more than 50% of the wastewater treatment plants are not adequately operating.

If it is assumed that EC countries enforce standards to ensure that the total volume of emissions does not exceed 1978 levels, the pollution control expenditures should increase sevenfold for Greece, fourfold for Spain and would more than double for Italy. The increase in France is much less. Table 1.59 gives capital expenditures for sewage treatment plants in France, Greece and Italy.

The Organization for Economic Cooperation and Development (OECD) expects that the southern European OECD countries with relatively fast economic growth will be confronted by severe regional pollution problems at the time when they have just established their environmental machinery. These regional environmental problems should be integrated into the economic development (Gerelli et al., 1987).

1.6.   The state of the environment: projections

1.6.1.   The Blue Plan scenarios

The main objective of the Blue Plan scenario is to assess the consequences of various Mediterranean development strategies on the Mediterranean environment. In addition to the environment, the other dimensions are: the management of the different areas; the national development strategies; the Mediterranean population and its movements; and the international economic context. On the basis of these dimensions five coherent sets of hypotheses and scenarios were formulated (Table 1.60).

Two types of scenarios were defined: trend scenarios; and alternative scenarios. Trend scenarios describe evolutionary processes which do not radically depart from trends observed up to now. Alternative scenarios are characterized by a more goal-oriented attitude on the part of Mediterranean governments both at the national and the international level.

The economic engine of the trend scenarios is the expansion of an international market characterized by American-Japanese economic and technological predominance. Individual Mediterranean countries adapt to this "Pacific" predominance.

The three trend scenarios differ from one another according to the extent to which the above pattern develops. In the worse trend scenario, international economic growth remains weak, especially because the dominant partners in the world economy are unable to coordinate their policies in the political, financial and macroeconomic areas. Consequently, the problem of Third World debt in particular would remain acute. On the contrary, in the moderate trend scenario a better coordination of economic policies between the EC, the United States and Japan makes it possible to achieve comparatively stable economic growth. The reference trend scenario is situated between these two contrasting scenarios.

With respect to the environment, the three trend scenarios would lead to governmental efforts parallel to the economic possibilities. In all cases the strongest economic and technological partners also press for the adoption of certain environmental standards.

The main feature of the two alternative scenarios is the greater self-assertion on the part of the Mediterranean countries. This is facilitated by the formation of a multipolar world structure in which western Europe, the United States, the countries of the former U.S.S.R., Japan and perhaps one or two other countries or groups of countries assert themselves. Europe is politically more assertive, but playing a different role in the two alternative scenarios.

The alternative reference scenario A-1 has overall relationships between Mediterranean countries bordering the basin. Together they create a region of harmonious development with optimal exchanges and agreement on migratory flows of inhabitants.

The alternative "integration" scenario A-2 has a more "regional" concept of these relationships. Groups of countries reach economic cooperation, with maximal exchange and migration within these groups, while maintaining

certain barriers between the groups.
In the alternative scenarios, national planning and environmental policies are much more goal-oriented, or incorporated within the decision-making processes and developmental plans. For example, systematic preference is given to nonpolluting or limited-pollution manufacturing processes, biological treatment processes, water-saving irrigation methods and "systemic" solutions rather than purely mechanical ones (UNEP/WG.171/3, 1987).

1.6.2.   Predictions with the scenarios

The scenarios stress the close links which exist between national development and the state of the environment resulting from environmental strategies.
It was possible in a number of cases to identify the negative environmental effects of economic development on the Mediterranean Sea. For example, large agricultural intensification caused a drop in productivity of the sea because of the large quantities of discharged nutrients and the resulting eutrophication and nearshore degradation.

All trend scenarios lead to a direct and indirect deterioration of the environmental condition of the Mediterranean Sea. This holds for both the scenarios with comparatively weak economic growth and strong population pressures and the scenario with strong economic growth but insufficient attention to the protection of the environment. The alternative scenarios correct or even reverse these trends, sometimes at a cost justified by the gravity of these threats. Efforts undertaken to combat urban and industrial pollution start to bear fruit in some countries but often lack consistency.

Most of the threats to the coast have repercussions for the sea such as the impact on preferred reproduction zones for marine species. This threat to the reproduction of species, aggravated by overfishing in the trend scenarios and by the degradation of the marine environment by land-based pollution, justifies the most rigorous environmental protection policies of the alternative scenarios, as recommended by UNEP (UNEP/WG.171/3, 1987).

1.6.2.1. Population growth

The coastal urban population could grow up to 140 million (alternative scenario) and 172 million (worse trend scenario) in the year 2025. In the first scenario the connectivity to urban sewage system would be 80% in the north of the basin and 70% in the south and east. In the worse trend scenario the connectivity in the south and east would not exceed an average of 45%. The volume of waste would be larger in the alternative scenario, but domestic pollution would be 20 to 30% lower.
The strong growth of fertilizer consumption is likely to lead to big increases in the discharge of nitrogen and phosphorus into the sea: a factor of 3 for the worse trend scenario and a little over 2 for the alternative scenario. The decrease from 3 to 2 is due to management of fertilizer application and erosion control.

1.6.2.2. Industrial development

Industrial activities pose a serious and growing environmental threat, perhaps less from major plants whose emissions can be more easily controlled than for the small- and medium-sized plants which could proliferate along the coast. A large number of industrial installations are already located on the coast and this trend will continue. Only vigorous regulation of installations, preventive actions and clean technologies would make it possible to counteract the exponential growth of pollution.
Between 1985 and 2025 the iron and steel industry is likely to level off in the north of the Mediterranean at a little over 30 million tons per year, with the decline in Spain, France and Italy being compensated by increases in Greece

and the former Yugoslavian states. In the south and east of the basin the output in 2025 will be at least 50 million tons per year. Control of pollution in the traditional iron and steel industry is possible, but at a relatively high cost (20 to 25% of the overall investment).

The production of cement in northern countries will remain below 90 million tons per year. The heavy dust emissions of cement works can be reduced from more than 3 kg per ton of cement to less than 0.5 kg per ton by using antipollution devices.

Ammonia production capacity on the southern and western shores (3.5 megatons per year) already exceeds that of the northern shores and should continue to grow. Pollution control is possible depending on the resources allotted to it.

1.6.2.3. Oil pollution

Pollution associated with maritime oil transport should slightly increase, since overall traffic would not grow much. A decrease in the short and medium term is not expected, because of delays in renewing fleets with more up-to-date vessels and lack of reception facilities required for compliance with the MARPOL 1973/1978 obligation concerning oil release in the Mediterranean. The alternative scenarios take into account a more rapid construction of these installations.

In particular, the risk of accidental pollution from chemical products will increase the most with strong economic growth and increased trade (moderate trend scenario) and less in the alternative scenarios with vigorous controls.

Thermal waste is an important problem linked to growing electrification and the location of thermal power stations on the coast. These stations will create localized zones of higher temperatures which will affect fauna and flora, whose sensitivity to temperature in the Mediterranean is known. The alternative scenarios assume coordination at regional levels regarding the number of planned facilities. Concern for the long-term future implies particular attention to the global problem of the warming up of the surface of the Earth by 1.5 to 4.5° C on account of the "greenhouse effect", due to the carbon dioxide release and industrial gases. It could cause the sea level of the Mediterranean to rise by 40 to 120 cm, respectively, with considerable economic and ecological consequences for several cities and deltas. The Nile Delta in particular is already subject to a severe degree of erosion and advancing inundation.

1.7. Measures and investment requirements

In this section we will consider several kinds of pollution or polluting activities and the investment requirements for connected measures.
Table 1.61 gives an estimate of the pollution-control expenditure indices for 1990 for several countries. Assuming that the emissions are kept at the 1978 level, France has to spend 1.6 times more money on environmental protection in the year 1990 as in 1978 (constant 1978 prices). Italy has to multiply its expenditures by a factor 2.4, Spain by 4, and Greece has to spend seven times more money than in 1978.

The European Regional Development Fund Environmental Program has calculated the funding needs for the four southern countries of the Community (see Table 1.62). The financial needs have been subdivided according to the typology of actions. They refer to projects for which investments were formulated. For Greece and Italy, money is especially needed for water projects (wastewater treatment plants and sewer networks); for Spain the main item is air pollution and erosion. It should be noted that not all the regions have submitted detailed proposals and therefore the figures reported give a low estimation of the financial requirement. The total funding need for the 3 Mediterranean countries was 891 million ECU with most of it required in Italy (Gerelli et al., 1987).

1.7.1. Waste

The disposal and utilization of urban and industrial waste poses great problems. A common practice is discharge to the sea, often untreated. Treatment before the discharge can reduce the environmental problems of the sea. The selection of treatment methods for municipal and industrial wastes needs to be based upon treatment efficiency and environmental effectiveness (WHO/UNEP, 1982).

Uncontrolled tipping and illegal dumping are widespread in the Mediterranean area, endangering the environment and public health. Insufficient awareness on the part of operators and the public is the root cause of the problem. The focus of priority action should be:
- urban waste:
    . to create a greater awareness of the importance of proper waste management among the public at large and the responsible authorities,
    . to combat illegal dumping and to create more effective waste collection, waste treatment and waste disposal systems,
    . to clean up polluting disposal sites,
    . to encourage more efficient management of waste, which should be reused as compost, fuel or secondary raw material;
- industrial waste:
    . to encourage "in-plant" treatment; closed loop operations; changes in products and processes,
    . to promote nonpolluting and low-waste technologies,
    . to plan for the establishment of a market in waste and waste exchanges,
    . to encourage the construction of treatment centers, especially for toxic and dangerous wastes,
    . to promote recycling;

- sewage sludge:
  - risk-free use of sewage sludge in agriculture should be promoted,
  - the management, control, safe disposal or marketing of sewage sludge should be organized in conjunction with programs to build treatment stations;
- agricultural wastes:
  - to promote problem-free reuse on land of agricultural waste,
  - to encourage appropriate treatment and/or conditioning units to make this possible (DocTer, 1988)

It is considered worthwhile to provide incentives for setting up combustion/gasification and ethanol-production lines so that agricultural waste can be used to provide energy. Problems exist in collecting and transporting the waste and the unreliability and high cost of the technology for processing it. A plant to treat pig excrement, producing biogas, costs in Italy 250,000-300,000 lire per m^3 digestor (Chiesa, 1988).

In the long term, it is necessary to develop technologies which generate less waste or waste that is easier to reuse or dispose of (DocTer, 1988).

1.7.2. Sewage treatment

A critical factor is the selection of the type of sewage treatment, i.e. secondary (biological treatment) or primary (sedimentation). Of the two, biological treatment is by far the more efficient, generally removing about 50-80% of the organic load compared to about 30% by primary treatment. Table 1.63 and 1.64 give the reductions that can be reached with several sewage treatment methodologies.

In southern Europe the capacity of treatment plants is rising (Sobemap, 1982), but the level of treatment remains low in many areas due to inadequate operation. The majority of the sewage, therefore, still enters the environment in an untreated form (CEC, 1987).

In 1978 the EC countries invested an average of 0.071% of their GDP in treatment plants. This was 0.052% in France, 0.016% in Italy and 0.0005% in Greece. The net pollutant load discharged into the environment decreased by the construction of new sewage treatment plants, using more advanced technologies and by improving the efficiency of existing works. Operational improvements are by far the least costly to implement.

The most recent figures for Italy indicate that 60% of the population is served by sewage treatment plants in the late 1980s (OECD, 1991).

1.7.3. Oil pollution of the Mediterranean Sea

A large part of the oil pollution from tankers is caused by operational spills (ballasting/deballasting and tank washings: 500,000 tons per year) and by bilge and bunkering discharges (tankers and nontankers: 50.000 tons per year). These spills can be reduced by new technologies and procedures. The most important are (EEC, 1982):

Load-on-Top (LOT):
In this procedure all the oily water is brought together in special cargo tanks or slob tanks. Gravity separates the water from the oil in about 2 to 5 days. In the Mediterranean many routes are too short or through waters too rough to allow effective use of this procedure.

Crude Oil Washing (COW):
Crude oil washing is a process whereby part of the crude cargo is used to wash selected tanks during cargo discharge. Crude washing has proven to be more effective than water washing for sludge removal, because crude acts as a solvent, dissolving sludge and sediments and allowing their movement out of the tank. The application of the COW procedure requires considerable tanker modifications and generally has been made or is under execution only for tankers over 70,000-80,000 DWT and of recent construction.

Segregated Ballast Tank (SBT):
The tankers designed according to this technique have a sufficient number of tanks reserved exclusively for ballast (i.e. segregated ballast). The use of segregated ballast tankers reduces the tanker carrying capacity, and for this reason it has been avoided wherever possible by tanker operators.

Reception Facilities:
In the ports and terminals of the Mediterranean Sea, reception facilities subject to the provisions of Regulation 10 and 12 of MARPOL 1973/1978 are necessary. The ballast, storage and treatment capacity of these facilities must be sufficient for ships that are not equipped with SBT or COW techniques.
The MARPOL Convention contains obligations for the ships and also for the ports. The oily water reception facilities of the ports are additional to the oil separating systems on board. They must have the ability to receive bilge waters and other oily waters that the ships have to discharge.

Gravity type separators followed by secondary treatment (induced air flotation, flocculation-flotation, sand filtration or filtration/coalescent processes) can achieve an oil content of 10 to 15 mg/l. The induced air flotation process requires low investment and operating costs and is simple to operate and maintain (EEC, 1982).
Marson (1980) describes a plant built at Sidi Kerir on the Mediterranean coast of Egypt. The plant uses the corrugated plate method of gravity oil removal. The effectiveness of this method is high, as the free-oil content of the ballast water is reduced, after purification, to a value of 1.0 to 1.5 mg/l. The sale of the collected oil contributes to the costs and no oil pollution was detected in the adjacent inshore waters of the Mediterranean.

For the Mediterranean countries of the EC, Renson (1988)

gives the following estimated costs for the construction of reception facilities in ports: France: $1,550,000; Greece: $6,150,000; Italy: $46,350,000; and Spain: $1,370,000.

1.7.4.   Industrial pollution policy

The various Mediterranean countries should have stricter waste disposal practices. Relevant regulations are often insufficient, and enforcement will depend on the quality of the inspectors.
Identification and supervision of hazardous industries is a priority. The Mediterranean countries need better systems for the treatment of toxic wastes.
For a successful implementation of environmentally friendly industrial policies, the collaboration of the enterprises is necessary. Treatment installations are sometimes expensive (steel works) but can also be profitable due to collection of recyclable materials. The development of information exchanges on clean technologies stimulates their adaptation. Incentives for the application of existing industrial treatment techniques should be enlarged and operator training should increase in this area (UNEP/MAP, 1988).

1.7.4.1.   Criteria for heavy metals

A major element of environmental policies is the establishment of criteria for polluting substances. In order to reduce the level of a pollutant in seawater to a concentration that is not harmful to marine organisms and to the ecosystems, it is necessary to limit the release of pollutants into the marine environment both in quantity discharged per unit time and as concentration of the pollutant in the liquid effluent. This requires that the concentration in the marine environment (environmental quality criteria) must be below a concentration which does not cause significant harm ("minimum risk concentration"). The "minimum risk concentration" can be derived from the lowest effective concentration at which the most sensitive marine organism is affected. Reducing the lowest effective concentration by a safety factor, usually a factor of 5 to 10, gives an estimate of the "minimum risk concentration" (UNEP/WG. 160/9, 1987).

As examples, the cases of cadmium (Cd) and mercury (Hg) will be discussed below.

1.7.4.2. Cadmium

The lowest apparent concentration of cadmium which causes an effect to the most sensitive species (marine algae) is given as approximately 1 µg ionic Cd/l. Applying a safety factor of 5 results in a "minimum risk concentration" of 0.2 µg ionic Cd/l.

The effluent can often be diluted by jet diffusers in the sewage outfall by a factor of 1,000 in the mixing zone adjacent to the outfall of the pipeline. Therefore a maximum concentration of 0.2 mg Cd/l in such effluent can be tolerated. A further condition is that the cadmium-containing waste water is released into the open sea providing maximum dilution.

Lagoons or semi-enclosed bays with limited exchange to the open sea cannot be chosen as release sites for new plants. In the case of existing plants, the turnover time and recharge of the water contained in the semi-enclosed water body receiving the discharge must be determined. The effluent concentration and the amount of cadmium discharge in such a semi-enclosed water body must be reduced accordingly. The cadmium concentration in sediments and resident biota in an area at a distance of 5 kilometers from the outfall should not increase more than 50% above background levels which are to be determined before the waste discharges from the new plant begins. In the case of an existing plant, concentration of total cadmium in sediments and biota should decrease with a half-life of 5 years until levels less than 50% above background are reached. The background levels should be determined in an unpolluted, ecologically similar area (UNEP/WG.160/9, 1987).

1.7.4.3. Mercury

The application of a safety factor of 4 to the effective concentration of 20 ng Hg/l of the most sensitive phytoplankton species results in 5 ng Hg/l. This can therefore be taken as the "minimum risk concentration".
Although the seawater concentration of 10 ng Hg-T/l (Table 1.7) may be considered a typical level for uncontaminated Mediterranean seawater this lower effective concentration is not contradictory because the total mercury (Hg-T) concentration in seawater will not be equal to the "bioeffective concentration". Most of the mercury in seawater is generally not in a bioavailable form. The "bioeffective concentration" of coastal seawater can therefore be estimated to be less than 10% of the Hg-T, i.e. 1 ng Hg/l. In addition, not all mercury discharged in waste waters will be in a bioavailable form, because some of the mercury will react with components contained in the waste and in the marine environment.
Using a dilution factor of 1,000 in the mixing zone adjacent to the outfall of a pipeline with jet diffusers, a maximum concentration of 5 µg Hg/l in such effluent could be allowed.
The total amount of mercury per unit time to be discharged should also be limited. Such limits are normally linked to production or mercury processing capacity. In the case of the chlor-alkali electrolysis industries, the application of the best technical means available makes it possible to limit discharges of mercury in new industrial plants to less than 0.5 grams per ton of installed chlorine production capacity using the recycled brine process.
Several studies on the release of mercury from chlor-alkali plants have shown that mercury concentrations in sediments and sensitive biota return to background at a distance of about 20 kilometers from the release point. So, multiple mercury releases into the same marine environment within a

range of 10 kilometers must be considered in the total amount to be released per unit of time.

It was estimated that the half-life of the concentration of mercury in sediments and biota is about 5 years. This means that the mercury concentration after release into the environment decreases by half every 5 years until levels are reached which do not exceed background levels by more than 50% (UNEP/WG.160/8, 1987).

The contracting parties of MED POL adopted a maximum concentration of 50 µg Hg/l for all effluent discharges before dilution into the Mediterranean Sea in April 1987. They agreed to ensure that outfalls for new discharges of mercury into the sea would be designed and constructed in such a way as to achieve a suitable effluent dilution in the mixing zone so that the mercury concentrations in biota and sediments at a distance of 5 kilometers from the outfall structures will not be more than 50% above background levels. Existing discharges will be adjusted to this objective within a period of 10 years (UNEP/WG.160/13, 1987).

1.7.5.   Management of nature sites

With regard to forests the cooperation of the Mediterranean countries is focused on the following areas, involving: upkeep and testing of stable multipurpose farm-forest grazing systems; management and protection of watersheds; multipurpose forest management; combating diseases specific to Mediterranean trees; and combating of fires.

The rich genetic heritage of the Mediterranean regions, of both wild species and cultivated or domestic varieties, is seriously threatened. The application of the Barcelona Convention protocol on "specially protected areas" and the work of the Salambo Regional Activity Center (Tunisia) should enhance the protection of coastal and marine regions. In cooperation with the International Union for the Conservation of Nature and Natural Resources (IUCN), it is essential to extend action to all the Mediterranean-climate land ecosystems in the region, particularly through the expansion and improvement of the biosphere reserve network, the creation of biotope reserves and the adoption of a regional conservation strategy. The conservation of outstanding sites and Mediterranean landscapes should bolster this effort. The participation of local populations in the management of protected areas is essential and offers an opportunity for exchange of experience.

1.7.6.   Management of marine living resources

Overfishing is a major threat for the Mediterranean ecosystem. Optimization of the fishery efforts requires effective international cooperation and information about existing fish stocks (demersal and pelagic species) and their migration and reproduction cycle.
Above all, priority should be given to concerted action between countries exploiting the same resource, and the formulation of measures to limit fishing activity and ensure distribution of this resource, together with supervision of the effective implementation. Legislation on the use of the coastal zone and habitat enhancement (artificial reefs) and national management and development plans for fisheries should be harmonized as far as possible.
The multispecies nature of the fishery in the Mediterranean poses certain difficulties in developing uniform measures for the protection of the resource. Measures that are favorable for some species may not be so for others. Experimental closing and opening of different areas and seasons and license restrictions may be used beneficially for the protection and management of the overexploited trawl and coastal resources in the Mediterranean.
A legal mesh size of 40 mm, required in trawlers as opposed to 28 mm used in Greece, is essential for the protection of demersal resources. This kind of objective requires financial support for the fishermen, i.e. cheap loans for buying new nets, etc. The use of illegal dynamite for fishing shows that enforcement is very difficult.

1.7.7.   Policy recommendations

In order to prevent a further downgrading of the Mediterranean environment, environmental policies should be strengthened with the adoption of a more goal-oriented action focusing in particular on:

- the strengthening of physical planning and programs and, if necessary, the formulation and publication of "national environmental protection plans" with deadlines for achievement of objectives;

- the establishment of coastal "charters", including the active participation of local institutions, socio-professional organizations and the population;

- the study of employment possibilities for young people in the field of environmental protection and the more effective economic use of natural resources;

- the training of environmental experts able to link scientific research, regulations and enforcement and the implementation of new developmental activities;

- raising the awareness of elected representatives and the local authorities and national agencies working in the area of development or physical planning of environmental issues.

Without greater awareness on the part of the public, it will be futile to expect a rapid and smooth evolution towards sustainable development in the Mediterranean basin as a whole. More systematic and consistent efforts would therefore have to be undertaken to:

- develop general education concerning the Mediterranean environment with the help of teaching materials focusing on the realities and problems of the region;

- disseminate objective information to the public about the possibilities and constraints of the local and regional environment in which they live, directed at various age groups and stressing the fact that one generation takes over from another;

- encourage national and local associations for environmental protection and landscape conservation, underscoring in particular tangible action and evidence of results (UNEP/MAP, 1988).

1.7.8.   Overall investment requirements

In order to implement the above formulated actions, an extensive investment needs survey was conducted with officials of the four Mediterranean EC countries. The total required investment is $21,796 million (U.S.) with the most needed for Italy ($17,220 million). The investment needs are in the area of policy actions, institutional actions and technical assistance, and for treatment facilities. The details of the investment needs are given in the appendix.

1.8.   Conclusions

a) The Mediterranean Sea is strongly linked to the economic and social life of the region. The narrow coastal zone is the place where most of the increasing human activities are concentrated. The result is a physical degradation of the environment and pollution by toxic and/or nontoxic substances.

b) Petroleum hydrocarbons in the ecosystem (about 716,000 tons per year) pose a serious environmental threat. Especially the eastern part of the basin is relatively contaminated by pelagic tar from tanker deballasting. The other major substances are heavy metals and nutrients. Natural concentrations of heavy metals in the basin are already high; together with metals from anthropogenic sources they accumulate in several food chains.

c) The major nutrient supplies originate from the Atlantic Ocean influx, river discharges and from wastewater discharges. The areas of enhanced productivity are located within the vicinity of the above sources. In some parts of the Mediterranean (Adriatic Sea, Saronikos Gulf, Lake of Tunis, Gulf of Lions) there are severe eutrophication problems, also affecting the tourist industry.

d) The benthic community suffers from oxygen deficiency as a result of algal bloom die-offs and decomposition. Recovery from an episodic benthic mortality takes place only gradually. When these events occur with increasing frequency, the ecosystem diversity is eventually lost.

e) Shellfish waters can be contaminated by domestic sewage discharges, outflows from polluted rivers and drainage from polluted areas. Domestic sewage contaminates shellfish-growing waters by pathogenic microorganisms.

f) Urbanization and tourist development of many beaches had a large impact on the population of the loggerhead turtle, the monk seal and several other animals, depriving them of their traditional breeding areas.

g) A noticeable part of the pollution entering the Mediterranean derives from sources located on land via atmospheric releases and deposition. Pollutants of major concern are heavy metals and metalloids (such as lead, cadmium, mercury, arsenic and tin); petroleum hydrocarbons; and chlorinated hydrocarbons.

h) The Mediterranean soil is a scarce resource, long exploited and nonrenewable. Activities such as agriculture, urbanization, infrastructure developments, and industrial installations often infringe upon some of the more fragile areas. Increased erosion reduces land productivity and causes downstream flooding and habitat siltation.

i) As pointed out by the Blue Plan scenarios, the expected environmental degradation can only be halted by rigorous environmental protection policies, management, planning, and large investments. There are three main ways of reducing the pollution to the sea: river basin management; sewage treatment plants; and construction and operation of port reception facilities for oily waste waters.
These activities should be strengthened by public awareness campaigns, directed at all sectors of economic life.

j) The reestablishment of an environmental equilibrium, and not necessarily the reconstruction of the former state, should be achieved first. Accumulation of nondegradable pollutants is an irreversible phenomenon. The disappearance of certain phylogenetically and ecologically fragile species leads to an impoverishment of the common natural heritage.

References

AUBERT, M. and AUBERT, J., Effets pathologiques de l' eutrophisation marine. In: Eutrophication in the Mediterranean Sea: receiving capacity and monitoring of long term effects. MAP Technical Reports Series No. 21, pp. 81-89. UNEP, Athens 1988.

BARISICH, A., The protection of the sea: a European Community policy. International Journal of Estuarine and Coastal Law, 1987, 2, no. 1. London.

BELLAN, G., Effects of pollution and man-made modifications on marine benthic communities in the Mediterranean: a review. In: Mediterranean marine ecosystems (eds. Moraitou-Apostolopoulou, M. and Kiortsis, V.), 1983 pp. 163-194. Plenum Press, New York and London.

BELLAN-SANTINI, D. and LEVEAU, M., Eutrophication in the Golfe du Lion. In: Eutrophication in the Mediterranean Sea: receiving capacity and monitoring of long term effects. MAP Technical Reports Series No. 21, pp. 107-121. UNEP, Athens 1988.

BENZHITSKY, A. G. and POLIKARPOV, G. G., Surface hydrocarbons in the Mediterranean. 1976. Okeanologiya 16.

BORREGO, J. J., DE VICENTE, A. and ROMERO, P., Study of the microbiological pollution of a Málaga littoral area. VI ICSEM/IOC/UNEP Workshop on pollution of the Mediterranean. Session IV: relations between marine pollution and public health. Cannes, 2-4 December 1982.

BUAT-MÉNARD, P., Assessing the contribution of atmospheric transport to the total pollution load of the Mediterranean Sea: facts and models. In: Strategies and advanced techniques for marine pollution studies: Mediterranean Sea (eds. Giam, C. S. and Dou, H. J.-M.), 1986 pp. 187-199. Springer, Berlin-Heidelberg.

BURNS, K. A., VILLENEUVE, J.-P. and FOWLER, S. W., Fluxes and residence times of hydrocarbons in the coastal Mediterranean: how important are the biota? Estuarine Coastal Shelf. Sci. 1985, 20, 313-330.

CEC, Community's fourth Environmental Action Program (1987-1992). Draft for a resolution of the Council of the European Communities. COM(86) 485 final. Brussels, October 1986.

CEC, The state of the environment in the European Community 1986. Brussels-Luxembourg 1987.

CHIESA, G., Residue and waste. 1988. Naturopa, Council of the EC, 58E, Strasbourg.

CORINE PROGRAM for gathering, coordinating and ensuring the consistency of information on the state of the environment and natural resources in the European Community. Progress of work. June 1986.

DOCTER, European environmental yearbook 1987. 1988. DocTer International U.K., London.

EEC, Feasibility study on deballasting facilities in the Mediterranean Sea. October 1982.

EMELYANOV, E. M. and SHIMKUS, K. M., Geochemistry and sedimentology of the Mediterranean Sea. 1986. D. Reidel Publ. Comp. Dordrecht.

EUR/ICP/CEH 054, Health effects of methylmercury in the Mediterranean area. Athens 1986. World Health Organization Regional Office for Europe, Copenhagen 1987.

FATTAL, B. and SHUVAL, H. I., Epidemiological research on the relationship between microbial quality of coastal seawater and morbidity among bathers on the Mediterranean Israeli beaches. 1986. Environmental Health Laboratory, Hebrew University-Hadassah Medical School, Jerusalem.

FERRARA, R. and MASERTI, B., Mercury levels in the atmosphere of the Tyrrhenian area in the Mediterranean basin. 1988. Rapports et Procès-Verbaux des Réunions. Commission Internationale pour l'Exploration Scientifique de la mer Méditerranée. Vol. 31, no. 2, p. 31.

FOWLER, S. W., Assessing pollution in the Mediterranean Sea. In: Pollutants and their ecotoxicological significance (ed. Nürnberg, H. W.), 1985 pp. 269-287. John Wiley & Sons Ltd. New York.

GERELLI, E. et al., Regional economic and environmental development, southern European countries. PROGNOS, Basel 1987.

GESAMP (IMCO/FAO/UNESCO/WMO/WHO/IAEA/UN/UNEP Joint Group of Experts on the Scientific Aspects of Marine Pollution), Principles for developing coastal water quality criteria. UNEP Regional Seas Reports and Studies No. 42. UNEP, 1984.

GESAMP, Atmospheric transport of contaminants into the Mediterranean region. UNEP Regional Seas Reports and Studies No. 68. UNEP, 1985.

GOLIK, A., WEBER, K., SALIHOGLU, I., YILMAZ, K. and LOIZIDES, L., Decline in tar pollution in the Mediterranean Sea. 1988. Rapp. Comm. int. Mer Médit., 31, 2 p. 164.

HENRY, P.-M., The Mediterranean: a threatened microcosm. Ambio, 1977, 6, 300-307.

ICP/CEH 001 m06, Correlation between coastal water quality and health effects. Long-term program for pollution monitoring and research in the Mediterranean Sea (MED POL Phase II). WHO Regional Office for Europe, Copenhagen 1986.

IMO/FAO/UNESCO/WMO/WHO/IAEA/UN/UNEP, Review of the state of the Mediterranean marine environment. GESAMP Working Group 26. UNEP, Athens, November 1987.

KUWABARA, S., The legal regime of the protection of the Mediterranean against pollution from land-based sources. 1984. Tycooly International, Dublin.

LE LOURD, P., Oil pollution in the Mediterranean Sea. Ambio, 1977, 6, 317-320.

LOGEMANN, D., Development programs threaten the nature around the Mediterranean Sea (Dutch). 1988. Natuur en Milieu 12, no. 10. Utrecht.

MAP BROCHURE, Overview of the Mediterranean basin (Development and environment). Mediterranean action plan. Blue Plan. First phase. 1986.

MARCHETTI, R., GAGGINO, G. F. and PROVINI, A., Red tides in the northwest Adriatic. In: Eutrophication in the Mediterranean Sea: receiving capacity and monitoring of long term effects. MAP Technical Reports Series No. 21, pp. 133-142. UNEP, Athens 1988.

MARSON, H. W., The successful treatment of tanker ballast water at a Mediterranean site. IAWPR specialized conference on Mediterranean pollution. 1980. Prog. Wat. Tech. 12, no. 1. Pergamon Press.

MASERTI, B. E., FERRARA, R. and PATERNO, P., Posidonia as an indicator of mercury contamination. Mar. Pollut. Bull., 1988, 19, 381-382.

MEDITERRANEAN COORDINATING UNIT and the Program Activity Center for Oceans and Coastal Areas of the United Nations Environment Programme, Mediterranean action plan. Booklet. Athens, September 1985.

MEDSAP, Protection of the environment in particularly threatened areas. Strategy and action plan for the protection of the environment in the Mediterranean region. Doc. MEDSAP/87/A2. Revision 3. 1987.

MEDWAVES, Issue No 4, January-March 1986. Coordinating Unit of the Mediterranean Action Plan, Athens.

MEDWAVES, Issue No 8, I/1987, Coordinating Unit of the Mediterranean Action Plan, Athens.

MUJERIEGO, R., BRAVO, J. M. and FELIU, M. T., Recreation in coastal waters: public health implications. In: Workshop on Marine Pollution of the Mediterranean, pp. 585-594. CIESM. Cannes 1982.

OECD, Environmental indicators. A preliminary set. OECD, Paris 1991.

OSTERBERG, C. and KECKES, S., The state of pollution of the Mediterranean Sea. Ambio, 1977, 6, 321-326.

PARKER, A. J., The Mediterranean Sea, a submarine museum. Unesco Courier, June/July 1988.

PICER, M., Levels and trends of the pollution of chlorinated hydrocarbons in sediments from the Mediterranean Sea. 1988. Rapp. Comm. int. Mer Médit., 31, 2 p. 150.

RENSON, M., Compte rendu de mission, Commission des Communautés européennes. XI/B/1, MR/rk. Brussels, 10 February 1988.

SALAMUN-VRABEC, J. and STIRN, J. Bioassay on productivity potential of typical water masses of the North Adriatic. Univ. Ljubljana, Inst. Biol. Year. Rep. 1982, 148-161.

SALIOT, A. and MARTY, J. C., Strategies of sampling and analysis for studying the hydrocarbon pollution at the water-atmosphere interface. In: Strategies and advanced techniques for marine pollution studies (eds. Giam, C. S. and Dou, H. J.-M.), 1986 pp. 157-186. Springer, Berlin-Heidelberg.

SHUVAL, H. I., Thalassogenic diseases. UNEP Regional Seas Reports and Studies No. 79. UNEP, 1986.

SICRE, M. A., HÔ, R., MARTY, J. C., SCRIBE, P. and SALIOT, A., Non-volatile hydrocarbons at the sea-air interface in the western Mediterranean Sea in 1983. VIIes Journ. Étud. Pollut. CIESM. Lucerne 1984.

SOBEMAP, Cost and efficiency of sewage treatment plants in the European Community. Final report. Volume 11: summary and conclusions. Brussels, September 1982.

SPIVACK, A. J., HUESTED, S. S. and BOYLE, E. A., Copper, nickel and cadmium in the surface waters of the Mediterranean. In: Trace metals in sea water (eds. Wong, C. S., Boyle, E., Bruland, K. W., Burton, J. D. and Goldberg, E. D.), 1983 pp. 505-512. Plenum Press, New York.

STACHOWITSCH, M., Mass mortality in the Gulf of Trieste: the course of community destruction. P.S.Z.N.I. Marine Ecology, 1984, 5, 243-264.

STERGIOU, K. I., Allocation, assessment and management of the cephalopod fishery resources in Greek waters, 1964-1985. 1988. Rapp. Comm. int. Mer Médit., 31, 2 p. 253.

STERGIOU, K. I. and PANOS, Th., Allocation of gadoid fishery in Greek waters, 1964-1985. 1988. Rapp. Comm. int. Mer Médit., 31, 2 p. 281.

STIRN, J., Eutrophication in the Mediterranean Sea. In: Eutrophication in the Mediterranean Sea: receiving capacity and monitoring of long term effects. MAP Technical Reports Series No. 21, pp. 161-187. UNEP, Athens 1988.

THE SIREN news from UNEP's Regional Seas Programme. No. 31. July 1986.

UNEP ACHIEVEMENTS. Brochure. UNEP Headquarters, Nairobi 1987.

UNEP/ECE/UNIDO/FAO/UNESCO/WHO/IAEA, Pollutants from land-based sources in the Mediterranean. UNEP Regional Seas Reports and Studies No. 32. UNEP, 1984.

UNEP/FAO, Baseline studies and monitoring of DDT, PCBs and other chlorinated hydrocarbons in marine organisms (MED POL III). MAP Technical Reports Series No. 3. UNEP, Athens 1986.

UNEP/IG.56/5, Report of the fourth ordinary meeting of the contracting parties to the Convention for the protection of the Mediterranean Sea against pollution and its related protocols. Genoa, 9-13 September 1985.

UNEP/IG.74/4, Proposed 100 coastal historic sites of common interest. UNEP, Athens 1987.

UNEP/IOC, Assessment of the state of pollution of the Mediterranean Sea by petroleum hydrocarbons. MAP Technical Reports Series No. 19. UNEP, Athens 1988.

UNEP/MAP, The Blue Plan, futures of the Mediterranean basin. Executive summary and suggestions for action. Sophia Antipolis 1988.

UNEP/UNESCO/FAO, Eutrophication in the Mediterranean Sea: receiving capacity and monitoring of long term effects. MAP Technical Reports Series No. 21. UNEP, Athens 1988.

UNEP/WG.160/8, Assessment of the state of pollution of the Mediterranean Sea by mercury and mercury compounds and proposed measures. UNEP, Athens 1987.

UNEP/WG.160/9, Assessment of the state of pollution of the Mediterranean Sea by cadmium and cadmium compounds and proposed measures. UNEP, Athens 1987.

UNEP/WG.160/10, Assessment of the state of microbial pollution of shellfish waters in the Mediterranean Sea and proposed measures. UNEP, Athens 1987.

UNEP/WG.160/11, Assessment of the present state of pollution by petroleum hydrocarbons in the Mediterranean Sea. UNEP, Athens 1987.

UNEP/WG.160/13, Report of the fifth meeting of the working group for scientific and technical co-operation for MED POL. Annex VI. UNEP, Athens 1987.

UNEP/WG.163/4. Annex I, Directory of marine and coastal protected areas in the Mediterranean. Draft. UNEP, Athens 1987.

UNEP/WG.171/3, Preliminary report on Blue Plan scenarios. Sophia Antipolis, July 1987.

UNESCO/IOC, Global oil pollution. (Levy, E. M. et al.). UNESCO, Paris 1982.

WETTMANN, R., MOTZ, G. B., ELAND, M., GERELLI, E., CELLERINO, R. and PANELLA, G., Environment and regional development of the less favoured and industrial declining regions. Report to the Community programs division, DG XVI of the Commission of the European Communities. Final report. PROGNOS, Basel, October 1987.

WHO, Environmental health criteria 1, mercury. WHO, Geneva 1976.

WHO TASK GROUP, Recommended health-based limits in occupational exposure to heavy metals. WHO Techn. Rep. Ser. 647. 1980.

WHO/UNEP, Waste discharge into the marine environment. Principles and guidelines for the Mediterranean Action Plan. 1982. Pergamon Press.

ZARKANELLAS, A. J., The effects of pollution-induced oxygen deficiency on the benthos in Elefsis Bay, Greece. Mar. Environ. Res., 1979, 2, 191-207.

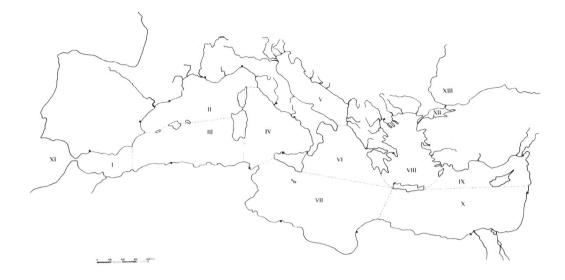

Figure 1.1   Regional entities of the Mediterranen proper and adjacent sea areas
[from: UNEP,RSRS no.32, 1984; p.8]

A. <u>Mediterranean proper</u>

	Regional sea	Bordering countries
I	Alboran	Spain, Morocco, Algeria
II	North-Western	Spain, France, Monaco, Italy
III	South-Western	Spain, Italy, Algeria, Tunisia -
IV	Tyrrhenian	Italy, France, Tunisia
V	Adriatic	Italy, Yugoslavia, Albania
VI	Ionian	Italy, Albania, Greece
VII	Central	Italy, Tunisia, Libya, Malta
VIII	Aegean	Greece, Turkey
IX	North-Levantin	Turkey, Cyprus, Syria, Lebanon
X	South-Levantin	Lebanon, Israel, Egypt, Libya

B. <u>Adjacent areas</u>

	Regional sea	Bordering countries
XI	Atlantic	Spain, Morocco
XII	Sea of Marmara	Turkey
XIII	Black Sea	Turkey, USSR, Rumania, Bulgaria

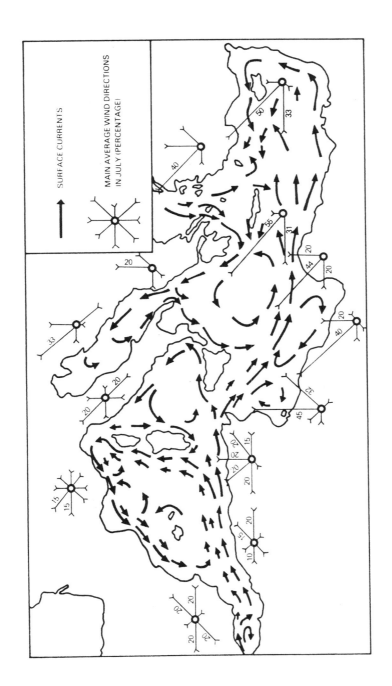

Figure 1.2 Surface currents and main winds in the Mediterranean in summer [AMBIO, Vol 6, No 6, 1977]

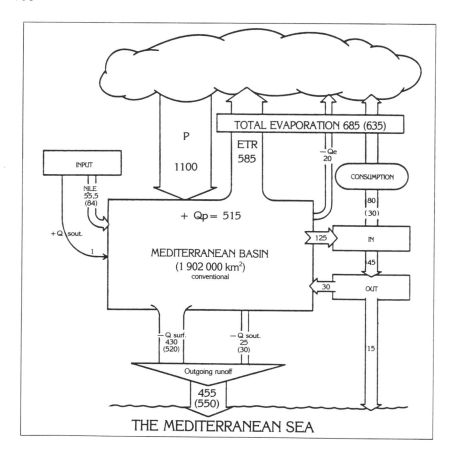

Figure 1.3 Mediterranean Basin water balance
[UNEP/WG.171/3, 1987]

(Current mean flow, figures in brackets = previous flow rates)

Unit : billion m³/annum
P : Rainfall.
ETR : Real evapotranspiration
+Qp : Runoff potential (= effective rainfall)
−Qe : Runoff loss via evaporation
+Q sout : Underground input
−Q surf : Outgoing surface runoff → sea
−Q sout : Outgoing underground runoff → sea

Outgoing runoff : Mediterranean ⟶ Atlantic   = 50 500 x $10^9 m^3$/year
Incoming runoff : Atlantic ⟶ Mediterraean = 53 000 x $10^9 m^3$/year
i.e. a positive annual balance for the Mediterranean Sea of 2 500 x $10^9 m^3$/year

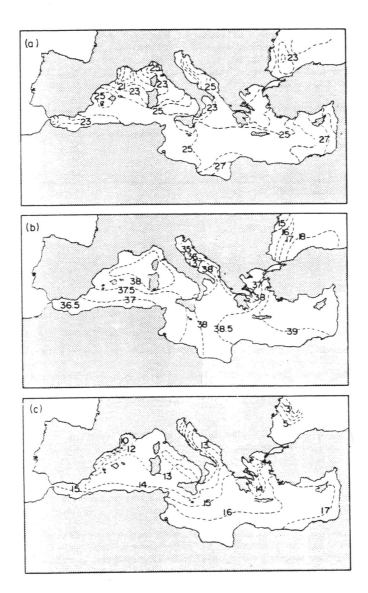

Figure 1.4 (a) Summer isotherms in the Mediterranean sea
(b) Salinity in the Mediterranean sea (average values)
(c) Winter isotherms in the Mediterranean sea
[WHO/UNEP, 1982]

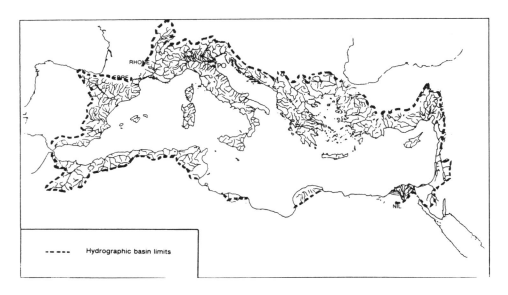

Figure 1.5   Hydrographic Basins   [MAP, 1986]

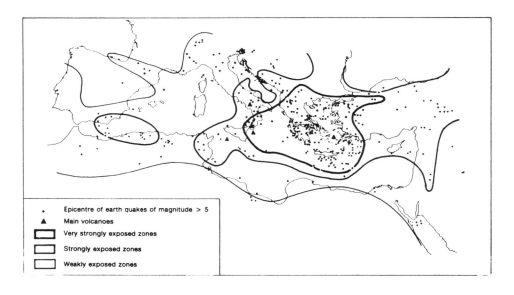

Figure 1.6   Seismicity   [MAP, 1986]

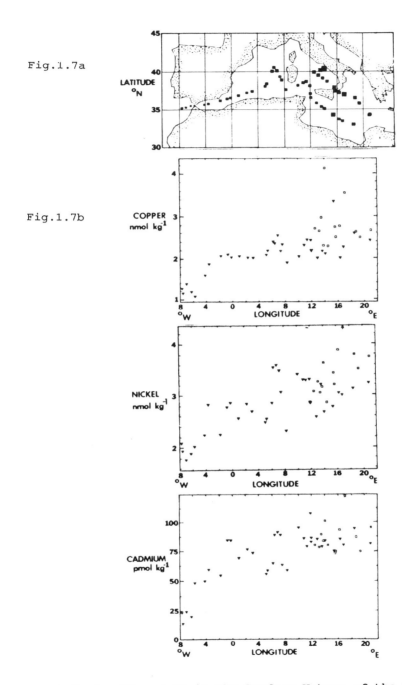

Fig.1.7a

Fig.1.7b

Figure 1.7 Cu, Ni and Cd in the Surface Waters of the Mediterranean
a) concentration of metal plotted vs. latitude
b) relative metal concentration. Length of symbol perimeter proportional to concentration   [Spivack et al., 1983]

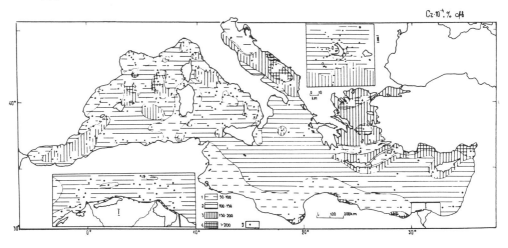

Figure 1.8

The distribution of Cr, in ppm, in the upper bottom sediment layer of the Mediterranean Sea (calculated on a CFB): 1-4 - content in ppm, 5 - the monitoring stations are shown with a dot. I - distribution of Cr in the upper sediment layer of the outer delta of the Nile; II - the same for the Santorini Volcano area. The shaded lines are the same for insets I and II, as for the main map.   [Emelyanov and Shimkus, 1986]

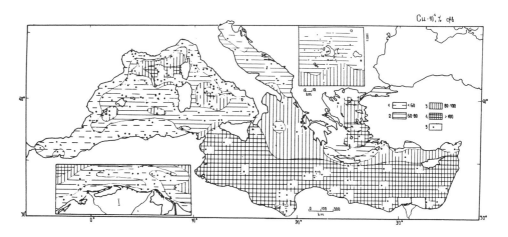

Figure 1.9

The distribution of Cu in the upper bottom sediment layer of the Mediterranean Sea (calculated on a CFB) in ppm: 1-4 - content in ppm, 5 - the monitoring stations are shown with a dot. (Insets: I and II are the same as for Cr).   [Emelyanov and Shimkus, 1986]

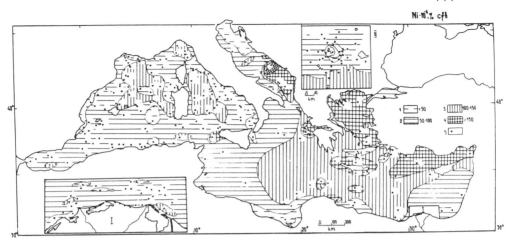

Figure 1.10

The distribution of Ni in the upper sediment layer of the Mediteranean Sea (calculated on CFB), in ppm; 1-4 - content in ppm, 5 - the monitoring stations are shown with a dot. Conventional signs are the same, as for Cr.   [Emelyanov and Shimkus, 1986]

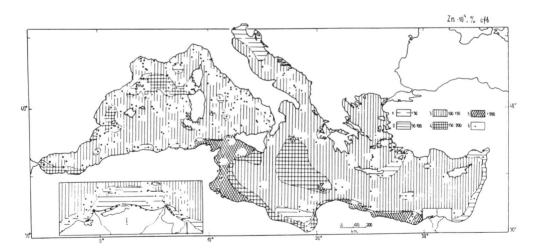

Figure 1.11

The distribution of Zn in the upper bottom sediment layer of the Mediterranean Sea (calculated on a CFB) in ppm: 1-5 - content in ppm, 6 - the monitoring stations are shown with a dot. Inset I is the same as for Cr.   [Emelyanov and Shimkus, 1986]

Fig.1.12a

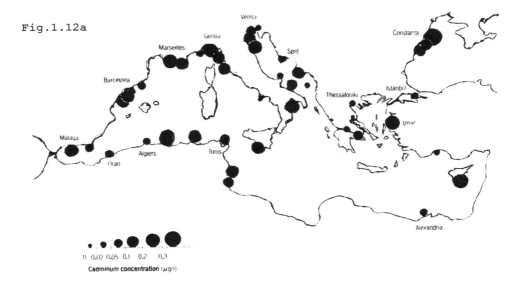

Fig.1.12b

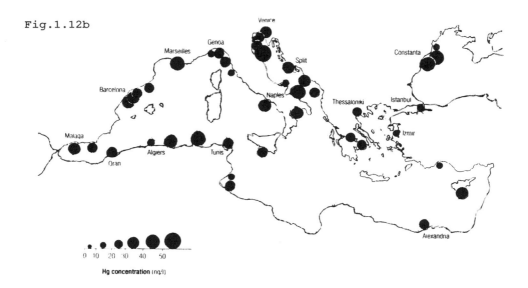

Figure 1.12   a) Cadmium concentrations in coastal waters
               of the Mediterranean [from: CEC, 1987]

               b) Mercury concentrations in coastal waters
               of the Mediterranean [from: CEC, 1987]

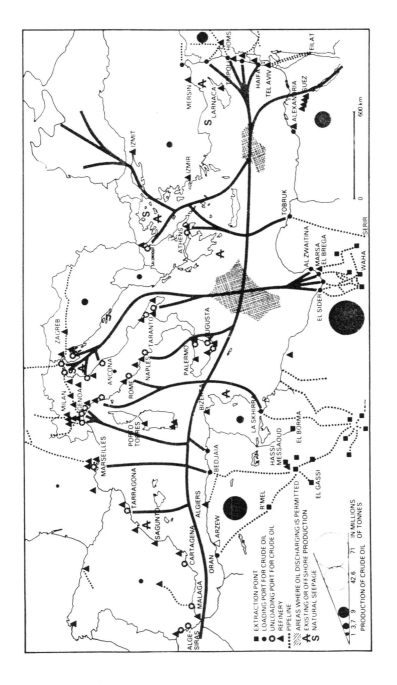

Figure 1.13 Location of the different sources of petroleum pollution in the Mediterranean Sea [AMBIO, Vol 6, No 6, 1977]

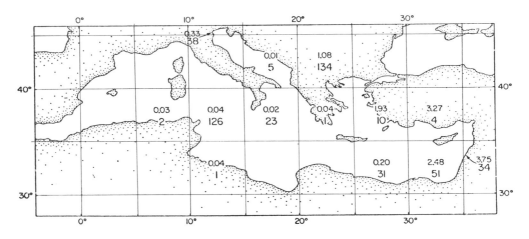

Figure 1.14　　Dissolved + dispersed petroleum residue concentrations for 5°x5° squares of latitude and longitude in the Mediterranean Sea (Upper value in ug/l), lower value number of samples) as tabulated by Unesco, 1982, for 459 samples　　[UNESCO/IOC, 1982]

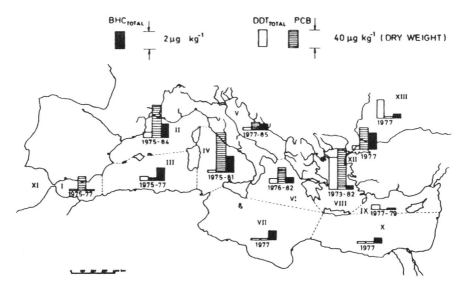

Figure 1.15　　DDT_{total}, BHC_{total} and PCB in sediments from the Mediterranean Sea sampled from 1973 to 1985　　[Picer, 1988]

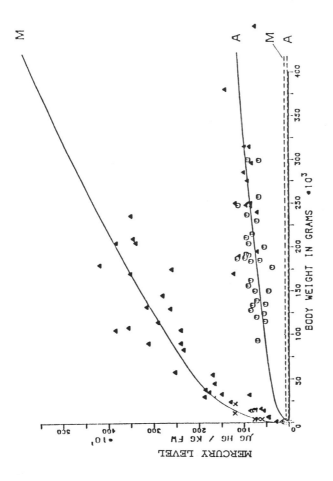

Figure 1.16 Total mercury concentrations in _Thunnus thynnus_ from the Strait of Gribaltar (☉), Tyrrhenian Sea (▲) and Spanish coast (×). The continuous line shows total Hg concentrations calculated by a model; intermittent line shows inorganic Hg concentration calculated by a model. M: prediction for Mediterranean tuna, A: prediction for Atlantic tuna. [UNEP/WG. 160/8, 1987]

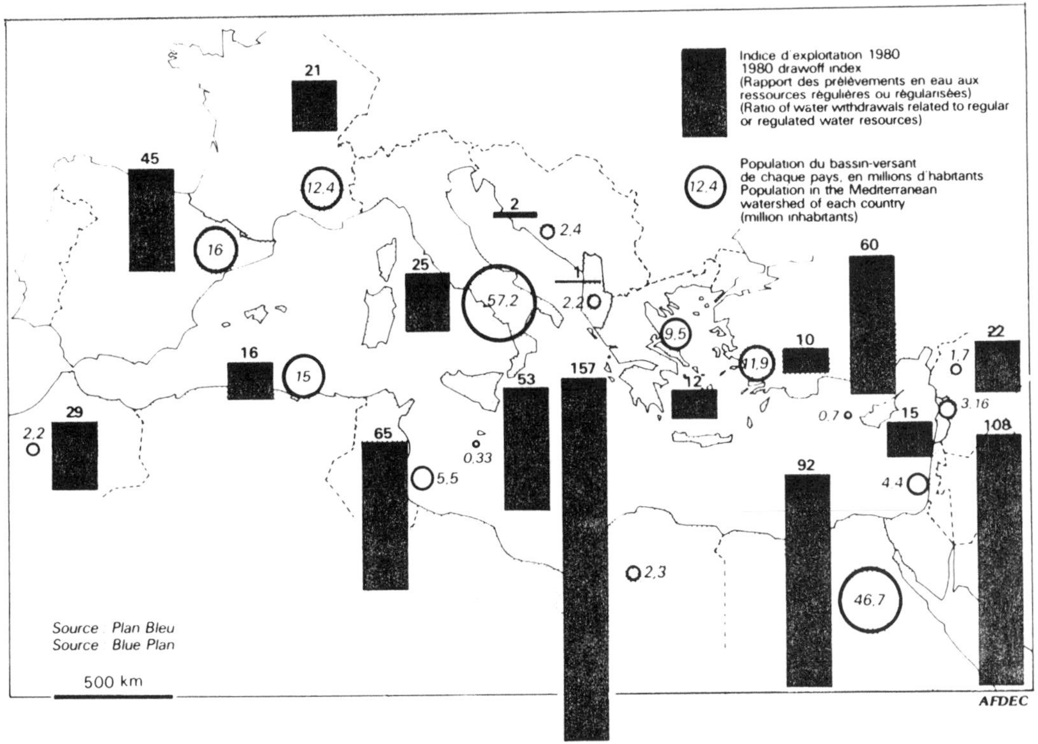

Un fort indice d'exploitation des ressources en eau nécessite le recours à des ressources non conventionnelles : nappes fossiles non renouvelées, recyclage d'eaux usées, multiples utilisations de l'eau, réduction des pertes, dessalement de l'eau de mer...
A high water drawoff index calls for utilization of non conventional resources : non-renewable fossile resources, recycling of used water, multiple water uses, reduction of losses and seawater desalination.

Figure 1.17 Water drawoff in the Mediterranean watershed (annual withdrawal in % of resources) [UNEP/MAP, 1988]

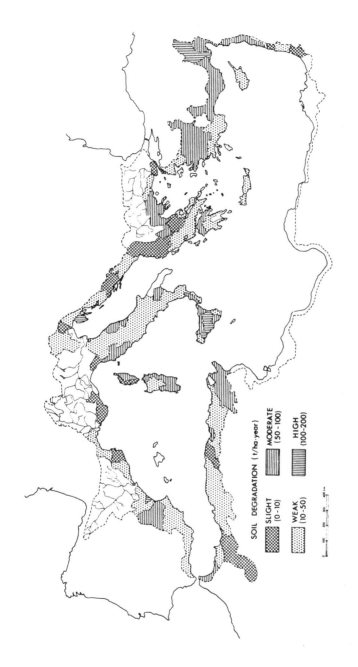

Figure 1.18 Distribution of erosion potentials within the Mediterranean watershed basin   [from: UNEP, RSRS no.32, 1984]

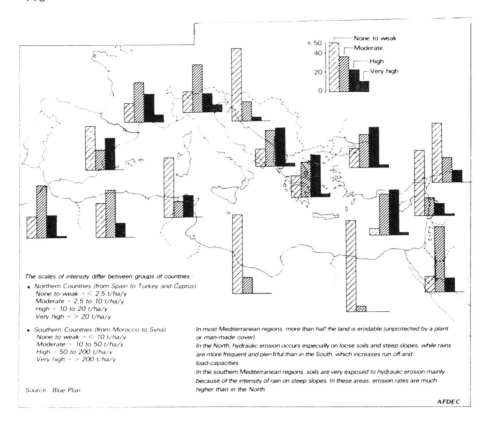

Figure 1.19 Intensity of hydraulic erosion of soil in the Mediterranean watershed, 1980 (as % of total watershed area) [UNEP/MAP, 1988]

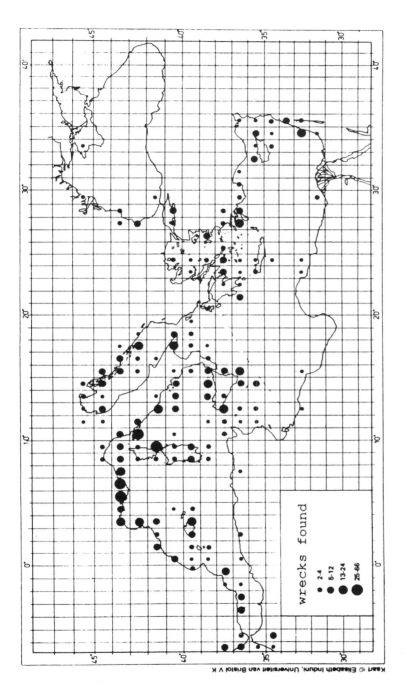

Figure 1.20   Historical shipwrecks   [Parker, 1988]

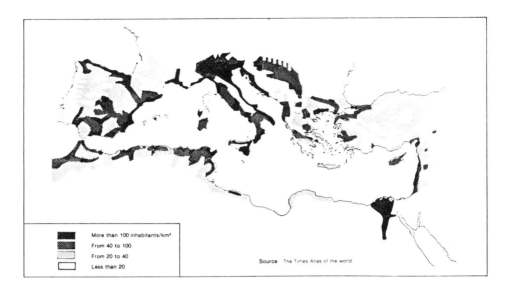

Figure 1.21   Population density   [MAP, 1986]

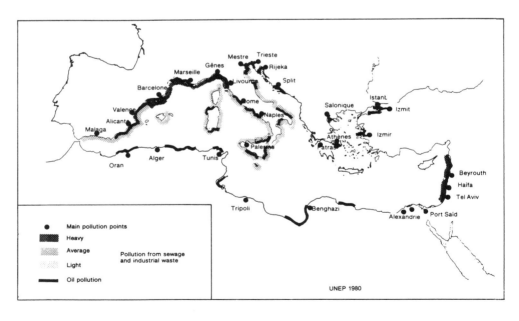

Figure 1.22   Land-based pollutions   [MAP, 1986]

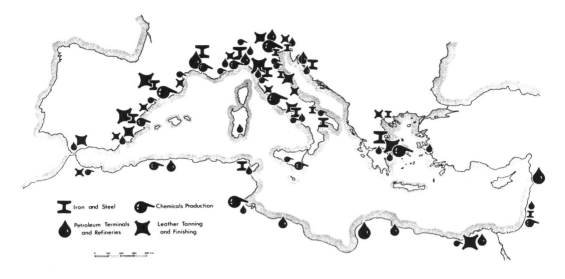

Figure 1.23　　Location of major industrial areas along the Mediterranean coastline
[from: UNEP,RSRS no.32, 1984]

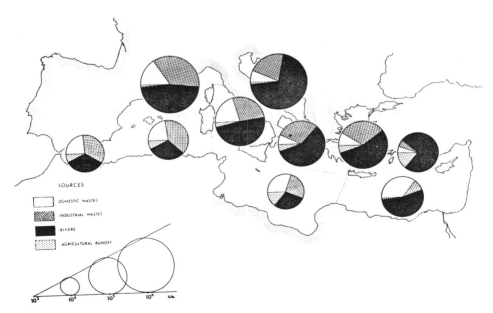

Figure 1.24　　Distribution of annual total loads of BOD that originates from different sources of wastes in the Mediterranean (in tons/ year)
[from: UNEP,RSRS no.32, 1984]

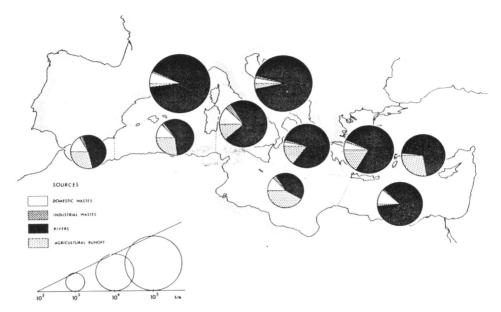

Figure 1.25   Distribution of annual total loads of phosphorus that originates from different sources of wastes in the Mediterranean (in tons/year)   [from: UNEP,RSRS no.32, 1984]

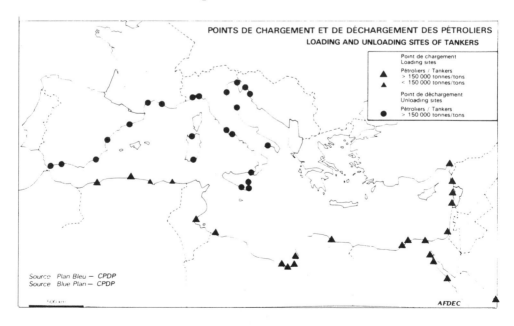

Figure 1.26   Oil terminals   [UNEP/MAP, 1988]

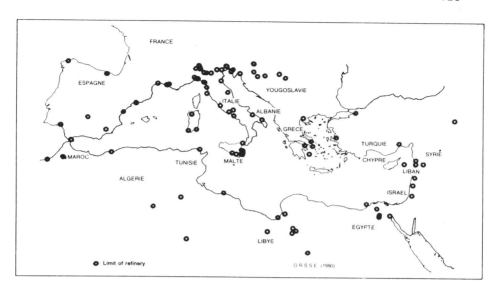

Figure 1.27   Refineries   [MAP, 1986]

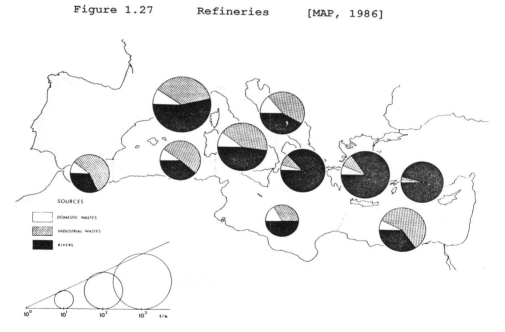

Figure 1.28 Distribution of annual total loads of chromium that originate from different sources of wastes in the Mediterranean (in tons/year) [from: UNEP, RSRS no.32, 1984]

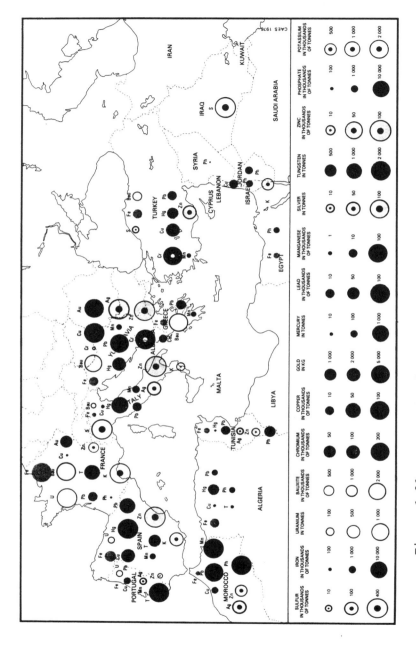

Figure 1.29 Production of minerals in the Mediterranean basin [AMBIO, Vol 6, No 6, 1977]

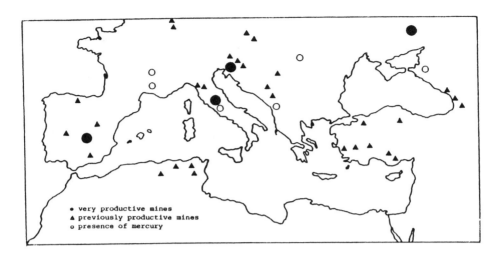

Figure 1.30 Locations of active and inactive mercury mines in the Mediterranean   [UNEP/WG.160/8, 1987]

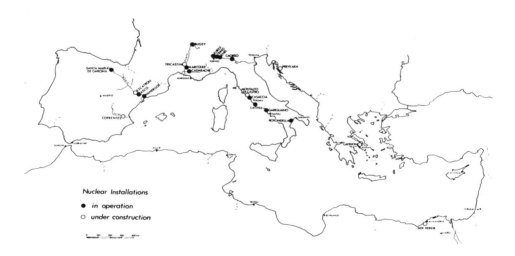

Figure 1.31 Location of nuclear power plants in the Mediterranean basin
[from: UNEP, RSRS no.32, 1984]

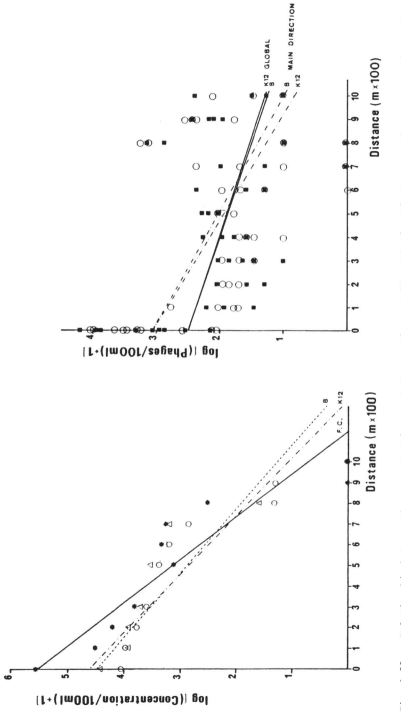

Fig. 1.32 Relationship between the concentrations and the distance to the pollution source. Synchronic samples.
✱——✱ Faecal Coliforms: y = 5.5288 - 0.0048 x. r = 0.9141.
○----○ Coliphages B: y = 4.4964 - 0.0033 x. r = 0.7638.
△—·—△ Coliphages K12: y = 4.5636 - 0.0034 x. r = 0.7580.
[Borrego et al, 1982]

Figure 1.33 Relationship between the concentrations of coliphages B and K12 and the distance from the pollution source. Asynchronic samples.
○——○ Coliphages B
■——■ Coliphages K12
[Borrego et al, 1982]

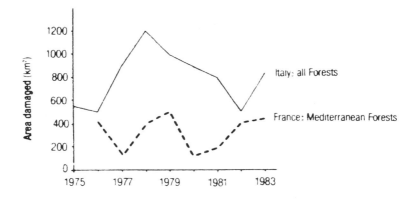

Figure 1.34   Extent of forest damage by fires in the Mediterranean region, 1975-1983 [from: Gerelli, 1987]

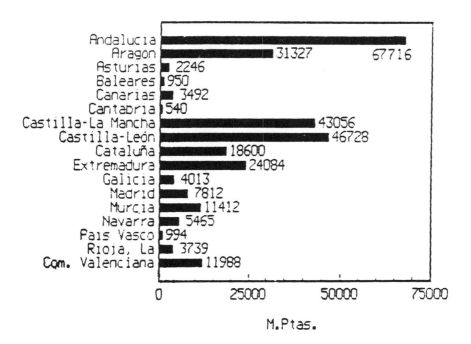

Figure 1.35   Economic losses due to soil erosion in Spain   [from: Gerelli, 1987; p.130]

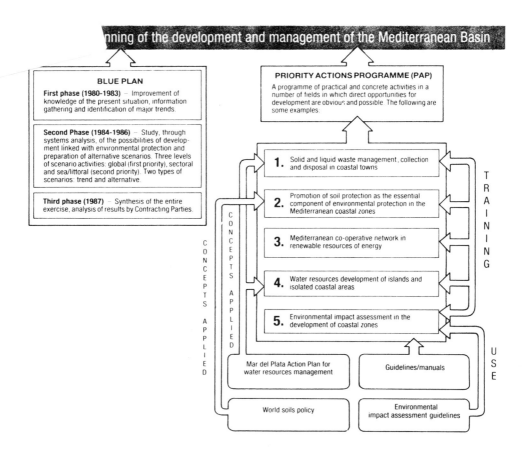

Figure 1.36   The Mediterranean Action Plan
[UNEP Achievements, 1987]

Table 1.1 General statistics of the Mediterranen, Black and Caspian Seas and the Atlantic Ocean [Emelyanov and Shimkus, 1986]

Geographic data	Mediterranean Sea	Black Sea	Caspian Sea	Atlantic Ocean
Sea area, 10^3 km^2	2965	413.5	154.1	93363
Sea volume, 10^3 km^3	3754	537	26.35	312576
Average depth, m	1500	-	170	3926
Maximum depth, m	5121	2215	768	9219
Coast-line length, m	-	4200	2152	-
Drainage system area (B) 10^3 km^2	3700	1864	-	34340
B-to-L* ratio	1.3	3.7	4.3	0.37

* L - basin area

Table 1.2 Morphometric data of the Mediterranean Sea [Emelyanov and Shimkus, 1986]

Basin or Sea	Area $:10^3$ km^2	Volume of water 10^3 km^3	Average depth m	Maximum depth m
1. Algerian-Provencal Basin	595.8	1016.9	1707	2887
2. Tyrrhenian Sea	258.8	407.3	1574	3719
The Western Basin	854.6	1424.2	1667	3719
3. African-Sicilian Sill	209.8	65.0	310	1730
The Central Basin	756.1	1302.8	1723	5120
4. Sea of Marmara	10.9	3.5	324	1264
5. Aegean Sea	188.5	90.0	477	2594
6. Adriatic Sea	139.3	33.0	237	1224
The Eastern Basin	552.1	988.6	1791	4486
The Mediterranean Sea	2501.5	3842.0	1536	5121

Table 1.3 Eutrophication of the Mediterranean [UNEP/UNESCO/FAO, Map TRS No. 21, 1988]

	PO_4-P (umol/l)	NO_3-N (umol/l)	NH_4-N (umol/l)	NO_2-N (umol/l)	chlorophyll a (ug/l)	phytoplankton (cells-/l)	C production (g/m^2Y)
mean concentration	< 0.03	< 0.1	< 0.5	< 0.1	< 0.5	10^4-10^5	64
eutrophied coastal waters	> 0.15	> 0.2	> 1.0	> 0.2			
higly eutrophied waters	> 0.30	> 0.5	> 2.5	> 0.5			
directly polluted waters		> 35.0	> 20.0				
Saronikos Bay					0.5-1.0	10^5-10^7	
Western part of the Saronikos Bay		1.0	5.0	1.0			
Elefsis Bay					0.5-1.0	10^4-10^7	
Lake of Tunis					> 100	10^7	
North Adriatic	0.12				> 1.0	10^7	
Northwest Adriatic	0.30				> 50	10^7-10^8	
Gulf of Trieste	0.15	2.5	3.3		1.26		
Kastela Bay						10^7	244
Etang de Berre						10^6-10^9	

Table 1.4 Primary production of phytoplankton in the Mediterranean Sea [Emelyanov and Shimkus, 1986]

Sea areas	in g/cm^2/day (0-50m)		calculating on a biomass basis of mg/m^3/day for 0-10m layer (from Kondratjeva, 1964)		
	on chlorophyll basis	on ^{14}C	on chlorophyll basis	on ^{14}C	by alcohol method
Aegean Sea	0.448	-	378	-	59
Levant Sea	-	-	-	25-55	21
Ionian Sea	0.273	-	231	-	16-67
Adriatic Sea	0.323	-	271	-	157
Tyrrhenian Sea	-	-	-	-	95
Coastal areas:					
1) May-June	-	0.15-0.35	-	14-58	-
2) October-November	-	0.04-0.08	-	-	-
Algerian-Provencal Basin	-	-	-	865	-
Tunisian Gulf	-	-	-	103	-
Algerian coast	-	-	-	1243	-
Gibraltar	-	-	-	1247	-
Atlantic Ocean (near Spain, Morocco)	-	0.19	50	-	-
Black Sea	-	-	-	58-100	106-1330

Table 1.5 Cadmium concentrations in the Mediterranean sediments (ug g^{-1} DW) [UNEP/WG.160/9, 1987]

Region	Method	Cd conc.	Reference
II			
-Var lagoon, France	$HF-HClO_4-HNO_3$	3.7	Chabert and Vicente, 1981
-Coastal lagoon,	<63um	10-32	De Leon et al., 1983
-Spanish coast	Conc. HNO_3	0.1-0.3	Peiro et al., 1983
-River Ebro Delta	HNO_3	0.12-0.37	Obiols et Peiro, 1981
-River Ebro Delta	HNO_3	0.04-2.1	Obiols et al., 1985
-Etang Salses-Leucate	<63 um	5.5	Buscail et al., 1985
-Etang Bages-Sigean	<63 um	6	Buscail et al., 1985
-Etang de Thau	<63 um	4	Buscail et al., 1985
-River Rhone Delta	HNO_3-HClO_4	0.25-5	Added et al., 1981;
-River Rhone Delta		0.3- 0.5	Span et al., 1985
-Marseille	<200um $HCl-HNO_3$	1.3-3	Arnoux et al., 1981
-Cannes	<63um $HNO_3-H_3PO_4-HCl$	1.8-7	Ringot, 1983
-Gulf of Nice	HNO_3-HCl	0.7-2.4	Flatau et al., 1983
-Italian Estuaries	HNO_3-HCl	0.21-0.55	Breder et al., 1981
	HNO_3	0.7-1.7	Frignani and Giordani, 1983
III			
-Portman	HNO_3-H-peroxide	up to 10.4	De Leon et al., 1985
-Castellon-Guardamar	"	ND-0.5	"
IV			
-Offshore sediments	HNO_3	0.5-2.5	Frignani and Giordani, 1983
V			
-River Po delta	HNO_3	0.16-1.7	Fascardi et al., 1984
-Gulf of Trieste		0.3-5.3	Majori et al., 1979
-Gulf of Venice	HNO_3	0.1-3.1	Angela et al., 1981
-Bay Mali Ston, Yugoslavia		0.1-0.2	Vukadin et al., 1985
-Northern Adriatic	-	0.05-5.6	Donazzolo et al., 1984a, 1984b
-Offshore sediments	HNO_3	0.80-1.2	Frignani and Giordani, 1983
VI			
-Patraikos Gulf, Greece	$HF-HNO_3\ HClO_4$	-	Varnavas and Ferentinos, 1983
-Gulf of Catania	HNO_3	2.2-4.6	Castagna et al., 1982
-Offshore sediments	HNO_3	0.6-1.1	Frignani and Giordani, 1983
VIII			
-Thermaikos-Kavala, Greece	63um HNO_3	0.6-1.1	Fytianos and Vasilikiotis, 1983
-Thermaikos Gulf, Greece	45um HNO_3	0.40-2.5	Voutsinou-Taliadouri, 1983
Industry	0.55	(0.45-1.15)	Voutsinou-Taliadouri
Axios river (1983)	2.5	(0.45-8.5)	and Varnavas 1986
Axios river (1985)	3.7		
Aliakmon river	0.4		
-Pagassitikos Gulf, Greece	45um HNO_3	0.4	Voutsinou-Taliadouri and Varnavas 1986
-East Aegean offshore	45um HNO_3	0.4	Voutsinou-Taliadouri, 1983
-Izmir Bay	$HCl-HNO_3$	0.2-40	Uysal and Tuncer, 1985
-Guelbahce Bay	$HCl-HNO_3$	1.4-14	Uysal and Tuncer, 1985
X			
-Haifa Bay	<250um	0.4-2.5	Krumholz and Fleischer, 1985
-Alexandria		2.8	El Sokkary, 1979
-Abu Kir Bay, Egypt	HNO_3	2	Saad et al., 1981
-Damietta estuary, Egypt	HNO_3	0.16-2	Saad and Fahmy, 1985
-Western Harbor, Alexandria	HNO_3-HClO_4	7-64	Saad et al., 1981
XIII			
-Black Sea, Nearshore	HNO_3	1.3-4.8	Pecheanu, 1983
Offshore		2.8	Pecheanu, 1983
Mediterranean		0.1-2.3	UNEP, 1978

Table 1.6 Some cadmium levels in the Mediterranean and in other selected areas which with the present state of knowledge may be considered typical [UNEP/WG.160/9, 1987]

Mediterranean:		
Air:	open sea	$0.4 - 2$ ng Cd m^{-3}
	cities	$30 - 200$ ng Cd m^{-3}
	total deposition	$10 - 50$ ng Cd cm^{-2} year^{-1}
Sea water:	open-sea	$5 - 150$ ng Cd l^{-1}
	coastal	up to 2000 ng Cd l^{-1}
Sediments:	background	0.15 ug Cd kg^{-1} DW
	polluted lagoon	up to 50 ug Cd kg^{-1} DW
Plankton:	open-sea	2000 ug Cd kg^{-1} DW (400 ug Cd kg^{-1} FW)
	coastal	10000 ug Cd kg^{-1} DW (2000 ug Cd kg^{-1} FW)
Crustaceans:		50 ug Cd kg^{-1} FW
Molluscs:	<u>Mytilus</u> (coastal)	40-2000 ug Cd kg^{-1} FW
Fish:	<u>Mullus barbatus</u>	20-50 ug Cd kg^{-1} FW
	M. <u>surmuletus</u>	150 ug Cd kg^{-1} FW
Non-Mediterranean:		
Air:	open sea	$0.02 - 2.5$ ng Cd m^{-3}
	precipitation	$0.004 - 1.2$ ng Cd l^{-1}
total deposition		
	open ocean	5 ng Cd cm^{-2} year^{-1}
	North Sea	$20 - 85$ ng Cd cm^{-2} year^{-1}
	Baltic Sea	$13 - 20$ ng Cd cm^{-2} year^{-1}
Sea water:	open sea	$10 - 70$ ng Cd l^{-1}
Crustaceans:	ICES area	$20 - 1000$ ug Cd kg^{-1} FW
Molluscs:	<u>Mytilus</u>, ICES area	$5 - 1060$ ug Cd kg^{-1} FW
Fish:	ICES area	$20 - 60$ ug Cd kg^{-1} FW
	Oslo Commission	1000 ug Cd kg^{-1} FW

Table 1.7 Mercury levels in the Mediterranean and in other seas which may be considered "typical" at the present state of knowledge taking into consideration many reservations   [UNEP/WG.160/8,1987]

Mediterranean:

Air:	coastal	2 - 3 ng "gaseous" Hg m^{-3}
		1 % particulate Hg
	urban	10 - 20 ng "gaseous" Hg m^{-3}
	chlor-alkali plant	up to 73 ng "gaseous" Hg m^{-3}
Sea water:	open sea	7 - 25 ng Hg-T l^{-1}
	coastal	up to 100 ng Hg-T l^{-1}
Sediments:	open sea	0.01 - 0.03 mg Hg kg^{-1} DW
	coastal	up to 45 mg Hg kg^{-1} DW
Plankton:	open sea	15 - 560 ug Hg-T kg^{-1} DW
		(3 - 120 ug Hg-T kg^{-1} FW)
Crustaceans:		20 - 300 ug Hg-T kg^{-1} FW
	area II+IV	1000 - 1100 ug Hg-T kg^{-1} FW
Molluscs:		70 - 200 ug Hg-T kg^{-1} FW
	area IV+V	250 - 870 ug Hg-T kg^{-1} FW
Fish	pelagic	100 - 300 ug Hg-T kg^{-1} FW
	area IV	300 - 400 ug Hg-T kg^{-1} FW
	tuna	850 - 1700 ug Hg-T kg^{-1} FW
	M. barbatus	55 - 215 ug Hg-T kg^{-1} FW
	area II+IV	590 - 1450 ug Hg-T kg^{-1} FW
	var. spp.	10 - 815 ug Hg-T kg^{-1} FW

non-Mediterranean:

Air:	open-sea, Atlantic Northern Hemisph.	1 - 3 ng "gaseous Hg" m^{-3}
	Southern Hemisph.	0.5-2.5 ng "gaseous Hg" m^{-3}
	over remote land	2 - 9 ng "gaseous Hg" m^{-3}
Sea water:	open-sea	2 - 14 ng Hg-T l^{-1}
	coastal	8 - 12 ng Hg-T l^{-1}
Plankton:	open-sea	100 - 1100 ug Hg-T kg^{-1} DW
		(50 - 500 ug Hg-T kg^{-1} FW)
Crustacean:	brown shrimp (N. Atlantic)	20 - 390 ug Hg-T kg^{-1} FW
Molluscs:	M. edulis (N. Atlantic)	20 - 130 ug Hg-T kg^{-1} FW
Fish:	Herring	20 - 240 ug Hg-T kg^{-1} FW
	cod	30 - 480 ug Hg-T kg^{-1} FW
	hake/haddock	20 - 130 ug Hg-T kg^{-1} FW
	plaice (N. Atlantic)	20 - 500 ug Hg-T kg^{-1} FW

Table 1.8   Hydrocarbons in the surface microlayer
[UNEP/IOC, Map TRS No.19, 1988; 34]

Area	Year	Concentrations (in ug l^{-1})		Technique	Reference
off-Monaco	1981	6.0 - 11.4 23 - 61 4.3 - 4.9	(n-alk) (UCM)* (arom.)	GC GC GC	Burns and Villeneuve, 1983
North Western basin	1981	1.0 - 13.48 8.11- 22.1 0.26- 0.35	(n-alk) (UCM) (arom.)	GC GC UV-fl.	Ho et al., 1983
	1983	0.55	(n-alk)	GC	Sicre et al., 1985
Central Western basin	1981	0.69 6.8 0.70	(n-alk) (UCM) (arom.)	GC GC UV-fl.	Ho et al., 1983
	1983	0.96	(n-alk)	GC	Sicre et al., 1985
South Western basin	1981	0.57 0.25 - 5.15	(n-alk) (arom.)	GC UV-fl.	Ho et al., 1983
	1983	1.67 - 1.86	(n-alk)	GC	Sicre et al., 1985
Alboran Sea	1983	1.4	(n-alk)	GC	Sicre et al., 1985

* UCM : unresolved complex mixture.

Table 1.9 Concentrations of dissolved/dispersed petroleum residues in the Mediterranean (n = number; c = arithmetic mean; s = standard deviation; G.M. = geometric mean)
All values in ug/l   [UNEP/WG.160/11,1987]

Region	Normal statistics			Log-transformed data	
	n	c	s	n	M.G.
Mediterranean					
< 0.4 µg/l	465	2.0	5.0	462	0.33
	219	0.07	0.08	215	0.04
> 0.4 µg/l	247	3.7	6.4	247	2.0
Aegean Sea	134	1.3	0.79	134	1.1
Eastern region					
< 0.4 µg/l	133	5.5	8.3	131	1.5
	29	0.04	0.06	27	0.03
> 0.4 µg/l	104	7.0	8.8	104	4.0
Central region					
< 0.4 µg/l	176	0.17	0.42	175	0.06
	156	0.07	0.07	155	0.05
> 0.4 µg/l	20	1.0	0.86	20	0.77
Western region	22	0.02	0.01	22	0.02
Baffin Bay	104	0.11	0.12	93	0.09
Indian Ocean	45	60.1	92.7	36	8.9
Japan	1666	0.31	1.21	1640	0.13
North Am. East Coast	80	0.11	0.10	71	0.09
North Sea	90	0.02	0.12	9	0.15
South China Sea	272	0.20	0.28	256	0.10
Strait of Malacca	14	0.11	0.12	10	0.13

Table 1.10   Petroleum hydrocarbons in benthic sediments
[UNEP/IOC, MAP TRS No. 19, 1988; 40/41]

Area	Concentrations (in ug g^{-1})			
French coast				
(Fos sur Mer to Monaco) (1979)				
Côte Bleue	13 -952	aliphatics + aromatics		
Les Embiez	69 -93	"		"
Monaco	51 -77	"		"
Spanish coast (1980-1982)	0.6 -2.3	(D/w)	$C_{14}-C_{20}$	(GC)
off Valencia 3-10m (0-5cm)				
off Alicante	0.1 -5.8	(D/W)	$D_{14}-C_{24}$	(GC)
off Delta del Ebro	0.3 -1.1	(D/W)	$C_{15}-C_{24}$	(GC)
off other Catalan rivers	0.07 -0.56	(D/W)		(GC)
- river mouths and cities				
Ter river mouth (10-60m)	0.5 - 1.9	(D/W)	n-alkanes	(GC)
(3 samples)	1.8 - 9.8		UCM	
	5.1 -10.1		aromatics	
off Barcelona (10-80 m)	1.3 -17.0	"		"
(9 samples)	24.5 -52.8			
	3.1 -66.8			
off Tarragona (17-95 m)	0.9 - 5.0	"		"
(6 samples)	4.8 -77.1			
	7.8 -21.2			
Ebro Delta (10-100 m)	0.4 - 3.2	"		"
(5 samples)	1.3 -12.9			
	0.6 -15.2			
off Valencia (10-100 m)	0.8 - 1.0	(D/W)		(GC)
(5 samples)	3.8 -12.3			
	4.8 -26.0			
off Benidorm (10-100 m)	0.8 - 0.9	(D/W)		(GC)
(2 samples)	1.9 - 4.0			
	2.8 - 5.5			
Western Mediterranean (1000 m)	1.2 - 1.6	"		UCM
	0.6 - 2.3			aromatics
Italy, Taranto,				
Mar Piccolo, 1983	1.3-45 (av. 14.73)	(D/W)		"
8 stations (1-10m depth)				
Yugoslavia,	1.0-18.9			Fluorescence
Split, 1984				
Cyprus,				
Larnaca Bay, 1983	0.114-0.135 (4 samples)			"
(90 m) 1984	0.442-1.301 (4 samples)			"
Limassol Bay 1984	0.308-0.417 (2 samples)			"
(18-90m)				
Turkey, Iskenderun Bay	0.04-0.68 (av. 0.24 ug g^{-1})			PAHs (GC)
1980-82, 10-9m depth				
Turkey				
Candarli Bay, 1983-84	0.0043-0.375 (fw)			Fluorescence
Aliaga, 1983-84	0.0175-0.025 (fw)			"
Saros Bay, 1983	1.0 (fw)			"
Izmir Bay, 1983	0.047 (fw)			"
Southern Aegean Coast, 1983	0.1575 (fw)			"

Table 1.11 Distribution of hydrocarbons in the ecosystem compartments (in 10^3 tonnes year^{-1} [UNEP/IOC, MAP TRS No.19, 1988; 45]

Beach tar	100
Surface microlayer	0.018
Floating tar	8.8
Surface water (0-5 m)	30
Subsurface water	72
Sediment flocculent layer	230
Sediments	120
Biomass	0.220
Atmosphere	155
Total	716

Table 1.12 Summary assessment of the microbiological quality of shellfish in the Mediterranean, according to the EEC criteria (MED POL VII sampling stations with at least 10 shellfish analyses per year) [UNEP/WG.160/10, 1987]

Year	Stations surveyed	Stations with satisfactory shellfish
1976	15	4 (27%)
1977	12	0 (0%)
1978	21	11 (52%)
1979	20	6 (30%)
1980	21	15 (71%)
1981	0	0
Overall	89	36 (40%)

Table 1.13 Heavy metals ($\mu g\ g^{-1}$ dry weight[a]) in *Mytilus* from different regions of the Mediterranean Sea. Values given are ranges [Fowler, 1985]

Region	Cd	Cu	Zn	Pb	Hg
North-west Mediterranean (Ligurian Sea)	0.4–5.9	2.4–154	97–644	2.4–117	0.18–0.96
Adriatic (Gulf of Trieste)	1.4–1.7	6.2–9.8	87–137	3.8–15	0.28–1.3
Aegean (Saronikos Gulf)	0.06–0.08	4.5	12–87		0.06–0.2
(Turkey)	6.6–12	36–64	336–452	83–110	0.89–1.1
South-west Mediterranean (Algeria)	0.3–6.5		7.2–71		0.25–0.63

[a] Where necessary values were converted using a wet/dry weight ratio of 6.

Table 1.14 Average cadmium concentrations in Mediterranean marine organisms in ug/kg^{-1} FW [UNEP/WG.160/9, 1987]

Species	No. of samples	Mean	Standard Deviation
Engraulis encrasicolus	81	34	25
Merluccius merluccius	27	63	34
Mugil auratus	10	47	85
Mullus barbatus	318	34	28
Mullus surmuletus	218	140	83
Thunnus alalunga	38	23	6.5
Thunnus thynnus	111	38	43

Table 1.15 Recent data on cadmium concentrations (ug Cd kg^{-1} FW) in selected seafood and other marine species from the Mediterranean [UNEP/WG.160/9, 1987]

Species	date	mean S.D	location	reference
M. gallopr.	Aug. 1984	770 + 120*		Veglia & Vaissière, 1986
	Dec. 1984	730 + 150*		
	Apr. 1985	1900 + 230*		
	July 1985	710 + 360*		
	1983/1985	1160 + 650*	Koper St.5	Tusnik & Planinc, 1986
		1390 + 400*	Piran St.23	
		1310 + 310*	Piran St.27	
		1390 + 430*	Piran St.35	
	July/Nov 1985	70	Valencia/Cast.	Hernandez et al., 1986
	1985	170 + 160	Aegean	Uysal and Tuncer, 1985
Corbula gibba	1984/85	210 + 45		Nat. Co-ord. Yugoslavia, 1986
Alcyomium palmatum		200 + 30		
Haliclona		30 + 40		
Pecten jabobeus		1015 + 60		
Ostrea edulis		730 + 40		
Chlamys opercularis		1180 + 60		
Ostrea edulis		100		Nat. Co-ord. Tunisia, 1986
Mullus barbatus		140		
Pagellus erythrinus		115		
Trachurus trachurus		135		
Palaemon serratus	July/Nov 1985	70	Valencia/Cast.	Hernandez et al., 1986
	M	81		
	F	42.5		
S. pilchardus		46	Valencia/Cast.	Hernandez et al., 1986
	1985	120 + 90	Aegean	Uysal and Tuncer, 1985
M. barbatus	M	16.6	Valencia/Cast.	Hernandez et al., 1986
	F	17.9		
M. surmuletus	M	8.4	Valencia/Cast.	Hernandez et al., 1986
	F	18.8		
S. scomber	1985	130 + 110	Aegean	Uysal and Tuncer, 1985
T. thynnus	F	280	Valencia/Cast.	Hernandez et al., 1986

* dry weight

Table 1.16   Overall averages of levels of mercury in composite samples of molluscs [UNEP/WG.160/8, 1987]

area	species	n	mean	range	
II	Mytilus galloprovin.	37	70	15 - 400	
III	Perna perna	192	76	20 - 370	
IV	M. galloprovincialis	59	240	25 - 1260	(!)
V	M. galloprovincialis	26	870 (!)	25 - 7000	(!)
VI	M. galloprovincialis	12	75	35 - 145	
VII	Lithophaga lithophaga	5	165	80 - 290	
VIII	M. galloprovincialis	175	105	5 - 920	(!)
IX	M. galloprovincialis	4	37	20 - 50	
	Donax trunculus	42	210	35 - 910	(!)
XI	M. galloprovincialis	3	190	20 - 290	
XII	M. galloprovincialis	3	160	140 - 170	

(!): value above 500 ug Hg-T kg^{-1} FW

Table 1.17   Mean levels in samples of crustaceans from the Mediterranean   [UNEP/WG.160/8, 1987]

area	species	n	mean	range
II	Nephrops norvegicus	129	1080 (!)	350 - 3000
IV	Nephrops norvegicus	86	1110 (!)	60 - 2900
VI	Nephrops norvegicus	7	290	190 - 360
VIII	Penaeus kerathurus	10	175	75 - 475
	Carcinus mediterraneus	13	215	115 - 345
IX	Penaeus kerathurus	7	20	10 - 50
XII	Parapenaeus longirostris	3	300	270 - 350

(!): value above 500 ug Hg-T kg^{-1} FW

Table 1.18   Averages of mercury concentrations in fishes (ug Hg-T kg^{-1} FW) according to UNEP sampling areas   [UNEP/WG.160/8, 1987]

area	species	n	mean	range
II	Engraulis encrasicholus	37	140	20 - 300
	Mullus barbatus	262	590 (!)	15 - 5600 (!)
	M. surmuletus	5	260	70 - 510 (!)
	Sarda sarda	14	1000 (!)	290 - 2300 (!)
	Thunnus thynnus	176	1100 (!)	20 - 6290 (!)
	Xiphias gladius	1	150	
III	M. surmuletus	204	90	30 - 230
IV	E. encrasicholus	44	157	65 - 380
	M. barbatus	195	1440 (!)	60 - 7050 (!)
	Thunnus alalunga	8	215	90 - 336
V	M. barbatus	6	190	100 - 390
VI	E. encrasicholus	11	145	55 - 270
	M. barbatus	13	190	45 - 330
	T. alalunga	8	275	60 - 400
VII	M. barbatus	11	165	30 - 280
	Trachurus mediterraneus	5	345	80 - 955 (!)
VIII	Merluccius merluccius	10	315	60 - 840 (!)
	Mugil auratus	16	350	85 - 2500 (!)
	M. cephalus	3	165	70 - 300
	M. barbatus	127	175	15 - 1400 (!)
	T. thynnus	7	370	70 - 890 (!)
	Tr. mediterraneus	3	340	320 - 365
	X. gladius	8	280	85 - 755 (!)
IX	Boops salpa	3	10	5 - 15
	Boops boops	5	135	40 - 430
	Mugil auratus	39	170	1 - 5600 (!)
	M. barbatus	6	55	2 - 90
	M. barbatus	168	140	30 - 475
	M. surmuletus	13	35	1 - 80
	Upenaeus moluccensis	7	200	100 - 430
	Dentex dentex	6	385	220 - 480
	D. gibbosus	12	140	100 - 180
	Epinephelus aeneus	4	250	100 - 400
	M. merluccius	6	150	31 - 260
	Pagellus acarne	7	190	70 - 340
	Pagellus erythrinus	112	205	55 - 805 (!)
X	Saurida undosquamis	143	135	40 - 650 (!)
	Sphyraena sphyraena	7	165	80 - 245
	Tr. mediterraneus	48	95	10 - 415
	U. moluccensis	120	440	40 - 1120 (!)
XI	M. surmuletus	5	150	15 - 380
	T. thynnus	1	550 (!)	
XII	M. merluccius	3	815 (!)	780 - 850 (!)
	M. barbatus	3	215	210 - 230
	P. erythrinus	3	220	210 - 225
	Tr. mediterraneus	3	345	340 - 350

(!) = levels above 500 ug Hg-T kg^{-1} FW

Table 1.19 Mercury concentrations (ug Hg-T kg^{-1} FW) in eggs and liver of Mediterranean birds [UNEP/WG.160/8, 1987]

	primary consumer			secondary consumer			tertiary consumer		
	n	mean	SD	n	mean	SD	n	mean	SD
Selvagens, Madeira									
eggs							24	400	± 185
liver							3	2440	± 460
Mistras									
eggs							6	1580	± 1000
liver							3	2180	± 2000
S. Gilla									
eggs				7	610	± 365	6	7760	± 4740
liver	2	5760		14	18800	± 13080	7	39420	± 19680
Elba									
eggs				25	585	± 345	16	2140	± 680
liver				4	1340	± 160			
Comacchio									
eggs	3	160	± 20	32	295	± 110	29	770	± 630
liver				4	2320	± 1680			
Marano									
eggs	10	150	± 150	21	440	± 110	22	2040	± 700
liver				8	1880	± 440	3	8480	± 8580
Linosa									
eggs							5	1300	± 380
liver							5	17240	± 19840
Dagonada									
eggs							2	1060	
liver							5	14960	± 10180
Danube, delta									
eggs	4	60	± 20	21	155	± 80	29	820	± 400

Note: the data have been converted into fresh weight by dividing dry weight by a factor of 5.

Table 1.20   Hydrocarbons in biota samples from the Spanish Mediterranean coast (ug/g dry weight)   [UNEP/WG.160/11]

Species	Area(year)	Saturate fraction (UCM)	Aromatic fraction (crude oil eq)
Mytilus (10 samples)	Palamós	106-190	-
	Barcelona	500-3200	-
	Ebro Delta	8-216	-
Mullus sp.	Palamós	12.6	4.4
(muscle) (14 samples)	Barcelona	22.2	9.3
	Ebro Delta	5.8	11.1
Merluccius sp.	Palamós	1.5	1.7
(muscle) (14 samples)	Barcelona	0.2	3.9
	Ebro Delta	0.2	2.4
Trachurus sp.	Palamós	11.2	4.2
(muscle) (14 samples)	Barcelona	1.4	10.9
	Ebro Delta	5.4	3.7
Engraulis sp. (muscle) (19 samples)	Barcelona	7.7	7.8

Table 1.21   Comparison of pesticide concentrations (ug/kg F.W.) in *Mullus barbatus* on Israeli Mediterranean shores with other Mediterranean areas in 1976
[UNEP/FAO, MAP TRS No 3, 1986]

Sampling area	fresh or dry weight basis	DDTs	PCB	BHC
Barcelona	F.W.	243.90	1,291.9	-
Venice	D.W.	89.0	225.8	22.7
Malta	F.W.	18.9	72.2	2.87
Turkey	F.W.	5.44	0.40	-
Israel	F.W.	30.9	71.5	53.43

Table 1.22 Water: Supply and Demand in the Mediterranean
Drainage Basin, 1980 [from: UNEP/WG.171/3, 1987]

COUNTRY	(a) Estimated population (millions)	SUPPLY (resources)		DEMAND			RATIO	
		(b) total water resources Gm/year	(c) stable or stabilized resources Gm/year	(d) water distributed (drawoffs) Gm/year	(e) net consumption Gm/year		(f) supply exploitation index in relation to (b) d/b	(g) demand exploitation index in relation to (c) d/c
SPAIN	15,5	28,3	24,1	12,7	~ 6,9		45	53
FRANCE, MONACO	11,2	74	~39	12,7	2,2		17	33
ITALY	57,3	187	50	56,2	22		30	112
MALTA	0,37	0,03	0,03	0,023	0,02		77	77
YUGOSLAVIA	6,9	62	~12	1,5	0,3		2,5	<12
ALBANIA	2,7	21,3	~7	0,2	0,04		1	<3
GREECE	9,6	62,9	~10	6,9	3,6		11	<70
TURKEY	11,0	77	~23	7,5	3,8		10	33
CYPRUS	0,6	0,9	0,4	0,54	0,3		60	135
SYRIA	1,0	4,4	~2,8	2,5	1,5		57	~90
LEBANON	3,1	4	3,3	0,6	0,4		15	18
ISRAEL	3,9	1,0	0,3	1,4	0,8		140	465
EGYPT	41,2	57,3	55,8	45	41		79	81
LIBYA	3	0,6	0,23	1,0	0,65		167	435
TUNISIA	6,3	3	2,9	~1,2	0,65		40	41
ALGERIA	14,6	13,2	~3,1	2,8	~1,4		21	90
MOROCCO	1,7	4,0	1,5	1,0	0,6		25	67
Σ	190	572	—	154	86			

Table 1.23 Atmospheric Concentrations of Pb and Cd over the Mediterranean Basin (ng m^{-3}) [from: GESAMP no.68, UNEP, 1985]

	Pb range	Pb mean	Cd range	Cd mean	n of samples	References
- Eastern and Central (1979)	2-25	14	--	--	5	Chester et al. (1981)
- Tyrrhenian Sea (1979)	3-16	10	0.07-1.6	0.40	9	Chester et al. (1984)
- Central and Tyrrhenian Sea						
(1980) -	10-98	50	0.2-6.0	2.10	19	Seghaier (1984)
(1982) -	3-39	15	0.2-2.4	0.9	16	Buat-Menard et al., unpubl. data
- Western Basin						
Phycemed 1 Cruise (1981)	3-58	33	0.1-5.5	1.4	13	Seghaier (1984)
Phycemed 2 Cruise (1983)	4-54	27	0.4-3.2	1.6	15	Buat-Menard et al., unpubl. data
- Alboran Sea (1981)	5-78	49	0.3-7	1.5	7	Seghaier (1984)
- Coastal Regions						
Marseille (1977-1979)		305		5.9	200	Viala et al. (1979)
Monaco (1978)		171		4.5	30	Seghaier (1984)

Table 1.24 Mercury concentrations (ng m^{-3}) in aerosol in the Western Mediterranean (Phycemed 81), around Sicily (Etna 80) and in the North Atlantic (EF: enrichment factor rel. to aluminium) [UNEP/WG.160/8, 1987]

Etna 80		Phycemed 81		North Atlantic	
mean	EF	mean	EF	mean	EF
0.1	560	0.24	910	0.065	450

Note: $EF = \dfrac{\text{(element conc./Al conc.) in sample}}{\text{(element conc./Al conc.) in earth's crust}}$

Table 1.25 Concentrations of hydrocarbons present in the atmosphere in gaseous and particulate forms. Concentrations are given in ng m^{-3}. n-Alk.= n-alkanes; PAH = polycyclic aromatic hydrocarbons; THC = total hydrocarbons. [Saliot and Marty, 1986]

Location of sampling	Gas. hydrocarbons			Part. hydrocarbons		
	n-Alk.	PAH	THC	n-Alk.	PAH	THC
Mediterranean Sea PHYCEMED I cruise leg GYL-ETR2	147	19.5	683.5	43.7	1.6	59.9
leg ETR2-ETE	65	15.1	476.1	10.8	1.8	30.1
Irish West coast	253			3.28		
Gulf of Guinea	30-281	7-18		1.5-14	0.1	
Tropical North Atlantic				4-50		
Tropical North Pacific	0.33			0.02-0.16		
Jasin site	94					
Indian ocean Cape Grim	162					

Table 1.26   Land use, 1984 (thousands of hectares)
[from: UNEP/WG.171/3, 1987]

	TOTAL	ARABLE LAND	PERMANENT CROP	PERMANENT CROP	FOREST	OTHER
SPAIN	49 940	15 620	4 920	10 640	15 625	3 135
FRANCE	54 563	17 468	1 344	12 385	14 603	8 763
ITALY	29 402	9 100	3 133	4 930	6 410	5 829
MALTA	32	12	1			19
YUGOSLAVIA	25 540	7 018	740	6 379	9 294	2 109
ALBANIA	2 740	592	121	400	1 038	589
GREECE	13 080	2 948	1 026	5 255	2 620	1 231
TURKEY	77 076	24 500	2 911	9 000	20 199	20 466
CYPRUS	924	365	67	93	171	228
SYRIA	18 406	5 104	550	8 319	498	3 935
LEBANON	1 023	210	88	10	82	633
ISRAEL	2 033	345	92	818	116	662
EGYPT	99 545	2 310	164		2	97 069
LIBYA	175 954	1 780	335	13 300	640	159 899
TUNISIA	15 536	3 167	1 520	3 024	555	7 270
ALGERIA	238 174	6 800	640	32 100	4 384	194 250
MOROCCO	44 630	7 815	516	12 500	5 200	18 599

Note : Figures refer to the whole country, not only the Mediterranean regions.

Table 1.27   Mediterranean specially protected areas
              [UNEP/WG. 163/4,1987]

LATITUDE	LONGITUDE	CODE	SPA SITE	COUNTRY
41.00	19.45	001	Diviaka Nature Reserve	Albania
41.75	19.47	001	Kune Nature Reserve	Albania
36.90	8.45	001	El Kala National Park	Algeria
36.75	3.50	001	Reghaia Nature Reserve	Algeria
34.87	33.55	001	Larnaka Lake Nature Reserve	Cyprus
34.50	32.95	001	Limassol Lake Nature Reserve	Cyprus
31.17	33.25	001	Bardaweel Nature Reserve	Egypt
30.75	29.20	001	Omayed Nature Reserve	Egypt
42.72	9.50	001	Bastia Fishery Reserve	France
42.58	8.70	001	Calvi Fishery Reserve	France
43.50	4.50	001	Camargue Nature Reserve	France
42.47	3.17	001	Cerbere-Banyuls Marine Reserve	France
43.32	5.17	001	Cote Bleue Regional Park	France
42.67	8.91	001	Ile Rousse Fishery Reserve	France
42.25	8.70	001	Piana and Porto Fishery Reserve	France
43.00	6.39	001	Port Cros National Park	France
42.75	9.33	001	Saint-Florent Fishery Reserve	France
42.35	6.22	001	Scandola Nature Reserve	France
42.08	8.62	001	Tuccia-Sagone-Cargese Fishery Reserve	France
35.26	23.48	001	Gorge Samaria (Lefki Ori) National Park	Greece
38.00	23.00	001	Pefkias-Xylokastron Aesthetic Forest	Greece
35.13	26.27	001	Vai Aesthetic Forest	Greece
32.62	34.92	001	Dor-Habonim Nature Reserve	Israel
33.07	35.60	001	Rosh Hanikra Marine Reserve	Israel
42.38	11.40	001	Burano Nature Reserve	Italy
40.27	14.93	001	Castellabate Fishery Reserve	Italy
41.63	12.53	001	Circeo National Park	Italy
45.70	13.70	001	Miramare Marine Reserve	Italy
42.46	11.25	001	Orbetello and Feniglia Nature Reserve	Italy
42.82	10.33	001	Porto Ferraio Fishery Reserve	Italy
38.70	13.20	001	Ustica Marine Reserve	Italy
32.67	21.67	001	El Kouf National Park	Libya
39.33	-0.37	001	Albufera de Valencia Regional Park	Spain
42.55	2.70	001	Castello de Ampurias Protected Landscape	Spain
41.15	0.50	001	Ebro Delta Regional Park	Spain
42.50	3.22	001	Medes Island Fishery Reserve	Spain
42.15	2.75	001	Pals Protected Landscape	Spain
42.50	2.70	001	San Pedro Pescator Protected Landscape	Spain
38.15	-0.45	001	Tabarca Marine Reserve	Spain
37.50	8.87	001	Galiton Marine Reserve	Tunisia
37.17	9.67	001	Ichkeul Natural Reserve	Tunisia
37.10	10.80	001	Zembra and Zebretta National Park	Tunisia
41.33	26.25	001	Gelibolu Peninsula National Park	Turkey
36.50	30.02	001	Olympos Seshore National Park	Turkey
44.92	13.75	001	Brioni Islands National Park	Yugoslavia
42.50	18.63	001	Kotor World Heritage Site	Yugoslavia
42.63	18.12	001	Lokrum Nature Reserve	Yugoslavia
42.78	17.37	001	Mijet National Park	Yugoslavia
43.03	17.45	001	Neretva Delta Nature Reserve	Yugoslavia

Table 1.28   100 Coastal historic sites of common interest [UNEP/IG.74/4, 1987]

COUNTRY

Albania (1)

- (Apollonia)
- (Dyrrachion)

Algeria

- Algiers
- Cherchell
- Hippone
- Tipasa*

Cyprus

- Famagusta
- Khiriotikia
- Kourion
- Limassol
- Paphos *
- Salamis

Egypt

- Abou Mîna *
- Cairo, Gîza, Memphis, Saqqara
- Necropolis of Alexandria

France

- Aigues Mortes
- Arles *
- Cucuruzzu and Filitosa (Corsica)
- Marseille : Old Port

Greece

- Athens
- Cnossos
- Corfu
- Delos
- Delphi
- Epidaurus and Nauplion
- Meteora (natural/cultural site)
- Mount Athos (natural/cultural site)
- Mycenae

- Olympia
- Paros and its quarries
- Phaestos
- Rhodes
- Thera (natural/cultural site)
- Thassos and its quarries
- Tiryns

Israel

- Acre
- Caesarea on the Sea

Italy

- Agrigente
- Amalfi
- Aquilea
- Bari
- Quarries of Carrara
- Nuraghic complex of Su Nuraxi, Barumini (Sardinia)
- Genoa
- Lecce
- Naples, Pompei, Herculanum and the Phlegrean fields
- Paestum/Poseidonia
- Palermo and Monreale
- Pisa
- Ravenna
- Rome and Ostia *
- Segesta
- Selinus
- Syracuse
- Tarquinia and the main Etruscan necropolises
- Trieste
- Venice and its lagoon

Lebanon

- Byblos*
- Sidon
- Tyre *

Libyan Arab Jamahiriya

- Cyrene *
- Leptis Magna *
- Sabratah *

Table 1.28   continued

Malta

- Ggantija *
- Hal Saflieni *
- Valletta *

Morocco

- Tetuan

Monaco

- Museum of Oceanography :cultural and natural

Spain

- Ampurias
- Barcelona
- Cordova *
- Granada *
- Ibiza : fortifications of Alt Vila
- Mallorca : Cathedral and Palacio de Palma
- Malaga
- Minorca : Toulas, talayots and navetas
- Murcia
- Seville
- Tarragona
- Valencia and Albufera

Syrian Arab Republic

- Ugarit /Ras Shamra

Tunisia

- Carthage and Sidi Bou Said *
- Jerba (cultural /natural site)
- El Jem *
- Kerkouan *
- Susa
- Tunis *

Turkey

- Antalya
- Aspendus
- Bursa
- Didyma
- Ephesus
- Halicarnassus
- Istanbul *
- Pergamum
- Priene
- Troy
- Xanthus

Yougoslavia

- Dubrovnik *
- Split *
- Trogir (natural/cultural site)
- Zadar (natural/cultural site)
- Kotor

---

* Sites on the World Heritage List

NOTE : (1) Albania has not ratified the MAP agreements

Table 1.29 Estimated annual pollution loads of the Mediterranean from land-based sources [from: UNEP, RSRS no.32, 1984]

Pollutant	Pollution loads originating in the coastal zone				Loads carried by rivers into the Mediterranean				Total Mediterranean Loads	
	Domestic t/a	Industrial t/a	Agricultural t/a	Sub-total t/a	Pollution t/a	Background t/a	Sub-total t/a	(range)	Pollution	Total(including background) t/a (range)
1. Volume:										
Total discharge $\times 10^9$	2	6	-*	(8)	(-)	420	420	(400-500)	(-)	430 (400-500)
2. Organic matter:										
BOD $\times 10^3$	500	900	100	1 500	1 000	(800)	1 800	(1200-2300)	2 500	3 300 (2700-3800)
COD $\times 10^3$	1 100	2 400	1 600	5 100	2 700	800	3 500	(2300-4700)	7 800	8 600 (7400-9800)
3. Nutrients:										
Phosphorus $\times 10^3$	22	5	30	57	260	40	300	(200-400)	320	360 (260-460)
Nitrogen $\times 10^3$	110	25	65	200	600	200	800	(600-1000)	800	1 000 (800-1200)
4. Specific organics:										
Detergents $\times 10^3$	18	-	-	18	42	0	42	(9-75)	60	60 (30-90)
Phenols $\times 10^3$	-	11	-	11	1	0	1	(0.5-1.8)	12	12 (6-18)
Mineral oil $\times 10^3$	-	120	-	120	(-)	0	(-)		(120)	(-)
5. Metals:										
Mercury	0.8	(7)	-	(8)	90	30	120	(40-200)	100	130 (50-200)
Lead	200	1 400	-	1 600	2 200	1 000	3 200	(2700-3800)	3 800	4 800 (4300-5400)
Chromium	250	950	-	1 200	1 200	400	1 600	(500-2700)	2 400	2 800 (1700-3900)
Zinc	1 900	5 000	-	6 900	14 000	4 000	18 000	(14000-22000)	21 000	25 000 (21000-29000)
6. Suspended matter:										
TSS $\times 10^6$	0.6	2.8	50	53	-	300	300	(100-500)	-	350 (100-600)
7. Pesticides:										
Organochlorines	-	-	-*	-	90	0	90	(50-200)	90	90 (50-200)
8. Radioactivity:										
Tritium Ci/a	-	400	-	400	2 100	(-)	2 100	(1600-3100)	2 500	(-)
Other radio-nuclides Ci/a	-	25	(-)	25	15	(-)	15	(10-25)	40	(-)

Legend:
- contributions from this source negligible
(-) insufficient data base for estimate
-* included in river assessment

Table 1.30 Industrial sectors and related pollutants considered for the pollution load assessment [from: UNEP,RSRS no.32, 1984]

	Food manufacturing	Beverage industry	Tobacco manufacturing	Rubber	Pulp and paper	Textile industries	Cement production	Leather tanning	Iron and steel basic industries	Motor vehicle production	Petroleum refineries	Crude oil terminals	Chemical industries total	Organic chemicals	Inorganic chemicals (fertilizer, chloralkali, etc)
1. Volume:															
total discharge	+	+	+	+	+	+	+	+	+	+	+	−	+	(+)	(+)
2. Organic matter:															
BOD	+	+	+	+	+	+	−	+	−	+	+	−	+	(+)	−
COD	+	+	+	+	+	+	−	+	−	+	+	−	−	(+)	−
3. Nutrients:															
Phsophorus	+	+	−	−	−	−	−	−	−	−	−	−	−	−	−
Nitrogen	+	+	−	−	−	−	−	+	+	−	−	−	−	−	(+)
4. Specific organics:															
Phenols	−	−	−	−	−	−	−	−	+	−	+	−	−	−	−
Mineral oil	−	−	−	−	−	−	−	−	+	+	+	+	−	−	−
5. Metals:															
Mercury	−	−	−	−	−	−	−	−	−	−	−	−	−	−	(+)
Chromium	−	−	−	−	−	(+)	−	+	−	+	−	−	−	−	(+)
Zinc	−	−	−	−	−	(+)	−	−	−	+	−	−	−	−	−
6. Suspended matter:															
TSS	+	+	−	−	+	+	+	+	+	+	+	−	−	(+)	(+)
7. Additional pollutants:															
Cyanides	−	−	−	−	−	−	−	−	+	−	−	−	−	−	−
Sulphides	−	−	−	−	−	−	−	−	−	−	+	−	−	−	−
Fluorides	−	−	−	−	−	−	−	−	−	−	−	−	−	−	(+)
Iron	−	−	−	−	−	−	−	−	+	−	−	−	−	−	(+)
Copper	−	−	−	−	−	(+)	−	−	−	−	−	−	−	−	−

Legend: "+" waste contributions of this pollutant evaluated and included in the assessment

"(+)" waste contributions of this pollutant only considered when directly reported from source

"−" waste contributions of this pollutant disregarded due to insignificance or absence of applicable estimates.

Table 1.31 Estimated annual pollution loads of the regional Mediterranean sea areas (all figures in tons per annum or percentages) [UNEP, RSRS no.32, 1984; p.23]

Sea area		I		II		III		IV		V		VI		VII		VIII		IX		X		TOTAL
Pollutant		t/a	%	t/a	%	t/a	%	t/a	%	t/a	%	t/a	%	t/a	%	t/a	%	t/a	%	t/a	%	t/a
1. Volume:																						
Total discharge ×10⁹		7	2	99	23	9	2	33	8	151	35	33	8	6	1	47	11	25	6	18	4	428
2. Organic matter:																						
BOD ×10³		90	3	950	29	120	4	370	11	800	25	230	7	70	2	330	10	140	4	150	5	3 250
COD ×10³		300	3	2400	28	400	5	1100	13	1700	20	600	7	300	3	950	11	550	6	300	3	8 600
3. Nutrients:																						
Phosphorus ×10³		7	2	126	35	9	3	29	8	85	24	23	6	7	2	33	9	19	5	20	6	358
Nitrogen ×10³		25	2	387	37	27	3	62	6	273	26	61	6	20	2	90	9	51	5	46	4	1 042
4. Specific organics:																						
Detergents ×10³		1.5	3	14.8	25	1.8	3	8.2	14	16.2	27	3.8	6	1.2	2	6.0	10	2.7	5	3.5	6	59.7
Phenols ×10³		1.2	10	3.9	31	0.6	5	1.0	8	1.6	13	1.5	12	1.1	9	0.9	7	0.2	2	0.4	3	12.4
Mineral oil ×10³		2	2	10	7	1	1	3	3	4	4	10	9	41	36	4	4	27	23	13	11	115
5. Metals:																						
Mercury		2	2	33	25	3	2	11	8	41	32	10	8	2	2	14	11	7	5	7	5	130
Lead		90	2	1360	28	120	2	630	13	1440	30	230	5	100	2	440	9	180	4	230	5	4 820
Chromium		100	4	1000	36	120	4	380	14	200	7	210	8	50	2	290	11	150	5	260	9	2 760
Zinc		300	1	5200	21	700	3	3000	12	8600	35	1600	6	500	2	2500	10	1100	4	1200	5	24 700
6. Suspended matter:																						
TSS ×10⁶		(-)	-	(-)	-	(-)	-	(-)	-	(-)	-	(-)	-	(-)	-	(-)	-	(-)	-	(-)	-	(-)
7. Pesticides:																						
Organochlorines		6.4	7	14.9	17	10.4	12	12.1	13	14.0	16	6.1	7	2.9	3	7.4	8	6.7	7	9.1	10	90
8. Radioactivity:																						
Tritium Ci/a		-	0	1100	44	-	0	120	5	1260	51	1	0	-	0	-	0	-	0	-	0	2 480
Other radio-nuclides Ci/a		-	0	16	42	-	0	14	37	7	18	3	3	-	0	-	0	-	0	-	0	38

Legend: (-) insufficient data base for estimate

Table 1.32   Organic load of domestic sewage discharged into the Mediterranean directly or through rivers, given in tonnes/year [AMBIO, Vol 6, No 6, 1977]

	BOD_5	Phosphorus content (P)	BOD_5/km coastline	P/km coastline
Spain	130 000	5 900	60	2.7
Northwestern basin	360 000	16 000	336	15.0
Italy	400 000	18 000	61	2.7
Yugoslavia	17 800	800	27	1.2
Malta	8 000	320	67	2.7
Greece	100 000	4 500	37	1.7
Turkey*	100 000	4 500	36	1.6
Cyprus	9 600	430	15	0.7
Syria	6 500	260	36	1.4
Lebanon	31 250	1 250	149	6.0
Israel	32 000	1 400	145	6.5

* The Black Sea coast of Turkey has not been included

Table 1.33   Organic load of industrial waste in the northwestern basin of the Mediterranean Sea [AMBIO, Vol 6, No 6, 1977]

	Inhabitant-equivalents (in millions)	BOD5* (tonnes/year)	BOD5 per km (tonnes/year)
Spanish section (including Ebro River)	5	100 000	342
French section (including Rhone River)	10	200 000	446
Italian section (including Arno River)	5.7	114 000	345
Total northwestern basin	20.7	414 000	

* Biological oxygen demand during a five-day period

Table 1.34   Weighed average oil content of European refinery effluents expressed as kg oil in effluent per 1000 tonnes of crude oil processed   [UNEP/MAP, TRS No.19, 1988]

Refinery location	before 1960	1960-1969	since 1969
Coastal	80	10.6	1.82
Inland	56	4.1	0.92

Table 1.35   Oil pollution of the Mediterranean [EEC, 1982; p.31/32]

ACTIVITY :	QUANTITY :
a) by maritime traffics:	
- tankers operational spills	500,000 t/y
- bilges and bunkering discharges	50,000 t/y
- tanks cleaning for repair	40,000 t/y
- accidental pollution	80,000 t/y
b) land-based discharges:	
- refining and petrochemical industries	30,000 t/y
- industrial machinery and motor vehicles	100,000 t/y
c) off-shore oil production	20,000 t/y
total:	820,000 t/y

Table 1.36   Inputs of petroleum hydrocarbons in the Mediterranean (10^3 tonnes per year) [UNEP/MAP TRS No.19, 1988]

Source	Estimate
Spilled oil from tankers, ballasting and loading operations, bilge and tank washings	330
Land based discharges, run-off Municipal Industrial	160 110
Atmospheric deposition	35
Total	635

Table 1.37   Outputs of hydrocarbons in the Mediterranean (10^3 tonnes per year) [UNEP/MAP TRS No.19, 1988]

Process	Estimate
Evaporation	125
Tar formation and stranding	100
Sedimentation	230
Biodegradation and biological uptake	180
Total	635

Table 1.38 Estimates of annual inputs of hydrocarbons from the atmosphere to the sea by dry and wet deposition for the two transects GYL-ETR2 and ETR2-ETE
[Saliot and Marty, 1986]

Sampling area	Hydrocarbon class	Annual inputs (mg m^{-2} yr^{-1}) wet deposition (rain scavenging)	dry deposition
GYL-ETR2	n-alkanes	1.22-12.2	0.69-6.9
	PAH	0.044-0.44	0.025-0.25
	total hydrocarbons	1.67-16.7	0.94-9.4
ETR2-ETE	n-alkanes	0.30-3.0	0.17-1.7
	PAH	0.05-0.5	0.03-0.3
	total hydrocarbons	0.84-8.4	0.47-4.7

Table 1.39 Comparison of Atmospheric and Riverine Inputs into the Mediterranean Sea
[GESAMP no.68, UNEP, 1985]

	Atmospheric Input y^{-1}	Riverine Input y^{-1}*
Pb	5000-30000 tons	2200- 3100 tons
Zn	4000-25000 tons	11000-17000 tons
Cr	200- 1000 tons	350- 1900 tons
Hg	20- 100 tons	30- 150 tons
^{137}Cs	980 Ci	32 Ci
^{238}Pu	0.45 Ci	0.12 Ci
^{239,240}Pu	20 Ci	0.46 Ci
^{241}Am	1.5 Ci	0.19 Ci

* The riverine inputs are based on measurements of the dissolved-phase transport only.

Note: It should be noted that fallout nuclides can have long residence times in drainage basins and therefore the riverine input may be more ained in the long term than the atmospheric input.

Table 1.40   Estimates on inputs of mercury in the Mediterranean [UNEP/WG.160/8, 1987]

Region	Originating in coastal zones				carried by		total
	domestic		industrial		rivers		
	t/year	% total	t/year	% total	t/year	% total	t/year
I	0.04	2	0.6	24	1.8	74	2.5
II	0.28	1	2.7	8	30	91	33
III	0.04	1	0.2	7	2.5	92	2.7
IV	0.12	1	1.1	10	9.5	89	10.7
V	0.08	>0	0.5	1	40	99	41
VI	0.03	>0	0.16	2	9.6	98	9.8
VII	0.03	2	0.16	9	1.5	88	1.7
VIII	0.05	>0	0.2	2	14	98	14.3
IX	0.01	>0	0.05	1	7	99	7.1
X	0.07	1	1.2	17	5.6	82	6.9
Total	0.75	0.6	6.87	5.4	121.5*	94	129.7

* in this amount 32 metric tons were considered as "background"

Table 1.41   Pesticide consumption by agriculture in the Mediterranean watershed
[from: UNEP, RSRS no.32, 1984; p.56]

Kind of Pesticide	Consumption (t active ingredient per year)			Total[5]	estimated area treated ($10^3 km^2$)
	Greece (1973)	Italy[3] (1975)	Spain[4] (1976)		
A. INSECTICIDES	1335.0	21205.0	6607.4	39683.7	647.1
1. Organochlorine compounds	85.5	2972.4	323.2	5684.3	216.8
1.1 DDT and related compounds	-	866.4	12.7	1970.7	29.1
1.2 BHC and lindane	-	1563.7	122.3	1943.3	126.6
1.3 Cyclodienes (aldrin, dieldrin, endrin, etc.)	-	-	99.2	291.3	15.4
1.4 Other organochlorine compounds	85.5	542.3	89.0	1444.0	44.4
1.5 Unspecified organochlorine compounds	-	-	-	35.3	1.3
2. Carbamates	493.0	2301.3	542.8	4061.4	87.6
2.1 Carbaryl	410.0	2110.3	495.6	3625.6	54.0
2.2 Other carbamates	83.0	191.0	47.2	368.8	32.2
2.3 Unspecified carbamates	-	-	-	67.0	1.4
3. Organophosphorus compounds	496.0	8733.1	1227.8	15432.9	325.6
3.1 Parathion	106.0	2594.6	59.2	2984.8	165.3
3.2 Malathion	109.0	997.0	237.6	1764.5	23.8
3.3 Diazinon	-	1119.4	39.5	1361.9	49.7
3.4 Other organophosphorus compounds	281.0	4022.1	891.5	8841.7	76.7
3.5 Unspecified organophosphorus compounds	-	-	-	480.0	10.1
4. Other insecticides	260.5	7198.2	4513.6	14505.1	17.1
B. FUNGICIDES	25323.5	114593.0	19567.1	187988.3	457.5
1. Copper compounds	2886.9	26109.4	2149.2	34379.2	135.3
2. Mercury compounds	-	-	5.0	9.9	27.1
3. Dithiocarbamates	1066.7	16698.4	1485.9	20962.1	195.0
4. Other fungicides[1]	21369.9	71785.2	15927.0	132637.1	97.1
5. Unspecified fungicides	-	-	-	-	3.0
C. HERBICIDES	489.2	5846.9	834.1	10130.3	133.6
1. Arsenic compounds	-	-	14.9	154.9	0.02
2. Phenoxy compounds	169.8	862.6	373.6	1990.2	37.9
2.1 2,4-D	131.4	673.2	373.6	1682.7	32.0
2.2 MCPA	38.4	189.4	26.5	283.5	5.5
2.3 2,4,5-T[2]	-	-	1.7	5.7	-
2.4 Other phenoxy compounds	-	-	18.3	18.3	0.4
3. Other herbicides	319.4	4984.3	445.6	7948.6	95.2
4. Unspecified herbicides	-	-	-	36.6	0.5
D. ALL OTHER PESTICIDES	-	6920.4	3367.3	14470.6	5.4
TOTAL	27147.7	148565.3	30375.9	252272.9	1243.6

[1] principally sulphur

[2] applied to non-cultivated areas such as highways and railway right-of-ways

[3] except Piemonte, Valle d'Aosta, Lombardia, Trentino Alto Adige and Umbrian regions

[4] Mediterranean watershed only

[5] including Cyprus (1976), Egypt (1975/76), Israel (1974), Lebanon (1973), Libya (1974), Syria (1976), Tunisia (1973/74), Turkey[4] (1976)

Table 1.42 Estimated annual radioactive discharges into the Mediterranean Sea (in curies per year) [AMBIO, Vol 6, No 6, 1977]

Total Activity from site or river	1977		1982		1987	
	Total activity except tritium	Tritium	Total activity except tritium	Tritium	Total activity except tritium	Tritium
Jucar River	—	—	3	50	3	50
Ebro River	4	270	7	770	10	1620
Vandellos Coast	8	200	11	450	14	700
Rhone River	—	—	—	—	—	—
Tyrrhenian Coast	6	100	6	100	19	450
Garigliano	6	5	6	5	6	5
Rotondella (Ionian Sea)	1	1	1	1	1	1
Adriatic Coast	—	—	—	—	6	500
Po River	6	1060	6	1060	11	1650
Yugoslavia Coast	—	—	—	—	6	300
Greece	—	—	—	—	3	150
Turkey	—	—	—	—	3	150
Israel	—	—	—	—	2	170
Egypt	—	—	—	—	4	200
	~30	~1630	~40	~2500	~90	~6000

Table 1.43 Characteristics of principal bacterial and viral fish- and shellfish-borne diseases in man [GESAMP No. 42, 1984]

	Etiological Agent	Principal aquatic food animals involved as source of infection	Sources of infection for aquatic food animal	Pathogenicity for aquatic food animal	Mode of transmission to man	Disease in man and most common manifestations
Bacterial infection	*Salmonella* spp. a) *S.typhi, S.paratyphi* b) other species (e.g., *S.typhimurium, S.enteritidis*)	fish or shellfish secondarily contaminated through polluted waters or through improper handling	a) human faeces and waters contaminated by human faeces b) human and animal faeces, polluted waters	nonpathogenic	ingestion of raw or insufficiently cooked contaminated fish or shellfish	a) typhoid and paratyphoid fever, septicaemia b) salmonellosis: gastroenteritis
	Vibrio parahaemolyticus	marine fish and shellfish	organism occurs naturally in the marine environment	may cause death of shrimps and crabs; experimentally pathogenic for fish	usually through consumption of raw or inadequately cooked fish or shellfish that has not been properly refrigerated	diarrhoea, abdominal pain
Bacterial intoxication	*Clostridium botulinum*	fermented, salted, and smoked fish	sediment, water, animal faeces	toxin can kill fish	ingestion of improperly processed fish or shellfish	botulism: neurological symptoms with high case-fatality rate
	Staphylococcus aureus	fish or shellfish secondarily contaminated through improper handling	man - nose and throat discharges, skin lesions	nonpathogenic	ingestion of fish or shellfish cross-contaminated after cooking	staphylococcal intoxication: nausea, vomiting, abdominal pain, prostration
Bacterial intravital intoxication [1]	*Clostridium perfringens*	fish or shellfish secondarily contaminated through polluted waters or through improper handling	polluted waters, human and animal faeces, sediment	nonpathogenic	ingestion of cooked fish or shellfish that has not been properly refrigerated	diarrhoea, abdominal pain
Bacterial skin infection	*Erysipelothrix insidiosa*	fish, particularly spiny ones (e.g. searobins, redfish) - organism is present in fish slime and meat		nonpathogenic	through skin lesions - usually an occupational disease	erysipeloid - severe inflammation of superficial cutaneous wounds
Viral Infection	virus of infectious hepatitis	shellfish	human faeces and water polluted by human faeces	nonpathogenic	ingestion of raw or inadequately cooked contaminated shellfish	infectious hepatitis

1) Intoxication by toxin produced in the body by bacteria present in heavily contaminated foods

Table 1.43   cont'd

	Etiological agent	Principal aquatic food animals involved as source of infection	Life cycle of parasite	Pathogenicity for aquatic food animal	Mode of transmission to man	Disease in man and most common manifestations
Parasitic infection - trematodes	Clonorchis sinensis (Chinese liver fluke)	freshwater fish-Cyprinidae family (e.g. carp, roach, dace)	1st intermediate host: snail 2nd intermediate host: fish Definitive host: man, dog, cat, other fish-eating mammals	muscle cyst infection	ingestion of raw or insufficiently cooked, infected fish (dried, salted or pickled fish may be involved)	clonorchiasis: signs and symptoms related to liver damage
	Opisthorchis felineus O.viverrini	freshwater fish-Pyprinidae family (e.g. whitefish, carp, tench, bream, barbel)	1st int.host: snail 2nd int. host: man, dog, fox, cat, other fish-eating mammals	muscle and subcutaneous cyst infection	ingestion of raw or insufficiently cooked, infected fish	opisthorchiasis: cirrhosis of the liver
	Heterophyes heterophyes	freshwater or brackish water fish	1st int. host: snail Def. host: man, dog, cat, other fish-eating mammals, fish-eating birds	encyst in muscles and skin	ingestion of raw or insufficiently cooked, infected fish (frequently salted or dried fish)	heterophyiasis: abdominal pain, mucous diarrhoea; eggs may be carried to the brain, heart, etc., causing atypical signs
	Metagonimus yokogawai	freshwater fish (e.g. trout, sweetfish, dace, whitebait)	1st int. host: snail Def. host: man, dog, pig, cat, fish-eating birds	encyst in gills, fin or tail	ingestion of raw or insufficiently cooked, infected fish	metagonimiasis: usually mild diarrhoea
	Paragonismus westermani P.ringeri (Oriental lung fluke)	freshwater crab and crayfish	1st int. host: snail 2nd int. host: crab, crayfish Def. host: man, dog, pig, wild carnivores	encyst in gills, muscles, heart, liver	ingestion of raw or insufficiently cooked, infected crabs or crayfish, or ingestion of water contaminated by metacercarise that have escaped from a crab or crayfish	paragonimiasis: usually chronic cough and haemoptysis from flukes localized in the lungs: flukes may invade other organs
- cestodes	Diphyllobothrium latum	freshwater fish (e.g. pike, trout, turbot)	1st int.host: copepod 2nd int. host: fish Def. host: man, dog, cat, pig, fox, polar bear, other fish-eating mammals	plerocercoid larvae infection of muscles and other organs	ingestion of raw or insufficiently cooked fish (frequently inadequately pickled fish)	diphyllobothriasis: disease may be mild or inapparant; may see signs of gastroenteritis, aneamia, weakness
- nematodes	Anisakis matina	marine fish (e.g. cod, herring, mackerel)		internal larvae infection	usually from ingestion of raw or partially-cooked, pickled or smoked herring	anisakiiasis: eosinophilic enteritis
	Anglostrongylus cantonensis	freshwater shrimp, land crab, possibly cartain marine fish	1st int. host: slug, land snail Def. host: rat Paratenic hosts: shrimp, land crab		ingestion of raw or inadequately cooked shrimp or crabs (sometimes pickled)	easinophilic meningitis

Table 1.44 Overall summary of hair mercury determinations as related to exposure [EUR/ICP/CEH 054, 1987]

Country	Dietary survey	Exposure conditions			Mercury levels in hair				Risk groups	
		Area	Seafood meals	Population sector	No. of samples	Concentration Range		No. of samples		
						Total mercury	Methylmercury	4-10 ppm Me Hg	10 ppm Me Hg	
Greece	n = 1500 simplified questionnaire	Non-polluted (fishermen)	4+/week	M	17	0.56 –38.0	0.16 –36.0	0	1
		Non-polluted	2 /week	F	–	–	–	–	–
				Ch.	87	–	0.09 – 4.15	1	0
		Polluted	2 /week	M	2	0.98 – 1.16	0.62 – 1.00	0	0
				F	19	0.69 – 2.60	0.25 – 1.42	0	0
				Ch.	15	0.48 – 5.10	0.37 – 1.59	0	0
Italy	n = 200 simplified questionnaire	Polluted (fishermen)	4+/week	M	7	3.58 –29.66	3.45 –25.31	2	4
		Polluted	2+/week	F	–	–	–	–	–
				Ch.	–	–	–	–	–
		Non-polluted control	0 /week	M	19	0.26 – 2.97	0.17 – 2.10	0	0
				F	–	–	–	–	–
				Ch.	–	–	–	–	–
Yugoslavia	n = 255 complete questionnaire	Non-polluted	0-7/week av. 2-3	M	–	–	–	–	–
				F	–	–	–	–	–
				Ch.	42	0.41 – 3.80	0.28 – 3.60	0	0
	n = 314	Polluted	0-7/week av. 3-5	M	–	–	–	–	–
				F	–	–	–	–	–
				Ch.	42	0.10 – 5.30	0.09 – 3.70	0	0
TOTAL	2269				250			3	5

Table 1.45 Tourism receipts (millions U.S. Dollars)
[UNEP/MAP, 1988]

	1970	1975	1979	1980	1981	1982	1983	
ESPAGNE	1 681	3 404	6 484	6 968	6 716	7 126	6 836	SPAIN
FRANCE	1 189	3 449	6 823	8 235	7 193	6 991	7 226	FRANCE
ITALIE	1 639	2 579	8 218	8 914	7 554	8 339	9 034	ITALY
MALTE	29	76	206	329	265	185	148	MALTA
YOUGOSLAVIE	274	768	825	1 115	1 350	844	929	YUGOSLAVIA
GRECE	194	644	1 662	1 734	1 881	1 527	1 176	GREECE
TURQUIE	52	201	281	327	381	370	408	TURKEY
CHYPRE	21	15	141	200	243	292	332	CYPRUS
SYRIE			136	156	175	150	110	SYRIA
ISRAEL	104	233	797	866	977	894	1 029	ISRAEL
EGYPTE	71	279	511	562	375	386	469	EGYPT
LIBYE	11	18	9	12	11	12	12	LIBYA
TUNISIE	61	280	521	605	581	545	573	TUNISIA
ALGERIE		51	43	48	152	197	202	ALGERIA
MAROC	136	434	388	397	440	354	417	MOROCCO

Table 1.46   Fishing tonnages in the Mediterranean
                  [MAP, 1986]

	1968	1980
SPAIN	83,197	146,606
FRANCE	39,857	46,800
ITALY	296,952	352,631
MONACO	–	–
MALTA	1,300	1,023
YUGOSLAVIA	30,061	34,968
ALBANIA	4,000	–
GREECE	55,815	75,745
TURKEY	33,287	41,405
CYPRUS	1,354	1,304
SYRIA	800	976
LEBANON	2,500	2,400
ISRAEL	6,864	3,702
EGYPT	13,560	19,939
LIBYA	5,000	4,803
TUNISIA	14,537	60,154
ALGERIA	18,200	38,878
MOROCCO	10,578	27,316
TOTAL	617,692	858,650

Table 1.47   Long-term effects of cadmium on marine biota   [UNEP/WG.160/9, 1987]

Conc. ug l^{-1}	species	effect	reference
	algae		
0.4	P. micans	no growth inhibition	Kayser and Sperling, 1980
0.5	Chlamydomonas	50% growth reduction	Mihnea and Munteanu, 1986
1.2	P. micans	growth inhibition	Kayser and Sperling, 1980
1	I. galbans	growth inhibition	Li, 1980
5	P. micans	growth inhibition	Prevot, 1980
10	P. micans	increased vacuolation and number lysosomes	Soyer and Prevot, 1981
2	S. faeroense	no growth inhibition	Kayser, 1982
10	S. faeroense	growth inhibition	Kayser, 1982
10	4 phytoplankter	growth inhibition	Fisher et al., 1984
25	4 diatoms	no growth inhibition	Fisher and Fround, 1980
25	S. costatum	growth reduction *	Berland et al., 1977
	crustaceans		
4.8	M. bahia	no effect	Nimmo et al., 1978
6.4	M. bahia	reduced reproduction	Nimmo et al., 1978
10	E. depressus	increased mortality	Mirkes et al., 1978
5	M. bahia	no effect	Gentile et al., 1982
10	M. bahia	reduced reproduction	Gentile et al., 1982
	molluscs		
0.5	M. galloprov.	increased cadmium body levels if fed with algae grown in 0.5 ug Cd l^{-1}	Mihnea and Munteanu, 1986
5	C. virginica	larvae, delayed develop.	Zaroogian and Morrison, 1981
10	C. margarita	larvae, reduced growth	Watling, 1982
5	M. edulis	no effect	Stromgren, 1982
10	M. edulis	reduced growth	Stromgren, 1982
400	S. officinalis	50% reduced hatching	Establier and Pascual, 1983
	fish		
5	C. heringa	larvae hatch earlier hatched smaller larvae	Ojaveer et al., 1980
5	P. platessa	reduced feeding and growth rate	Von Westernhagen et al., 1978
5	P. flesus	reduction in potassium and calcium level in blood	Larsson et al., 1981
10	F. heteroclitus	inhibition of fin regeneration	Weis and Weis, 1976
50	L. limanda	30% mortality	Westernhagen et al., 1980
50	T. adspersus	enzyme inhibition enzyme induction	MacInnes et al., 1977

* medium contains TRIS

Table 1.48   Number and extent of forest fires in Community member states, 1985 [CEC, 1987]

Member state	Number	Area affected by fire (ha)		Total area (ha)
		Forest	Farmland or natural land	
Italy	16,903	62,515	98,420	160,935
France	5,596	35,050	24,800	59,850
Greece	727	37,511	34,354	71,865
Spain	9,770	147,235	208,763	355,998
Portugal	5,459	81,475	54,095	135,570
Total	38,455	363,786	420,432	784,218

Note: data refer to the nine months to 30/9/1985 and are provisional estimates.

Table 1.49   Economic impacts of forest fires in Spain [Gerelli, 1987]

COMMUNIDAD	(A)	(B)	(C)	(D)
ANDALUCIA	662	22091	415,00	0,150
ARAGON	167	4073	379,50	0,880
ASTURIAS	265	3558	274,20	1,020
BALEARES	110	400	11,80	0,190
CANARIAS	72	1333	81,20	0,300
CANTABRIA	176	1886	149,35	0,840
CASTILLA - LA MANCHA	199	6655	142,75	0,140
CASTILLA Y LEON	965	26374	1258,90	0,820
CATALUNA	394	9412	582,97	0,630
EXTREMADURA	383	23699	298,96	0,810
GALICIA	2924	38888	1417,57	1,450
MADRID	116	494	16,32	0,120
MURCIA	24	91	2,14	0,004
NAVARRA	0	0	0,00	0,000
PAIS VASCO	238	1806	225,14	0,840
RIOJA, LA	59	1050	5,29	0,020
COMMUNIDAD VALENCIANA	470	22500	535,03	0,450

(A) : Total number of reported fires
(B) : Surface affected (Ha.)
(C) : Estimation of direct money loss
(D) : Percentage of Regional Agricultural income.

Table 1.50   The land-based sources protocol
[MEDWAVES No. 4, 1986]

WHAT DOES THE PROTOCOL STATE?

Article 4

1. This Protocol shall apply:
(a) To polluting discharges reaching the Protocol Area from land-based sources within the territories of the Parties, in particular:
directly, from outfalls discharging into the sea or through coastal disposal;
indirectly, through rivers, canals or other watercourses, including underground watercourses, or through run-off;
(b) To pollution from land-based sources transported by the atmosphere, under conditions to be defined in an additional annex to this Protocol and accepted by the Parties in conformity with the provisions of article 17 of the Convention.
2. This Protocol shall also apply to polluting discharges from fixed man-made off-shore structures which are under the jurisdiction of a Party and which serve purposes other than exploration and exploitation of mineral resources of the continental shelf and the sea-bed and its subsoil.

Article 5

1. The Parties undertake to eliminate pollution of the Protocol Area from land-based sources by substances listed in annex I to this Protocol.

(Annex I states that:
"The substances, families and groups of substances listed, have been selected mainly on the basis of their Toxicity; Persistence; Bioaccumulation".)

Article 6

1. The Parties shall strictly limit pollution from land-based sources in the Protocol Area by substances or sources listed in annex II to this Protocol.

(Annex II states that:
"The substances, families and groups of substances, or sources of pollution, listed, have been selected mainly on the basis of criteria used for annex I, while taking into account the fact that they are generally less noxious or are more readily rendered harmless by natural processes and therefore generally affect more limited coastal areas.

The present annex does not apply to discharges which contain substances listed in section A that are below the limits defined jointly by the Parties".)

Article 7

1. The Parties shall progressively formulate and adopt, in co-operation with the competent international organizations, common guidelines and, as appropriate, standards or criteria.
2. Without prejudice to the provisions of article 5 of this Protocol, such common guidelines, standards or criteria shall take into account local ecological, geographical and physical characteristics, the economic capacity of the Parties and their need for development, the level of existing pollution and the real absorptive capacity of the marine enviroment.

Article 8

Within the framework of the provisions of, and the monitoring programmes provided for in, article 10 of the Convention, and if necessary in co-operation with the competent international organizations, the Parties shall carry out at the earliest possible date monitoring activities in order:
(a) Systematically to assess, as far as possible, the levels of pollution along their coasts, in particular with regard to the substances or sources listed in annexes I and II, and periodically to provide information in this respect;
(b) To evaluate the effects of measures taken under this Protocol to reduce pollution of the marine enviroment.

Article 10

1. The Parties shall, directly or with the assistance of competent regional or other international organizations or bilaterally, co-operate with a view to formulating and, as far as possible, implementing programmes of assistance to developing countries, particularly in the fields of science, education and technology.

Table 1.51   Implementation workplan for land-based sources protocol   [UNEP]

1. USED LUBRICATING OILS	1986
2. SHELL-FISH AND SHELL-FISH GROWING WATERS	1986
3. CADMIUM AND CADMIUM COMPOUNDS	1987
4. MERCURY AND MERCURY COMPOUNDS	1987
5. ORGANOHALOGEN COMPOUNDS	1987
6. PERSISTENT SYNTHETIC MATERIALS WHICH MAY FLOAT, SINK OR REMAIN IN SUSPENSION	1988
7. ORGANOPHOSPHOROUS COMPOUNDS	1988
8. ORGANOTIN COMPOUNDS	1988
9. RADIOACTIVE SUBSTANCES	1989
10. CARCINOGENIC, TERATOGENIC OR MUTAGENIC SUBSTANCES	1989
11. PATHOGENIC MICROORGANISMS	1989
12. CRUDE OILS AND HYDROCARBONS OF ANY ORIGIN	1990
13. ZINC, COPPER AND LEAD	1990
14. NICKEL, CHROMIUM, SELENIUM AND ARSENIC	1990
15. INORGANIC COMPOUNDS OF PHOSPHOROUS AND ELEMENTAL PHOSPHORUS	1991
16. NON-BIODEBRADABLE DETERGENTS AND OTHER SURFACE-ACTIVE SUBSTANCES	1991
17. THERMAL DISCHARGES	1991
18. ACID OR ALKALINE COMPOUNDS	1992
19. SUBSTANCES HAVING ADVERSE EFFECT ON THE OXYGEN CONTENT	1992
20. BARIUM, URANIUM AND COBALT	1992
21. CYANIDES AND FLUORIDES	1993
22. SUBSTANCES, OF A NON-TOXIC NATURE, WHICH MAY BECOME HARMFUL OWING TO THE QUANTITIES DISCHARGED	1993
23. ORGANOSILICON COMPOUNDS	1993
24. ANTIMONY, TIN AND VANADIUM	1994
25. SUBSTANCES WHICH HAVE A DELETERIOUS EFFECT ON THE TASTE AND/OR SMELL CF PRODUCTS FOR HUMAN CONSUMPTION	1994
26. BIOCIDES AND THEIR DERIVATIVES NOT COVERED IN ANNEX I	1994
27. TITANIUM, BORON AND SILVER	1995
28. MOLYBDENUM, BERYLLIUM, THALLIUM AND TELLURIUM	1995

Table 1.52   Specially protected areas in the Med
[MEDWAVES No. 8, 1987]

THE SPA PROTOCOL: HIGHLIGHTS

The main provisions of the Specially Protected Areas Protocol are the following:

ARTICLE 1

1. The Contracting Parties to this Protocol shall take all aprropriate measures with a view to protecting those marine areas which are important for the safeguard of the natural resources and natural sites of the Mediterranean Sea Area, as well as for the safeguard of their cultural heritage in the region.

2. Nothing in this Protocol shall prejudice the codification and development of the law of the sea by the United Nations Conference on the Law of the Sea convened pursuant to resolution 2750 C (XXV) of the General Assembly of the United Nations, nor the present or future claims and legal views of any State concerning the law of the sea and the nature and extent of coastal and flag State jurisdiction.

ARTICLE 2

For the purposes of the designation of specially protected areas, the area to which this Protocol applies shall be the Mediterranean Sea Area as defined in article 1 of the Convention for the Protection of the Mediterranean Sea against Pollution; it being understood that, for the purposes of the present Protocol, it shall be limited to the territorial waters of the Parties and may include waters on the landward side of the baseline from which the breadth of the territorial sea is measured and extending, in the case of watercourses, up to the freshwater limit. It may also include wetlands or coastal areas designated by each of the Parties.

ARTICLE 3

1. The Parties shall, to the extent possible, establish protected areas and shall endeavour to undertake the action necessary in order to protect those areas and, as appropriate, restore them, as rapidly as possible.

ARTICLE 6

1. If a Party intends to establish a protected area contiguous to the frontier or to the limits of the zone of national jurisdiction of another Party, the competent authorities of the two Parties shall endeavour to consult each other with a view to reaching agreement on the measures to be taken and shall, among other things, examine the possibility of the establishment by the other Party of a corresponding protected area or the adoption by it of any other appropriate measure.

2. If a Party intends to establish a protected area contiguous to the frontier or to the limits of the zone of national jurisdiction of a State which is not a party to this Protocol, the Party shall endeavour to work together with the competent authorities of that State with a view to holding the consultations referred to in the preceding paragraph.

ARTICLE 9

1. The Parties shall, in promulgating protective measures, take into account the traditional activities of their local populations. To the fullest extent possible, no exemption which is allowed for this reason shall be such as:

a) to endanger either the maintenance of ecosytems protected under the terms of the present Protocol or the biological processes contributing to the maintenance of those ecosystems;

b) to cause either the extinction of, or any substantial reduction in, the number of individuals making up the species or animal and plant populations within the protected ecosystems, or any ecologically connected species or populations, particularly migratory species and rare, endangered or endemic species.

ARTICLE 10

The Parties shall encourage and develop scientific and technical research on their protected areas and on the ecosystems and archaeological heritage of those areas.

ARTICLE 11

The Parties shall endeavour to inform the public as widely as possible of the significance and interest of the protected areas and of the scientific knowledge which may be gained from them from the point of view of both nature conservation and archaeology. Such information should have an appropriate place in education programmes concerning the environment and history. The Parties should also endeavour to promote the participation of their public and their nature conservation organizations in appropriate measures which are necessary for the protection of the areas concerned.

ARTICLE 13

The Parties shall exchange scientific and technical information concerning current of planned research and the results expected. They shall, to the fullest extent possible, coordinate their research. They shall, moreover, endeavour to define jointly or to standardize the scientific methods to be applied in the selection, management and monitoring of protected areas.

ARTICLE 14

2. The parties shall designate persons responsible for protected areas. Those persons shall meet at least once every two years to discuss matters of joint interest and especially to propose recommendations concerning scientific, administrative and legal information as well as the standardization and processing of data.

ARTICLE 16

Changes in the delimitation of legal status of a protected area or the suppression of all or part of such an area may not take place except under a similar procedure to that followed for its establishment.

Table 1.53   Quality requirements for bathing water [E.E.C.]

Parameters		G (Guide)		I (Mandatory)		Minimum sampling frequency
Microbiological						
1 Total coliforms	(/100ml)	500	(80%)	10 000	(95%)	Fortnightly (1)
2 Faecal coliforms	(/100 ml)	100	(80%)	2 000	(95%)	Fortnightly (1)
3 Faecal streptococci	(/100 ml)	100	(90%)	-		(2)
4 Salmonella	(/l)	-		0	(95%)	(2)
5 Entero viruses	(PFU/10 l)	-		0		(2)
Physico-chemical		-				
6 pH		-		6 to 9 (0)		(2)
7 Colour				No abnormal change in colour (0)		(1)
8 Mineral oils	(mg/l)	-		-		(2)
		-		No film visible on the surface of the water and no odour		Fortnightly (1)
9 Surface-active substances reacting with methylene blue	(mg/l) (lauryl- sulfate)	≤ 0.3 -		- No lasting foam		(2) Fortnighty (1)
10 Phenols (phenol indices)	(mg/l) C_5H_5OH	≤ 0.3 -		- No specific odour		(2) Fortnightly (1)
11 Transparency	(m)	≤ 0.005 2		≤ 0.05 1 (0)		(2) Fortnightly (1)
12 Dissolved oxygen	(% satn)	80 to 120		-		(2)
13 Tarry residues and floating materials such as wood, plastic articles, bottles, containers of glass, plastic, rubber or any other substance. Waste or splinters.		Absence				Fortnightly (1)

(1) When a sampling taken in previous years produced results which are appreciably better than those in this Annex and when no new factor likely to lower the quality of the water has appeared, the competent authorities may reduce the sampling frequency by a factor of 2.
(2) Concentration to be checked by the competent authorities when an inspection in the bathing area shows that the substance may be present or that the quality of the water has deteriorated. Percentage values in ()= proportion of samples in which the numerical values stated must not be exceeded.
*PFU=Plaque Forming Unit - a method of counting viruses; one PFU may be regaded as one infective virus particle.

Extract from the E.E.C. Directive on the Quality of Bathing Water

Table 1.54 Limit values and time limits for discharges of cadmium
[UNEP/WG.160/9,1987]

Industrial sector (1)	Unit of measurement	Limit values which must be complied with as from	
		1.1.1986	1.1.1989 (2)
1. Zinc mining, lead and zinc refining, cadmium metal and non-ferrous metal industry	Milligrams of cadmium per litre of discharge	0.3 (3)	0.2 (3)
2. Manufacture of cadmium compounds	Milligrams of cadmium per litre of discharge	0.5 (3)	0.2 (3)
	Grams of cadmium discharged per kilogram of cadmium handled	0.5 (4)	(5)
3. Manufacture of pigments	Milligrams of cadmium per litre of discharge	0.5 (3)	0.2 (3)
	Grams of cadmium discharged per kilogram of cadmium handled	0.3 (4)	(5)
4. Manufacture of stabilizers	Milligrams of cadmium per litre of discharge	0.5 (3)	0.2 (3)
	Grams of cadmium discharged per kilogram of cadmium handled	0.5 (4)	(5)
5. Manufacture of primary and secondary batteries	Milligrams of cadmium per litre of discharge	0.5 (3)	0.2 (3)
	Grams of cadmium discharged per kilogram of cadmium handled	1.5 (4)	(5)
6. Electroplating (6)	Milligrams of cadmium per litre of discharge	0.5 (3)	0.2 (3)
	Grams of cadmium discharged per kilogram of cadmium handled	0.3 (4)	(5)
7. Manufacture of phosphoric acid and/or phosphatic fertilizer from phosphatic rock (7)		–	–

(1) Limit values for industrial sectors not mentioned in this table will, il necessary, be fixed by the Council at a later stage. In the meantime the Member States will fix emission standards for cadmium discharges autonomously in accordance with Directive 76/464/EEC. Such standards must take into account the best technical means available and must not be less stringent than the most nearly comparable limit value in this Annex.

(2) On the basis of experience gained in implementing this Directive, the Commission will, pursuant to Article 5 (3), submit in due course to the Council proposals for fixing more restrictive limit values with a view to their coming into force by 1992.

(3) Monthly flow-weighted average concentration of total cadmium

(4) Monthly average

(5) It is impossible for the moment to fix limit values expressed as load. If need be, these values will be fixed by the Council in accordance with Article 5 (3) of this Directive. If the Council does not fix any limit values, the values expressed as load given in column "1.1.1986" will be kept.

(6) Member States may suspend application of the limit values until 1 January 1989 in the case of plants which discharge less than 10 kg of cadmium a year and in which the total volume of the electroplating tanks is less than 1.5 m³, if technical or administrative considerations make such a step absolutely necessary.

(7) At present there are no economically feasible technical methods for systematically extracting cadmium from discharges arising from the production of phosphoric acid and/or phosphatic fertilizers from phosphatic rock. No limit values have therefore been fixed for such discharges. The absence of such limit values does not release the Member States from their obligation under Directive 76/464/EEC to fix emission standards for these discharges.

Table 1.55 Limit values and time limits for mercury discharges by industrial sectors other than the chlor-alkali electrolysis industry [UNEP/WG.160/8,1987]

Industrial sector (*)	Limit value which must be complied with as from: 1 July 1986	1 July 1989	Unit of measurement
1. Chemical industries using mercury catalysts:			
(a) in the production of vinyl chloride	0.1	0.05	mg l^{-1} effluent
	0.2	0.1	g t^{-1} vinyl choride production capacity
(b) in other processes	0.1	0.05	mg l^{-1} effluent
	10	5	g kg^{-1} mercury processed
2. Manufacture of mercury catalysts used in the production of vinyl chloride	0.1	0.05	mg l^{-1} effluent
	1.4	0.7	g kg^{-1} mercury processed
3. Manufacture of organic and non-organic mercury compounds (except for products referred to in point 2).	0.1	0.05	mg l^{-1} effluent
	0.1	0.05	g kg^{-1} mercury processed
4. Manufacture of primary batteries containing mercury	0.1	0.05	mg l^{-1} effluent
	0.05	0.03	g kg^{-1} mercury processed
5. Non-ferrous metal industry (**)			
5.1 Mercury recovery plants	0.1	0.05	mg l^{-1} effluent
5.2 Extraction and refining of non-ferrous metals	0.1	0.05	mg l^{-1} effluent
6. Plants for the treatment of toxic wastes containing mercury	0.1	0.05	mg l^{-1} effluent

(*) Limit values for industrial sectors other than the chlor-alkali electrolysis industry which are not metnioned in this table, such as the paper and steel industries or coal-fired power stations, will, if necessary, be fixed by the Council at a later sage. In the meantime, the Member States will fix emission standards for mercury discharges autonomously in accordance with Directive 76/464/EEC. Such standards must take into account the best technical means available and must not be less stringent than the most nearly comparable limit value in this Table.

(**) On the basis of experience gained in the implementation of this Directive the Commision will, pursuant to Article 6 (3), submit to the Council proposals for more stringent limit values to be introduced 10 years after the notification of this Directive.

The limit values given in the table correspond to a monthly average concentration or to a maximum monthly load.

The amounts of mercury discharged are expressed as a function of the amount of mercury used or handled by the industrial plant over the same period or as a function of the installed vinyl chloride production capacity.

Table 1.56   Interim environmental quality criteria for bathing waters in the Mediterranean Sea [UNEP/IG.56/5, 1985]

Parameter	Concentrations per 100 ml not to be exceeded 50% / 90% of the samples	Minimum number of samples	Analytical method	Intepretation method
Faecal coliforms	100 / 1000	10	WHO/UNEP Reference Method No. 3, "Determination of Faecal Coliforms in sea-water by the Membrane Filtration Culture Method", or WHO/UNEP Reference Method No. 22, "Determination of Faecal Coliforms in sea-water by the Multiple Test Tube Method".	Graphical or analytical adjustment to a log normal probability distribution

Table 1.57   Proposed interim criteria for shellfish waters in the Mediterranean [UNEP/WG.160/13, 1987]

Matrix	Parameter	Concentration
Shellfish	Faecal coliforms	< 300 per 100 ml flesh + intervalvular fluid or flesh in at least 75% of the samples.

Minimum sampling frequency	Analytical method	Interpretation method
every 3 months (more frequently whenever local circumstances so demand).	Multiple tube fermentation and counting according to MPN (most probable number) method. Incubation period: $37 \pm 0.5°C$ for 24 or 48 h followed by $44 \pm 0.2°C$ for 24 h.	By individual results, histograms or graphical adjustment of a lognormal-probability distribution.

Table 1.58　Estimated expenditure on environmental protection in Community member states, 1978　[CEC, 1987]

Member state	Total expenditure (million ECUs)	% GDP
Belgium	(290)	(0.4)[2]
Denmark	435	1.0[2]
France	2970	0.8
Germany	7854[1]	1.5
Greece	107	0.3
Ireland	(90)	(0.9)
Italy	n.d.	n.d.
Luxembourg	n.d.	n.d.
Netherlands	1412[2]	1.1
Portugal	20	0.2
Spain[3]	(175)	(0.2)
United Kingdom	(3608)	(1.4)
Total	(16,884)	(1.2)

Notes: ( ) - estimated   1 - revised data   2 - revised 1980 data   3 - partial data

Table 1.59　Capital expenditure on sewage treatment plants in Community member states, 1970-73 and 1979-82　[CEC, 1987]

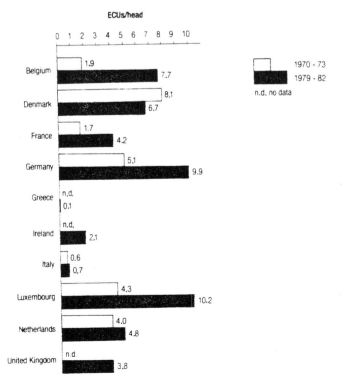

Table 1.60   Summary of the blue plan scenarios [UNEP]

	International economic context	Populations and their movements	National development	Management of Mediterranean areas	Taking the environment into account
Reference trend scenario T-1	reduced growth, no efficient European policy, trade benefits the strongest partner, oil prices are unstable	current growth trends continue, urbanisation heading for between 70% and 80%, tourism continues to grow	co-operation continues along its current main lines	attempts to distribute economic activities throughout the region	limited attempts to protect the environment
Worse trend scenario T-2	weak economic growth, strongest partner uses bilaterism, fixed oil prices	in the North as T-1, in the East and South: very high levels of population growth	national strategies have to adapt to a comparatively harsh international context	regional and physical planning face pressure from economic needs	no real action to protect the environment
Moderate trend scenario T-3	growth at the expences of the environment, a harsh international trade competition, increasing oil prices	the population growth rates are higher than in T-1, but lower than T-2, urban growth rates are lower than in T-1	better adjustment of individual strategies, better co-operation	attempts as in T-1 in a more determined fashion	a strenghtening of international agreements, and a greater respect of standards, conventions, etc.
Alternative reference scenario A-1	considerable growth, coherent development at the basin, intensif exchanges between North and South, regularized oil prices	for the North same rates as in T-3, lower than T-3 for the South and East	at domestic level development towards "self reliance"	enlightening regional policy, a better rural-urban balance and coast-hinterland balance	strong environmental protection and pollution control policies
Alternative integration scenario A-2	considerable growth, different new economic units at the basin, regularized oil prices	the same rates as in A-1	development strategies remain self-reliant but on a more regional basis	a reduced use of the coastline, a better rural-urban balance	somewhat lighter policies than in A-1

Table 1.61   Pollution control expenditure indices for 1990 (constant 1978 prices, 1978=100) [Gerelli, 1987; p.28]

Country	Constant emission scenario
Belgium	225
France	160
Germany	130
Netherlands	185
United Kingdom	110
Denmark	260
Norway	125
Sweden	130
Greece	(700)
Italy	240
Portugal	(1000)
Spain	(400)

Assumes that for 1990 total emissions are kept at 1978 levels except for motor vehicles, which may be lower. The upper end of the cost range of all cost estimates was assumed.

Table 1.62   Funding needs (1)   [Gerelli, 1987; p.134/135]

GREECE	million ECU		
. Water projects (waste water treatment plants, sewer network)	181.4		
. Air pollution	75.129		
. Research activities	0.194		
Total	256.723		

ITALY	M.ECU 1986	temp.	perm.
. Water	10.940	280.000	35.000
. Solid waste	1.059	20.000	3.000
. Solid conservation	10	--	--
. Protected areas	300	?	1.350
. Training	233	--	?
. Data collection and processing	82	1.380	520
. Other measures	72	?	?
Total	13.146	301.380	39.870

PORTUGAL	Millions ECU
. Water projects	532.15
. Solid waste	108.06
. Nature preservation	147.21
. Soil erosion	98.55
. Monitoring	18.3
Total	904.27

SPAIN	Million ECU
. Water projects	68.0
. Air	112.7
. Soil conservation	98.9
Total	279.6

(1) The exchange rates used refer to July 1987: Greek drachmas 155.259; Lit 1497.83; Spanish pesetas 143.369; Portughese escudos 161.791

Table 1.63 Typical nutrient compositions for various types of sewage effluents   [UNEP/MAP, TRS No 21, 1988]

TYPE OF EFFLUENT	Nutrient Concentration mg.dm^{-3} of N, P			
	N_{tot}	NH_4	NO_2+NO_3	P_{tot}
Raw Sewage	45	25	7	9
Primary Treatment	31	15	7	6
Secondary Treatment	25	2	13	6
Oxidation Ponds	11	7	1	6
PO_4-Precipitation	23	2	13	2

Table 1.64   Domestic sewage: estimated annual per capita loads and reduction due to treatment [UNEP, RSRS no.32, 1984; p.10]

Variable	Annual per capita sewage loads		Cumulative percentage reduction due to sewage treatment[a]		
	Units per capita and year	Min - Max	Screening/ grit chamber	Primary sedimentation	Biological treatment
1. Volume:					
Total discharge	m^3	30 - 200	0	0	0
2. Organic matter:					
BOD	Kg	10 - 25	0-10	10-30	50-80
COD	Kg	20 - 55	0-10	10-20	30-60
3. Nutrients:					
Phosphorus	Kg	0.5 - 1.1	0-10	10-20	10-30
Nitrogen	Kg	4	0-10	20-40	20-50
4. Specific organics:					
Detergents	Kg	0.4 - 1.0	0-10	0-10	30-70
5. Metals:					
Mercury	g	0.02 - 0.04	0-10	0-10	40-60
Lead	g	10 - 20	0-10	20-40	60-90
Chromium	g	10 - 30	0-10	20-40	50-90
Zinc	g	50 - 100	0-10	20-50	50-80
6. Suspended matter:					
TSS	Kg	20 - 30	0-10	50-70	70-95
VSS	Kg	15 - 20	0-10	40-60	70-95

a) all figures are percentage values based upon raw sewage concentrations

2. SPANISH MEDITERRANEAN ENVIRONMENT

Five autonomous regions, covering 12 provinces, are located in the Mediterranean region. They have undergone important changes over the past years. The process of environmental degradation has been accelerated due to increasing pressures of human habitation, including tourism and industrial activity. In many areas large amounts of pollutants are emitted, resulting in a lower environmental quality. These developments have made the Mediterranean region one of the most contaminated in Spain. Exposure to contaminants can result in health problems and affect economic processes (income from tourism).

The state of the environment (including water quality and land quality) are discussed in section 2.1, while the pressures on the environment (including discharges from municipalities, industries, rivers and ships) are outlined in section 2.2. The economic impacts and institutional aspects are described in sections 2.3 and 2.4, respectively.

2.1. State of the environment

2.1.1. Water quality

2.1.1.1. Heavy metals in marine waters and sediments

Figure 2.1 summarizes the results of a July 1987 survey of metal concentrations in seawater along the Mediterranean coast by the European Community (EC) (CEC, 1988). Forty Mediterranean beaches were investigated. Roquetas de Mar (Pl. Serena) and Mojácar were systematically polluted by mercury, lead and zinc. El Grau de Castellón, Mongat and Playa de Magalluf (Balearic Islands) were heavily polluted by cadmium, lead and zinc, whereas Castelldefels had high cadmium and zinc concentrations. Heavy zinc pollution was measured at Alicante (San Juan).

High values of lead in sediments have been found near Portman and the adjacent Cartagena area, due to the mining waste from the nearby lead and zinc ore washing plants (from 300 to 1,160 mg/kg), as shown in Figure 2.2A. The ore washing plant discharges in Portman Bay were stopped on March 30, 1990, after an agreement between the mining company and MOPT was reached. Near Barcelona an elevated concentration of lead was recorded (230 mg/kg) possibly due to industrial discharge and urban run-off.

The most significant mercury values are found between Barcelona (4.9 mg/kg) and the Besós River (3.1 mg/kg), as seen in Figure 2.2B.

The concentrations measured for cadmium were "normal" except near Portman (3.2 mg/kg), as a result of mining activity (MOPT, parte II, 1987).

De León et al. (1985) reported concentrations for cadmium up to 10.4 mg/kg at Portman (Table 2.1).

2.1.1.2. Heavy metals in marine organisms

Mercury (Hg) levels in marine organisms are generally elevated along the Spanish Mediterranean coast, as seen in Table 2.2. Accumulation of Hg was most evident in shrimps (Aristeus antennatus). In Palma de Mallorca the recorded range was 70-1,230 µg/kg wet weight. The striped mullet Mullus barbatus contained lower concentrations for corresponding locations, with Palma de Mallorca also the most polluted one.

From Valencia to Castellón even higher mercury concentrations in marine organisms were recorded, with maximum values of 1,921 and 713 µg/kg wet weight in striped mullets (Mullus barbatus) and shrimps (Parapenaeus longirostris), respectively (Hernández Hernández et al., 1987).

Elevated levels for lead (Pb) are observed in the mussel

Mytilus galloprovincialis (ranging from 100 to 125,000 µg/kg wet weight; 125,000 µg/kg at Portman and 26,680 µg/kg at adjacent Cartagena), as shown in Table 2.3. Lead concentrations in molluscs (e.g. Mytilus galloprovincialis) follow the lead concentrations in sediments, as shown previously.

Shrimps (Aristeus antennatus) contain lower levels of lead (0-760 µg/kg wet weight); an occasional maximum (19,000 µg/kg wet weight) in Cartagena is observed in the striped mullet Mullus barbatus.

A general increase in the concentration of cadmium (Cd) (Table 2.4) is observed in the three different organisms, in the order M. barbatus< A. antennatus< M. galloprovincialis. In the latter Cd ranged from 56 to 930 µg/kg wet weight. As in the case of lead, cadmium concentrations in molluscs are higher in places where sediments are also highly polluted. Similar results were also reported by De León et al. (1985) as shown in Table 2.1.

2.1.1.3. Organics in marine sediments and seawater

Elevated concentrations of insecticide dichlorodiphenyl trichlorethane (DDT) and polychlorinated biphenyls (PCBs) in marine sediments along the Spanish coast have been found, as shown in Figure 2.3. For both organics the highest level was recorded around Barcelona between the Llobregat and Besós rivers: 117 µg/kg dry weight of DDT and 524 µg/kg of PCBs. Although the application of DDT has been banned, it is still persistent in the natural environment.

A more detailed sampling of the sediments along the Catalonian coast (Table 2.5) shows that the most polluted zones are located near the Ebro Delta and heavily industrialized areas in Barcelona.

The Balearic region has lower concentrations of PCBs and DDTs than the Catalonian coastal area (Table 2.6). The most significant DDT and PCB values in the Baleares are found near Palma (5.16 µg/kg and 8.82 µg/kg, respectively) and

Ibiza (2.76 µg/kg and 8.95 µg/kg, respectively) harbors.

Seawater values of petroleum hydrocarbons along the Mediterranean coast are shown in Table 2.7. The highest values are observed in Alicante (maximum: 8.26 µg/l) and Portman (maximum: 6.5 µg/l).
The concentrations of aliphatic hydrocarbons (C_{14} to C_{24}) in the delta sediments of the Catalonian rivers are shown in Table 2.8. The deltas of the Tordera and Besós are the most polluted with 565 and 424 µg/kg, respectively.

2.1.1.4. Organics in marine organisms

Concentrations of organochlorine pesticides and PCBs in marine organisms collected along the Castellón and Valencia coast showed generally no relationship with the ambient concentration at the sampling locations. On the other hand, DDT was higher in the mussel Mytilus galloprovincialis (48 µg/kg wet weight) collected at Cullera near the mouth of the Júcar River, which passes through a large and important agricultural area (Figure 2.4). For the organs of the red mullet (Mullus surmuletus) and the striped mullet (Mullus barbatus) the decreasing order in pesticide content was generally: liver >digestive organs >muscle and for those of shrimps (Aristeus antennatus): liver >gonads >digestive organs >muscle. In tuna (Thunnus thynnus) the highest organochlorine content is found in muscle tissue compared to other organs (Figure 2.5).

The concentrations of p,p'-DDE, p,p'-DDD, p,p'-DDT and total DDTs in samples of marine organisms from the delta zone of the Ebro River are listed in Table 2.9. The most elevated level - 11,696 µg/kg dry weight - is observed in the liver of the hake (Merluccius merluccius). The concentration of total DDTs in the liver and gonads is about 17 times higher than in the soft tissue or muscle due to its greater lipid content.

The concentrations of PCB Arochlor 1254, PCB Arochlor 1260 and total PCBs in samples of marine organisms from the Ebro Delta are given in Table 2.10. The most elevated level — 26.5 µg/g dry weight — is observed in the liver of the hake (<u>Merluccius</u> <u>merluccius</u>). The concentration of total PCBs in the liver and gonads is around 12 times higher than in the soft tissue or muscle due to its higher lipid content.

Along the Catalonian region considerable levels of petroleum hydrocarbons of 500-3,200 µg/g dry weight (measured on the basis of the unresolved complex mixture (UCM) in the chromatograms of the saturated fractions) were found in <u>Mytilus</u> off Barcelona, as shown in Table 2.11. Levels in muscle tissue of mullet (<u>Mullus</u> spp.), hake (<u>Merluccius</u> spp.), scad (<u>Trachurus</u> spp.) and anchovy (<u>Engraulis</u> spp.) were 0.2-22.2 µg/g dry weight and 1.7-11.1 µg/g dry weight for saturated and aromatic hydrocarbons, respectively.

The concentrations of aliphatic hydrocarbons (C_{14} to C_{24}) in samples of marine organisms from the delta zone of the Ebro River (Table 2.12) range from 2.3 to 697.7 µg/g dry weight. The average concentrations in liver and gonads (341 µg/g dry weight) are 40 times higher than in soft tissue or muscle (8.7 µg/g dry weight). These results are comparable with the bioconcentration phenomena of DDT and PCBs.

2.1.1.5. Bacteria and viruses in the marine environment

The results of a July 1987 survey on bacteriological bathing-water quality conducted by the EC (CEC, 1988) for 40 Spanish Mediterranean beaches are shown in Figure 2.6. Of the 40 beaches examined, 25 complied with the EC guide number (recommended criteria limit), 13 exceeded the EC guide number but were in accordance with the required standard, while the remaining 2 (Peñíscola and Puerto de Alcudia) exceeded the required standard.

The water quality as measured by total and fecal coliforms

and fecal streptococci in 1986 along the coast of Murcia was generally in accordance with the EC guide numbers (Figure 2.7). The highest bacterial levels were observed at San Javier in the so-called Mar Menor, a saltwater coastal lagoon, with a high population density along its coastline. The bacterial levels are highest during the summer season, when water temperatures and the wastewater discharges are the highest (Consejería de Política Territorial y Obras Públicas, 1987).

Seawater from three beaches (extensively used for recreational purposes) in the urban area of Barcelona were analyzed for human viruses in June 1979. The three beaches studied were:
- Barceloneta which is the natural beach of Barcelona, a densely populated city of around 3,000,000 inhabitants. Not far from this beach are several outfalls of waste waters from Barcelona, of which around 70% is being treated;
- Castelldefels which is located 20 kilometers south of Barcelona near a town of 20,000 inhabitants. The Llobregat River is the main pollution focus in this coastal area;
- Badalona which is located nearly 5 kilometers north of Barcelona near a city of 300,000 inhabitants. The main polluting sources are the Besós River and some local sewage outfalls.

Results from samples taken 4 or 5 meters offshore at the three surveyed beaches are shown in Table 2.13. Viruses were recorded in approximately similar amounts at the three beaches in all samples analyzed, ranging from 0.12 to 1.60 MPN CU (expressed as the most probable number of cytopathogenic units) per liter.

To study the spread of human viruses in marine waters, their presence at different distances (2, 5, 20 and 1,000 meters) from one of the sewage outfalls in Badalona was analyzed. Results in Table 2.14 indicate that viruses spread far from the shore, since the number of viruses recovered from samples taken 2, 5, 20 and 1,000 meters from the outfall does not differ significantly. Even at a distance of 1,000 meters were 4 of the 9 samples positive.

Bacteriological data obtained in a routine survey (Juan Martínez, personal communication) from samples corresponding in space and time to the previous ones showed that there is not a clear correlation between the presence of human viruses of fecal origin and bacterial indicators of fecal contamination. Thus, all samples from Castelldefels, around 80% of the samples taken in Badalona and 50% of the samples at Barceloneta Beach indicate a satisfactory bacteriological quality according to World Health Organization and Spanish Public Health regulations. Nonetheless viruses were recovered in similar amounts in all samples taken at these three beaches.
The same survey also showed that the amount of both fecal coliform bacteria and fecal streptococci decreased between 100 and 1,000 times when measured from 2 to 1,000 meters from the wastewater outfall. There was not a comparable decrease in the virus concentration. This shows that the survival of viruses in seawater is longer than the survival of the indicator bacteria.

<u>Vibrio cholerae</u>, a pathogen that causes the gastrointestinal disease known as cholera, has been isolated from Lake Albufera - a brackish lagoon - near Valencia and from coastal waters near the lake discharges. <u>V. cholerae</u> is considered a member of the autochthonous bacterial flora in aquatic environments and its presence is not correlated with the commonly used coliforms as fecal indicators. <u>V. cholerae</u> also survives longer in the water. Garay et al. (1985) determined that the <u>V. cholerae</u> density reached 10^5/100 ml in the lake and 10^4/100 ml in one of the coastal sites in September of 1982 (Figure 2.8). In the lake water, especially during the warm season, all vibrios isolated were very often identified as <u>V. cholerae</u>.
The many different uses of the lake water (fishing, recrea-

tional and irrigation purposes) and the periodic discharge of these microorganisms into the Mediterranean Sea, where the salinity is not an inhibiting factor for coliforms, represent a constant risk to public health. Therefore the water and shellfish quality must be assessed not only on a basis of fecal coliforms, as is generally the case, but also on that of Vibrio cholerae.

The Besós River, located north of Barcelona, is a major cause of the existing fecal contamination of the coastal seawater in that area. Bosch et al. (1986) described the occurrence of viruses along the Besós River and its main tributary, the Ripoll River. The viral load varied between 1.5 and 215 MPN CU/10 l. The density reached a maximum at the point where the Besós empties into the sea. The heavy use of neighboring beaches for recreational activities requires frequent monitoring for the presence of pathogenic viruses and bacteria.

2.1.1.6. Freshwater environment

The water quality of most Spanish rivers decreases downstream (MOPT, 1985). The largest river, the Ebro (in Cataluña), shows a steady decline in the General Quality Index (ICG) from 84 at the upper most station at Palazuelos to 55 at Pina after which some restoration occurs as shown in Figure 2.9A. The Ebro River is one of the largest river systems of the Mediterranean basin, draining approximately one-sixth of the Iberian peninsula. The mean water flow (1.6×10^{10} m^3 per year) has decreased by approximately 14% since the early years of this century, as a result of the construction of dams for irrigation projects. The silt retention behind dams has halted the delta formation.

The Llobregat River shows a more rapid decline in the index from 80 at Figols to 45 at San Juan Despí due to the numerous pollution discharges when flowing downstream (Figure 2.9B).

The ICG includes 22 parameters which are: dissolved oxygen; suspended matter; pH; conductivity; biochemical oxygen demand; total coliforms; chlorides; sulfates; phosphates; calcium; magnesium; sodium; nitrates; detergents; cyanides; phenols; cadmium; copper; chromium IV; mercury; lead; and zinc. The ICG is interpreted as follows:

ICG	interpretation
100	excellent
100-85	very good
85-75	good
75-60	usable
60-50	bad, needs control
< 50	very bad

The water quality in the major rivers, as exemplified by the general parameters and heavy metals, is regularly tabulated by MOPT.

2.1.2. State of the coastal region

2.1.2.1. Erosion

The Mediterranean coastal area and interior regions are vulnerable to soil erosion due to climatological factors such as dry periods during the summer, heavy rains in autumn, the type of soil, insufficient drainage and repeated forest fires. Mining also causes degradation of soils. Erosion contributes to the river water pollution through run-off carrying considerable amounts of suspended solids, pesticides and nutrients that are subsequently discharged into the sea.

There are different degrees of erosion (MOPT, 1987):
- light, when soil has lost less than 25% of the A-horizon;
- moderate, when soil has lost up to 75% of the A-horizon;
- severe, when soil has lost A-horizon and up to 25% of the B-horizon.

The main problems in Almería and Granada are desertification and severe soil erosion affecting 72% and 51% of the territory, respectively, due to climatic conditions, steep slopes and progressive deforestation (Table 2.15). The latter can be reduced by an active replantation program.
Topography and sudden rainfalls in Murcia make it easier for floods to occur and as a consequence 48% of the surface area was found to be affected by severe erosion. Agricultural use of fertilizers and pesticides on an eroding soil surface affects the water quality in the aquifers in Murcia.

2.1.2.2. Forest fires

There has been a sharp increase in the number of fires and affected areas over the period 1974-1985 (DocTer, 1988). In the whole of Spain 12,837 fires took place in 1985 that resulted in 469,426 hectares of damaged surface, whereas in 1974 there were 3,980 fires that damaged 140,211 hectares (Table 2.16). The figures for 1985 are far higher than for previous years and are four times greater than the reforestation figures for 1984. This means that more trees are being lost through fire each year than are being planted. Spain's woodland will gradually disappear unless this trend is reversed.

Only a small percentage of fires can be attributed to natural causes or negligence and there is growing concern about the large number of fires started intentionally. Although there is no definite proof, it has to be assumed that many of the fires classified as being "of unknown cause" are started intentionally.

The prevention of forest fires is of major importance particularly since they accelerate erosion which is already a more serious problem in Spain than in any other European country.

2.1.2.3. Natural interest areas

The 1975 Ley de Espacios Naturales Protegidos (Law on protected natural areas), which is still in force, established four categories that receive varying degrees of protection:
- the highly protected Reservas Integrales de Interes Científico (integral reserves of scientific interest);
- the Parques Nacionales (national parks);
- Parajes Naturales de Interés National (natural sites of national interest);
- the Parques Naturales (nature parks).

Natural interest areas with special protection in the Mediterranean coastal provinces comprise more than 48,893 hectares in Cataluña, 3,200 hectares in Valencia, 13,000 hectares in Murcia and 28,875 hectares in Andalucía for a total of more than 93,968 hectares (Table 2.17). Natural interest areas that currently have no special protection but are to be considered protected areas encompass 27,941 hectares in Cataluña, 33,804 hectares in Valencia and 505 hectares in Andalucía for a total of 61,530 hectares (Table 2.18). The Balearic Islands have 61,915 hectares of potentially protected natural areas (Table 2.19).

2.1.3. Air pollution

Air pollution comes mainly from three sources: industry and energy production, domestic heating, and motor vehicles.
The areas of San Adrián del Besós, Badalona, Barcelona, Tarragona, Castellón, Valencia and Cartagena are the most seriously affected. Barcelona is the coastal province with the greatest atmospheric pollution (Table 2.20).
Vehicles emit large amounts of lead into the atmosphere because of the high lead content of gasoline. The Royal Decree of February 20, 1985, reduced the permissible lead content of petrol from 0.6 g/l to 0.4 g/l, in line with EC regulations.

Increased enforcement of the air pollution control laws began after Spain became a member of the EC (DocTer, 1988).

2.1.4.   Urban and industrial solid waste

The control of urban solid waste (10.6 million tons per year, in 1985) is not well developed. Uncontrolled dumping in Spain amounts to 46.5% of the entire disposal. Controlled dumping accounts for 34.8%, composting for 14.2%, while 4.5% is incinerated. The coastal region generates 3.9 million tons per year of which 36.7% is dumped uncontrolled, 26% dumped controlled, 28% composted and 9% incinerated. The coastal region has 29 of the 40 compost plants which produce 473,800 tons of compost every year accounting for 70% of the total compost in Spain. The central government is trying to encourage the local authorities to cooperate in waste collection and limit the uncontrolled dumping which can threaten the Mediterranean water quality.

Plastic polyethylene pellets are becoming an environmental problem in the Mediterranean coastal area. After the entrance into the EC, Spain has further increased the manufacturing of plastic products. Plastics wastes are found in large quantities on the beaches of Costa Dorada (Barcelona) and Costa del Sol (Algeciras-Almería). The plastic factories are mainly located around Barcelona (south of the city), Valencia and Andalucía (in Algeciras Bay). Careless disposal practices of these factories and losses of raw materials during sea shipment and unloading at ports are likely causes (Shiber, 1987) of this waste problem. In addition, in Almería (Campo de Dalías) there is plastic waste coming from the greenhouse-based agriculture which represents the largest European concentration of under-plastic cultures. Polyethylene accumulates during the winter months, when the beaches are not used and cleaned up regularly.

Data on hazardous wastes are not readily available. Discharge to surface waters and contamination of groundwater aquifers is becoming an increasing problem in the coastal region.

The region of Valencia produces 90,000 tons of industrial solid waste per year, comparable to the quantity of domestic waste, 190,000 tons of recyclable industrial waste per year, 220,000 tons of inert industrial waste per year and 132,000 tons of toxic industrial waste (heavy metals, aromatic hydrocarbons and asbestos) per year which is 21% of the total industrial solid waste (Martínez de la Vallina, 1987).

2.2. Pressures on the environment

2.2.1. Municipal discharges

The highest population densities occur in the province of Barcelona (606 inhabitants per km^2) followed by Alicante (202 inhabitants per km^2). A parallel trend is seen in the housing density (Figures 2.10A,B). High population densities put pressures on water use, sewage treatment facilities and adequate disposal of solid waste.

Tourism exerts strong environmental pressures along the Mediterranean coast, especially during the summertime. A total of 13.6 million travellers stayed in hotels in 1985, while 787 thousand travellers stayed at camping sites in the coastal provinces. The largest number of travellers is seen in the Balearic Islands (primarily hotel visits), while the highest camping density occurs in Gerona (Figures 2.11A,B). The total number of beds in hotels in 1985 was 552,000, there were 255,000 sleeping places in camping sites, 176,000 beds in apartments and 995,000 beds in vacation homes. The population of the coastal municipalities increases from 6.75 million in the winter to 12.67 million in the summer as shown in Figures 2.11C,D.

The coastal municipalities use 640×10^6 m^3 per year of water of which 500×10^6 m^3 of waste water (78%) is discharged into the Mediterranean. Two-thirds of the discharged load is being treated. During the summer the maximum waste flow increases to 2.5×10^6 m^3 per day of which 49% can be treated. In the winter the minimum flow is 1.35×10^6 m^3 per day of which 91% is treated. The percentages of the treated and untreated waste water discharged into the Mediterranean are shown in Figure 2.12.

Figure 2.13 shows the distribution of domestic loads along the coastline entering the sea, using a logarithmic scale. The input of organic matter as biochemical oxygen demand (BOD) from the Mediterranean settlements was 166×10^3 tons per year over the 1975-1984 period, whereas for suspended solids the value was 249×10^3 tons per year. Approximately one-third of the discharge comes from the Barcelona Province.

Until 1985 (MOPT, 1987) 93 treatment plants and 236 submarine outfalls had been constructed for some of the 203 coastal municipalities. Of the 93 plants, the largest 56 represent 92% of the capacity (21,400 m^3/day.plant), while the remaining 37 represent 8% of the capacity (2,700 m^3/day.plant). A later survey counted 147 sewage treatment plants. The median percentage of the sewage that was treated was 67% and ranged from 1 to 80% (Table 2.21). Of the 147 plants, 37% used biological, 0.7% physical-chemical and 62% primary sedimentation treatment.

Andalucía (MOPT, 1987) is a region in which 78% of the 76 treatment plants did not work. Of the remaining 22% about one-third does not work properly. The region of Andalucía has allocated 1,000 million ptas. for construction of 50 alternative plants (including lagoons) to serve 200,000 inhabitants.

The number of untreated marine discharges totals 757 of

which 236 are outfalls (Table 2.22). The highest number of direct discharges (215) is found in the province of Gerona. Málaga has the largest number of outfalls. Of the outfalls where data are available, 52% has a depth of less than 15 meters and 95% has a length of less than 1,000 meters (Table 2.23).

2.2.2. Industrial discharges

A total of 1,421 potentially polluting industries are located in municipalities along the Mediterranean coast of which 1062 (75% of the total) are in the Barcelona Province only (Table 2.24).

MOPT (1987) estimated an industrial discharge of the coastal provinces of 634 x 10^6 m^3 of waste water per year, 90.5 x 10^3 tons of BOD per year and 140 x 10^3 tons of suspended solids per year. An earlier survey by the Ministry of Industry and Energy (1981) estimated a load of 70 x 10^3 tons of BOD per year and 96 x 10^3 tons of suspended solids per year (Figure 2.14). The MOPT survey found the highest BOD discharge in Cataluña while the MIE detected it in the Baleares. In Murcia an additional load of 210.0 x 10^4 tons per year of suspended solids (primarily clays) originating from zinc and lead mining is discharged into the sea near Portman (MOPT, 1987). It should be noted that the ore washing plant discharges in Portman Bay were stopped on March 30, 1990, after an agreement was made between the mining company and MOPT. Figure 2.15 shows the industrial discharges per sector into the Spanish Mediterranean. Coal mining is the most polluting sector as far as BOD is concerned.

2.2.3. River discharges

The rivers discharge a load of 53 x 10^3 tons of BOD per year and 157 x 10^3 tons of suspended solids per year into the Mediterranean Sea, as shown in Figure 2.16.

The highest BOD discharge is through the Ebro and Besós rivers, whereas the highest suspended solids load is carried by the Vinalopó and Ebro rivers. Both watersheds are strongly affected by erosion. The maximum BOD discharge occurs during the high-water period in November, while the highest suspended solids discharges are observed during the high-water period from February to March (Figure 2.17).

2.2.4. Maritime traffic

The Spanish ports handle a growing amount of maritime traffic. The volume of total traffic increased from 227 million tons in 1980 to 237 million tons in 1985 (Figure 2.18). Most of these are petroleum products followed by bulk solids. The Mediterranean ports handled 198 million tons in 1985 or 83.5% of all Spanish maritime traffic. Tarragona and Algeciras are the largest harbors for petroleum-related products (Table 2.25).

Pollution by hydrocarbons (bilge water) and heavy metals (corroding ship hulls) are detected in Spanish harbors. The amount invested in Spanish harbors in 1985 reached a figure of 22,083 million ptas. with an increase of 14.7% over 1983. Of the above amount 72.3% was self-financed and the remaining through State subsidies.

The analysis of the main Spanish Mediterranean petroleum ports (EEC, 1982) has pointed out that adequate reception facilities for ballast waters exist in all of them except in two cases: the port of Cartagena (Escombreras), where a secondary treatment installation has been planned, and in the CAMPSA depot of Barcelona where an improvement of the existing facility is required.
All Spanish ports on the Mediterranean Sea must be provided with systems to collect and treat the bilge waters. The ports having oil terminals (Tarragona, Algeciras, Barcelona, Málaga, Cartagena, Castellón de la Plana), which already have oily water reception facilities, must be

provided with barges for the reception of bilge water from ships, to be unloaded and treated in the existing reception facilities. The other ports (Palma, Valencia, Alicante and Almería) must be provided with a barge for the reception of bilge water equipped with a storage tank and an oil separator.

The amount of investment required for the port reception facilities is around $10 million (U.S.).

2.3. Economic impacts

2.3.1. Economic impacts of water pollution on tourism

Marine pollution can affect tourism income as bathing in polluted seawater poses a health risk. Between 1981 and 1982 increases were noted in eye and throat infections for users of the Spanish beaches (Table 2.26). A strong correlation was found between chlorophyll-a concentrations and eye infections and allergies (Table 2.27).

Tourism income losses can amount to several tens of millions of dollars per year. The total morbidity rate of all bathers in 1982 was 28.8 per 100 bathers (Table 2.26). If we assume that each of the health effects occurred separately and that the tourists lost 2 days per illness, in which they spent only half as much, and that the average stay was 14 days, then one can calculate a loss of ($7,140 $\times 10^6$ * 14-2/2 / 14 * 28.8/100) 143 x 10^6 dollars per year.

Mujeriego et al. (1982) carried out an epidemiological study at 24 beaches in Málaga and Tarragona during the summer of 1979. A total of 20,219 persons were interviewed based on a fixed questionnaire. The following conclusions were drawn: seawater transparency and the presence of floatable materials are the two major factors that determine the aesthetic quality of coastal waters in the opinion of recreationists; the aesthetic quality was related to the microbiological quality of coastal waters; the most frequent health ailments observed among recreationists were skin infections, with a morbidity rate of 2 percent, followed by ear and eye infections with a morbidity rate close to 1.5 percent; intestinal infections had morbidity rates below 1 percent; the habit of immersing the head in the water when bathing was significantly associated with ear and eye infections; microbiological limits in terms of fecal coliforms did not seem to provide consistent public health protection; compliance with coliform limits did not guarantee the absence of health effects associated with fecal streptococci.

The authors carried out microbiological tests on 19 of the 24 studied beaches during the summer, but not necessarily in close association with the dates that the people were interviewed. They plotted regression lines between reported morbidity rates of ear infections at each specific beach and the median fecal streptococci concentration at that specific beach. The coefficient and correlation were positive and significant for these relations as shown in Figure 2.19.

2.3.2. Other economic impacts

Soil quality is strongly related to economic activity in the agricultural sector. The estimated economic impact per year in Mediterranean Spain due to erosion affecting a surface area of 2,817 hectares (resulting in crop reduction) equals $110,666 \times 10^6$ ptas (Table 2.28).
Forest fires affect 54,494 hectares and the resulting losses amount to $1,547 \times 10^6$ ptas (Table 2.29). This represents 0.004-0.63% of the regional agricultural income.

The estimated annual cost due to floods ranges from 12,078 to 32,143 million ptas (Calderón et al., 1987). In order to protect the soil from erosion due to floods main actions include planting, soil conservation and regeneration.

2.4. The institutional/legal framework

Environmental protection in Spain is governed by an administrative division of power at the functional and territorial level. In the central governmental administrative structure there are 13 institutions regulated by five different ministries, with substantial environmental jurisdiction. In the five Autonomous Regions with a Mediterranean coastline there are more than 20 departments or services with environmental jurisdiction over the sea or the coastal strip.

The transfer of environmental policies to the Autonomous Regions (CCAA) had both positive and negative results. The division of responsibilities is not always clear and has been changing over time, depending upon the available financial resources and willingness. The regional wastewater treatment projects were proposed when the regional Governments received the environmental jurisdiction to apply corrective measures. However, some CCAAs have fragmented the environmental issues in a similar way as the central Administration. These CCAAs were often obliged to do this, because fragmented transfer of obligations occurred. The transfer of "packages" included responsibilities, employees and materials (offices, buildings, equipment, etc.) from the already fragmented central Administration to the regional Administration.

Four ministries have the main environmental competence in the central Administration: the Ministry of Public Works and Urban Planning (MOPT); the Ministry of Agriculture, Fishing and Food (MAPA); the Ministry of Industry and Energy (MINER); and the Ministry of Health and Consumer Affairs (MISACO). The MOPT and the MAPA have most of the jurisdiction, as well as most of the national budget for environmental protection.

MOPT has the main environmental powers in the Spanish

central Administration. The MOPT's General Secretariat for the Environment (SGMA) has the responsibility for developing the basic legislation, distributing the national environmental budget between the local and regional administrations, and managing international issues. This last task has become more important in recent years, because of Spain's entry into the EC and the need to harmonize the Spanish legislation with the EC's legal framework. Additional powers of MOPT concern water, coastal and urban planning and management.

The protection and conservation of natural areas is regulated by ICONA (Nature Conservation Institute), a dependent institution of MAPA. They are in charge of forestry exploitation of publicly owned land. Wood production, however, can sometimes conflict with nature conservation. The MAPA also is charged with regulating marine conservation through the Spanish Oceanographic Institute.

The MINER regulates the main industrial environmental problems, especially industrial emissions. Its traditional position has been one of minimizing environmental restrictions on industrial activity. MINER also controls electricity-generating plants using coal, which release large amounts of sulfur dioxide.

The MISACO is responsible for the establishment of minimum quality standards for several natural resources, such as drinking and bathing water. MISACO has been restrictive in publishing existing quantitative environmental health data because of expected negative tourism reactions.

The fragmentation of environmental powers has negative effects on the establishment of environmental legislation. Legislation for control, usually proposed by MOPT, has sometimes serious difficulties in being approved, or is weakened before approval in long discussions between the relevant departments. This is currently happening, for instance, with the Coastal Law and also with the Environmental Impact Regulations of large projects, which have to be enacted soon in order to synchronize the Spanish legislation to that of the EC.

Of the 25 major environmental laws, summarized in Table 2.30, 10 are national and 15 are regional.

2.5. Existing/planned action and investments

The government is trying to solve the environmental problems: to protect natural resources and restore good quality conditions for water, land and air.
At the central administration level, MOPT spent 5,423 million ptas. (20% as subsidies), MAPA 20,476 million ptas. (77% as subsidies), MINER 623 million ptas. and MISACO 84 million ptas. in 1986 on environmental expenditures (MOPT, 1987). MOPT financed 50-100% of the construction of sewage treatment plants. It spent a total of 31.1×10^9 ptas. (1985 value) between 1975 and 1985.

Eleven of the most polluted areas requiring central administration assistance are listed in Table 2.31. The Spanish government is undertaking feasibility studies for a new generation of wastewater treatment plants at these locations, requiring large investments. These new programs deal with problems concerning wastewater treatment that cannot be solved by the regional authorities alone.

The environmental affairs are handled by the autonomous regional governments under specific laws established by the central government. Improvement of the current economic situation allows for the increase in investment in restoring the quality of the Mediterranean Sea and the coastal area.

All autonomous regions have started to build sewage treatment plants. As a consequence of these investments, general conditions in the coastal area have improved.

The autonomous region of Cataluña made a territorial division of 13 zones and has an approved plan for 7 of them, 5 of which are in the coastal area (Table 2.32). A total of 53 billion ptas. has been invested in general sanitation in Cataluña. The total forecasted need between 1983 and 1996 is 145 billion ptas. Between 1975 and 1985 10.1 billion ptas. (1985 values) were invested in 18 sewage treatment plants.

The autonomous region of Valencia is under pressure to invest in sewage treatment of the coastal area and in that of the basins of the Vinalopó, Serpis, Albaida, Júcar and Magro rivers. Table 2.33 shows the forecast for investments and actions taken by the coastal municipalities. The main projects to be started are the construction of collectors in the city of Valencia and of collectors in the Albufera region, which carry most of the industrial wastes. A total of 20 billion ptas. is required between 1986 and 1995 to complete construction of 9 outfalls and 25 plants with a total capacity of 354×10^3 m^3 per day. Between 1975 and 1985 7.3 billion ptas. (1985 value) were invested in 21 treatment plants with a total capacity of 440×10^3 m^3 per day.

The autonomous region of Murcia is mainly working on the recovery of the Segura River. Projects are being considered for the treatment of the waste waters discharged into the rivers. A total of 5.5 billion ptas. is required for environmental investments, of which 1.91 billion ptas. were needed between 1986-1987, 1.97 billion ptas. were required for 1988-1989 and the remaining 1.63 billion ptas. were needed for 1990-1991.

The investment supplied by the autonomous region of Murcia to the coastal municipalities in the year 1985 equaled 600 million ptas., which represents 50% of the total project costs of 10 sewage treatment plants and 4 other works (Table 2.34). Between 1975 and 1985 296 million ptas. were invested for the completion of 3 sewage treatment plants with a capacity of 39×10^3 m^3 per day.

The autonomous region of Andalucía has developed an initial investment plan with a cost of 1.0 billion ptas. until 1990 for construction of alternative treatment plants. These investments are required for the satisfactory functioning of 50 treatment plants serving 200,000 inhabitants. A second investment round of 2.3 billion ptas. serves to improve the quality of deteriorated littoral zones as required by the Beach Control Law (PROVISAPLA-1983). The plans include construction of 16 treatment plants and 2 outfalls (Table 2.35).

The autonomous region invested a total of 14.04 billion ptas. between 1984 and 1987 80% of which was allocated for drinking-water supplies and 20% was allocated for sewage treatment (mainly spent in the coastal region).

The Balearic Islands have 11 treatment plants and several private ones with a total capacity of 100,000 m^3 per day representing an investment of 1.8 billion ptas. Between 1975 and 1985 Palma de Mallorca had its sewage treatment plants enlarged to a capacity of 60,000 m^3 per day at an investment of 0.92 billion ptas (1985 value). The second Palma South plant started operating in 1980. As a result, the untreated sewage discharged was reduced from 35,000 m^3 per day to 3,000 m^3 per day. The fecal coliform densities decreased from 110/100 ml in 1982 to 22.3/100 ml in 1985, an 80% decrease.

Future plans include the construction of 16 treatment plants and 8 outfalls for a total capacity of 51,500 m^3 per day or 242,000 inhabitants (Table 2.36).

The 60 treatment plants constructed in the Spanish Mediterranean at an investment cost of 22.5 billion ptas. (1985 value) cost on the average 375 million ptas. per plant. The 68 treatment plants approved in 1985 required a total investment of 47.6 billion ptas. (1985 value) or 700 million ptas. per plant.

A total of 180 beach projects and 74 rivermouth structures requiring 660 million ptas. and 250 million ptas., respectively, were planned in 1986 for the different coastal provinces (Table 2.37).

2.6.   Summary and conclusions

Heavy-metal accumulations in coastal waters and sediments are mainly caused by industrial and mine discharges. The mining waste is particularly a problem in the province of Murcia. The disposal problem of sediments rich in heavy metals remains after the stoppage of the ore washing plant discharges in Portman Bay on March 30, 1990. Industrial discharges are concentrated in the Barcelona Province. Food-chain accumulation has been observed for cadmium, lead and mercury.

The use of toxic pesticides and fertilizers greatly affects the freshwater and seawater quality. Pesticides are persistent in the environment. Their use has to be reduced or substituted by more biodegradable substances.

Direct discharge of wastes deteriorates the quality of the seawater. In addition, there are risks of bacterial and viral infections among bathers with subsequent economic loss. Tourism in the summer taxes the treatment facilities. More wastewater treatment plants are needed. In cases where the marine environment has a large absorption capacity, proper outfalls are foremost needed. The outfalls used for this purpose should be deeper and extend further from the coast.

When flowing downstream, most rivers show a decrease in water quality. A stricter river basin management has to be developed in order to reduce pollution. Such management should also consider reforestation as an important factor for the reduction in soil erosion and damage by flooding.

Other environmental issues to which attention needs to be given are: toxic waste disposal facilities; port reception facilities for ship wastes; coastal zone management; and wetland conservation projects.

References

BOSCH, A., LUCENA, F., GIRONÉS, R. and JOFRE, J., Survey of viral pollution in Besós river (Barcelona). J. Water Pollut. Control Fed., 1986, 58, 87-91.

CALDERON, E., ESTEBAN, A. and HERNANDEZ, I., Community program "ENVIREG". Draft final report, Spanish case. 1987.

CEC, 26.7.1987-1.8.1987 plages propres. Résultats de la campagne de sensibilisation. Brussels 1988.

CONSEJERIA DE POLITICA TERRITORIAL Y OBRAS PUBLICAS, Cuadernos informativos. Comunidad Autónoma de la Región de Murcia. 1987.

DE LEON, A. R., MAS, J., GUERRERO, J. and JORNET, A., Monitoring of heavy metals in superficial sediment and some marine organisms from the western Mediterranean coast. 1985. VIIes Journ. Étud. Pollut. CIESM, pp. 321-326.

DOCTER, European environmental yearbook 1987. 1988. DocTer International U.K., London.

EEC, Feasibility study on deballasting facilities in the Mediterranean Sea. October 1982.

GARAY, E., ARNAU, A. and AMARO, C., Incidence of *Vibrio cholerae* and related vibrios in a coastal lagoon and seawater influenced by lake discharges along an annual cycle. Appl. Environ. Microbiol., 1985, 50, 426-430.

GESAMP Working Group 26, Review of the state of the Mediterranean environment. UNEP, Athens 1987.

GUERRERO, J., DEYA, M. M., RODRIGUEZ, C., JORNET, A. and CORTES, D., Heavy metals levels in marine organisms from the Mediterranean Sea (Spanish coast). 1988. Rapp. Comm. int. Mer Médit., 31, 2 p. 155.

HERNANDEZ HERNANDEZ, F., MEDINA ESCRICHE, J. and PASTOR GARCIA, A., Contaminació del litoral per pesticides organoclorats i metalls pesats. In: El medio ambiente en la Comunidad Valenciana, pp. 248-253. Generalitat Valenciana, May 1987.

LUCENA, F., FINANCE, C., JOFRE, J., SANCHO, J. and SCHWARTZBROD, L., Viral pollution determination of superficial waters (river water and sea-water) from the urban area of Barcelona (Spain). Water Res., 1982, 16, 173-177.

MARTINEZ DE LA VALLINA, J. J., Residuos industriales tóxicos y peligrosos. In: El medio ambiente en la Comunidad Valenciana, pp. 332-335. Generalitat Valenciana, May 1987.

MINISTRY OF INDUSTRY AND ENERGY, Inventario nacional de focos industriales contaminadores de las aguas. Madrid 1981.

MOPT, Medio ambiente en España 1984. Dirección General del Medio Ambiente, 1985.

MOPT, Memoria. 1985. Dirección General de Puertos y Costas.

MOPT, Revista del Ministerio de Obras Publicas y Urbanismo. November 1986.

MOPT, Acciones para la protección de la zona del Mediterráneo 1975-1985. B-351, Parte II. 1987.

MOPT, Análisis de calidad de aguas 1985-86. Dirección General de Obras Hidraúlicas, April 1987.

MUJERIEGO, R., In: Coastal water quality control (MED POL VII). MAP Technical Reports Series No. 7, pp. 313-328. UNEP, Athens 1986.

MUJERIEGO, R., BRAVO, J. M. and FELIU, M. T., Recreation in coastal waters: public health implications. In: Workshop on Marine Pollution of the Mediterranean, pp. 585-594. CIESM. Cannes 1982.

PASTOR, A., HERNANDEZ, F., MEDINA, J., MELERO, R., LOPEZ, F.J. and CONESA, M., Organochlorine pesticides in marine organisms from the Castellón and Valencia coasts of Spain. Mar. Pollut. Bull., 1988, 19, 235-238.

SANCHEZ-PARDO, J. and ROVIRA, J., Hidrocarburos alifáticos, DDT's y PCB's en sedimentos marinos de la zona catalano-balear (Mediterráneo occidental)". Inv. Pesq., 1985, 49, 521-536.

SANCHEZ-PARDO, J. and ROVIRA, J., Hidrocarburos alifáticos, (C_{14} a C_{24}), DDT's y PCB's en sedimentos y organismos del delta del Ebro (Mediterráneo occidental). Inv. Pesq., 1985, 49, 637-651.

SHIBER, J. G., Plastic pellets and tar on Spain's Mediterranean beaches. Mar. Pollut. Bull., 1987, 18, 84-86.

UNEP/IOC, Assessment of the state of pollution of the Mediterranean Sea by petroleum hydrocarbons. MAP Technical Reports Series No. 19. UNEP, Athens 1988.

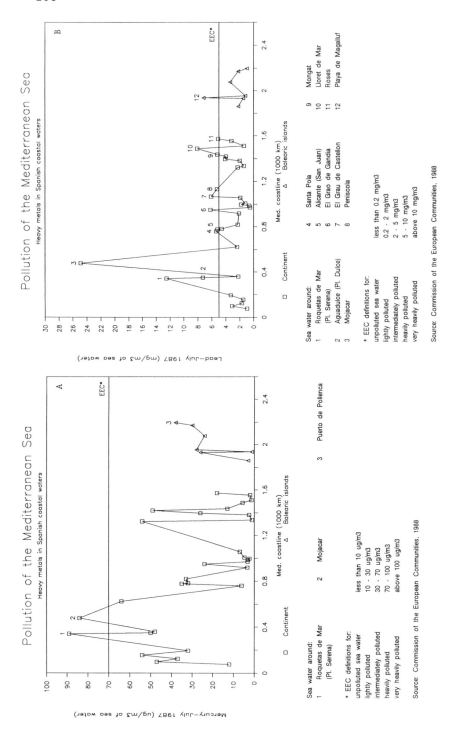

Figure 2.1 (A) Mercury, (B) Lead, (C) Cadmium and (D) Zinc concentrations along the Spanish coast

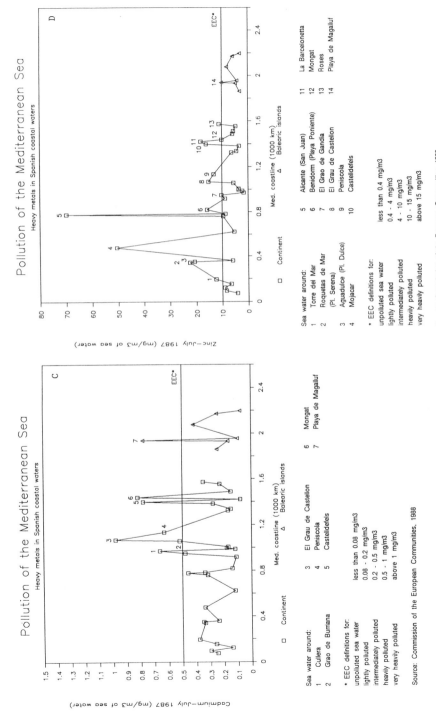

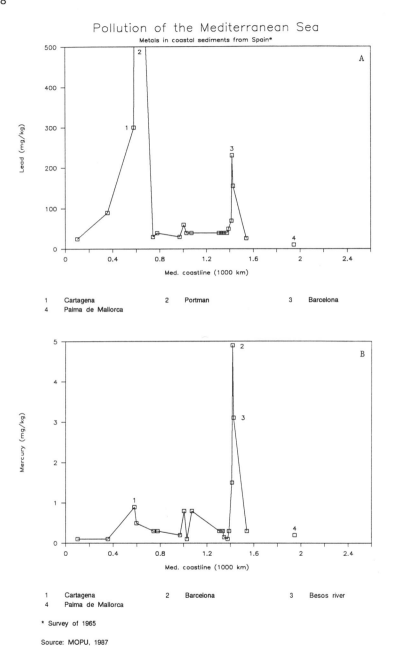

Figure 2.2 (A) Lead and (B) Mercury content in marine sediments along the Spanish coast

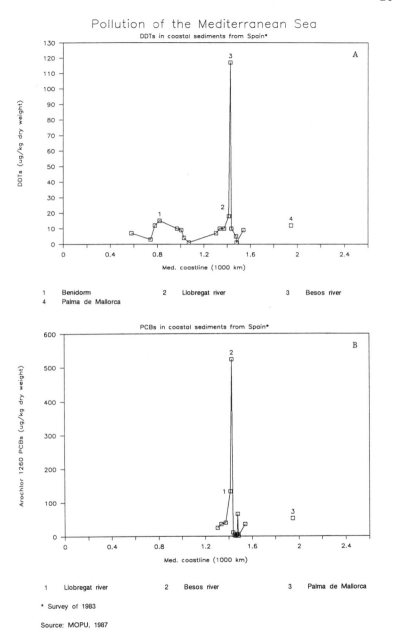

Figure 2.3 Organochlorines in sediments along the Spanish coast (samples taken at a depth of 10-100 mts, distance 2-5 kms and in the top 50 cm) (A) DDT and (B) PCB

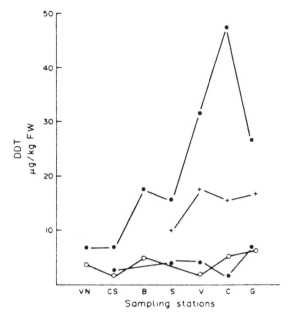

Figure 2.4 DDT content versus sampling stations for several species. VN = Vinaroz; CS = Castellón; B = Burriana; S = Sagunto; V = Valencia; C = Cullera; G = Gandia. -o- <u>Macropipus depurator</u>; -•- <u>Mytilus galloprovincialis</u>; -•- <u>Venus gallina</u>; -+- <u>Donax trunculus</u>. [Pastor, A., et al., 1988]

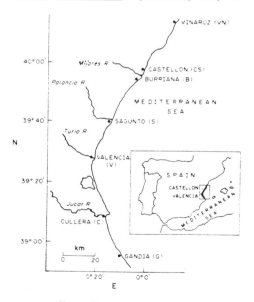

Map of the sampling stations

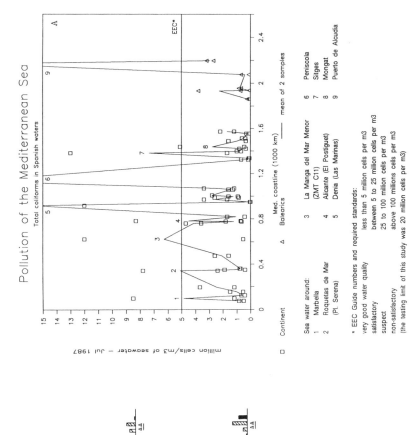

Fig. 2.5 Mean content of DDTs and PCBs in several tissues for the marine organisms: Mullus surmuletus (MS), Mullus barbatus (MB), Tunnus thynus (TT) and Aristeus antennatus (AA). [Pastor, A., et al., 1988]

Fig. 2.6 Bacteriological quality along the Spanish coast: (A) total coliforms, (B) faecal coliforms and (C) faecal streptococci

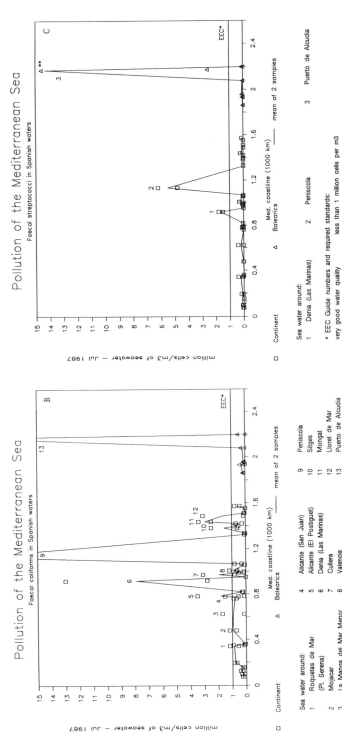

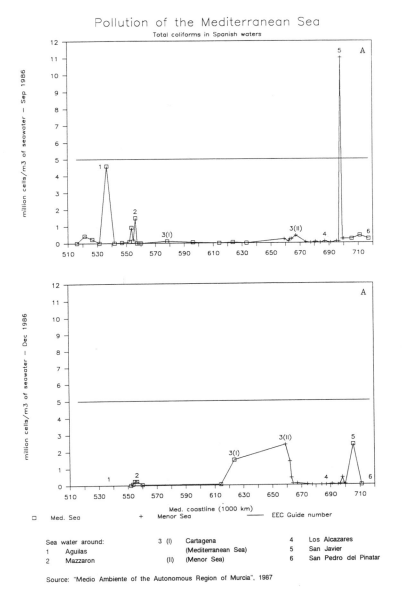

Figure 2.7 Bacteria concentration along the Murcia coast in September and December 1986
A: total coliforms; B: faecal coliforms; C: faecal streptococci

Pollution of the Mediterranean Sea
Faecal coliforms in Spanish waters

□ Med. Sea   + Menor Sea   — EEC Guide number

Sea water around:
1 Aguilas
2 Mazzaron
3 (I) Cartagena (Mediterranean Sea)
  (II) (Menor Sea)
4 Los Alcazares
5 San Javier
6 San Pedro del Pinatar

Source: "Medio Ambiente of the Autonomous Region of Murcia", 1987

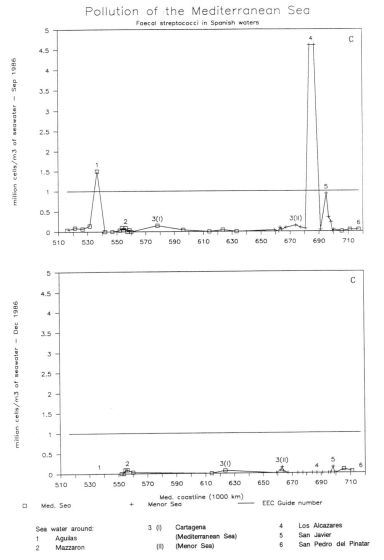

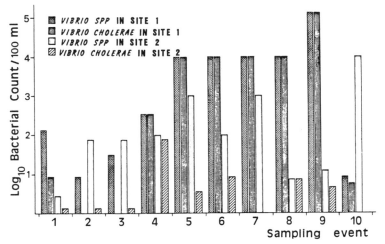

MPNs of *Vibrio* spp. and *V. cholerae* in sites 1 and 2.

Physicochemical parameters for sites 1 and 2

Sampling event	Date	Temp (°C)		pH		Salinity (‰)		Gates[a]
		Site 1	Site 2	Site 1	Site 2	Site 1	Site 2	
1	1/11/82	15	13	8.5	8	ND[b]	ND[b]	Open (1)
2	2/1/82	13	12	9.1	8.3	0.56	21.26	Open (3)
3	2/24/82	15	12	9.2	7.9	0.64	21.8	Open; B
4	5/3/82	20	18	8.2	8	0.7	32.7	Closed
5	6/1/82	23	22	9.1	8.2	ND	11.87	Open (4)
6	6/21/82	28	25	9	7.9	1.4	30.38	Closed
7	7/26/82	30.5	29	8.7	7.9	1.43	21.88	Closed
8	9/6/82	25.5	25	8.9	8.3	0.57	12.03	Open (6)
9	9/27/82	23	24	8.7	8	0.58	34.78	Open (3); B
10	12/7/82	10.5	14.5	8.2	8	0.34	28.5	Closed

[a] Values in parentheses represent the number of open gates. B, Sand bar between channel and sea.
[b] ND, Not determined.

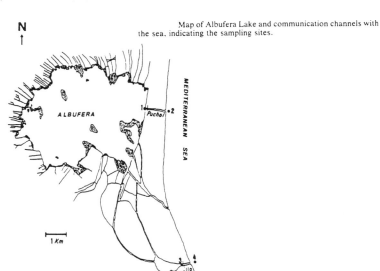

Map of Albufera Lake and communication channels with the sea, indicating the sampling sites.

Figure 2.8 Incidence of Vibrios in a coastal lagoon and seawater near Valencia   [Garay et al., 1985]

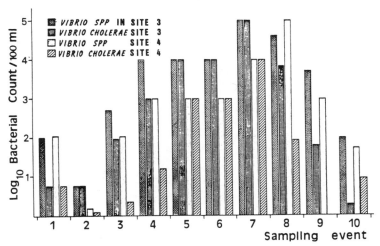

MPNs of *Vibrio* spp. and *V. cholerae* in sites 3 and 4.

Physicochemical parameters for sites 3 and 4

Sampling event	Date	Temp (°C)		pH		Salinity (‰)		Gates[a]
		Site 3	Site 4	Site 3	Site 4	Site 3	Site 4	
1	1/25/82	10	11	8.3	8.1	0.47	8.27	Open (1)
2	2/8/82	12.5	13.5	8	7.9	0.6	28.09	Open (1)
3	5/17/82	20	19	7.8	7.4	0.8	ND[b]	Closed
4	6/14/82	25.5	24	8.3	7.9	ND[b]	ND[b]	Closed
5	7/7/82	29	27.5	7.4	7.4	1.37	12.79	Open (1)
6	8/2/82	27	28	7.9	7.9	ND[b]	13.45	Open (1)
7	9/13/82	26.5	25.5	7.4	7.5	2.5	14.03	Open (2)
8	9/29/82	23	22.5	8.1	8.3	0.81	21.98	Open (2)
9	11/23/82	14	16	7.9	8	0.27	30.22	Closed
10	12/13/82	11.5	13	7.9	7.9	0.24	27.51	Closed

[a] Values in parentheses represent the number of open gates.
[b] ND, Not determined.

Figure 2.8   cont'd

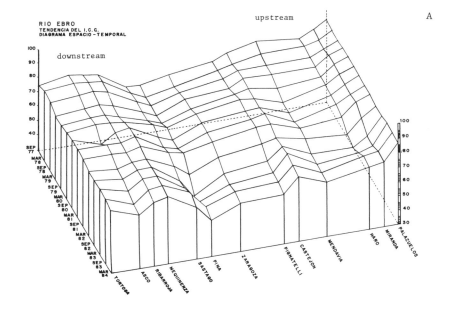

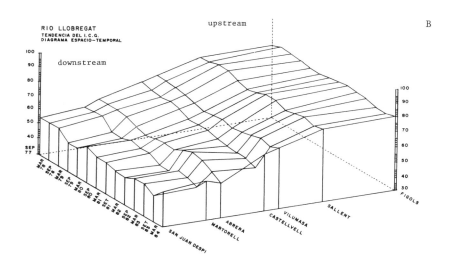

Figure 2.9   Trend in General Quality Index (I.C.G.) along (A) Ebro and (B) Llobregat Rivers [MOPU, 1984]

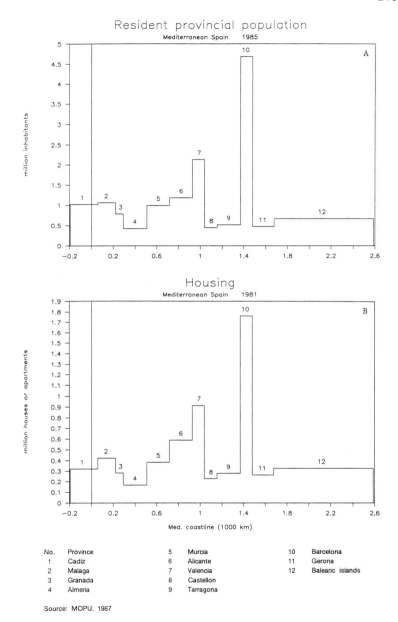

Figure 2.10   (A) Resident population and (B) housing in the different coastal provinces

220

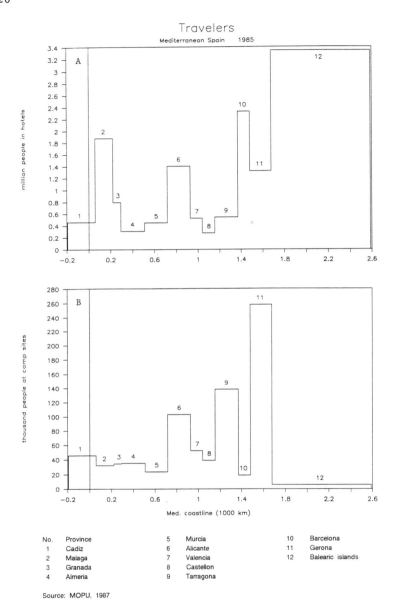

No.	Province				
1	Cadiz	5	Murcia	10	Barcelona
2	Malaga	6	Alicante	11	Gerona
3	Granada	7	Valencia	12	Balearic islands
4	Almeria	8	Castellon		
		9	Tarragona		

Source: MOPU, 1987

Figure 2.11   (A) Travelers in hotels, (B) Travelers in campings, (C) winter population of coastal municipalities and (D) summer population

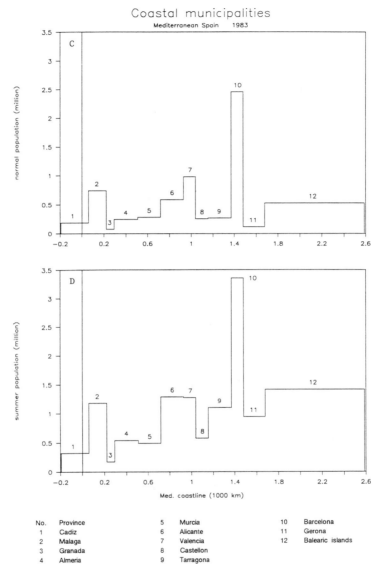

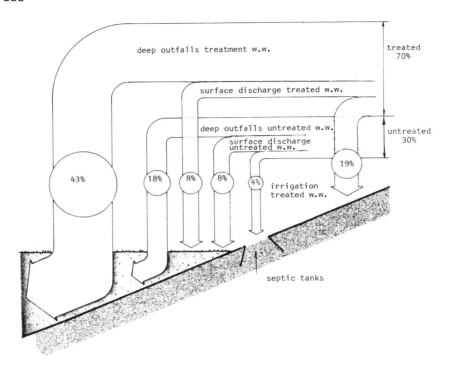

Figure 2.12   Discharge of treated and untreated wastewater into the Mediterranean Sea [MOPU, Feb. 1987; B-351 part II]

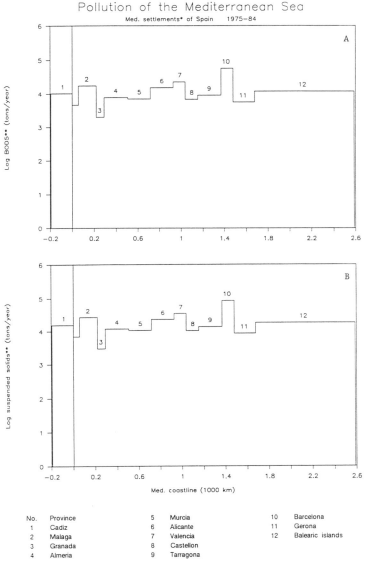

Figure 2.13 (A) BOD and (B) Suspended Solids dicharged by coastal municipalities in the different provinces

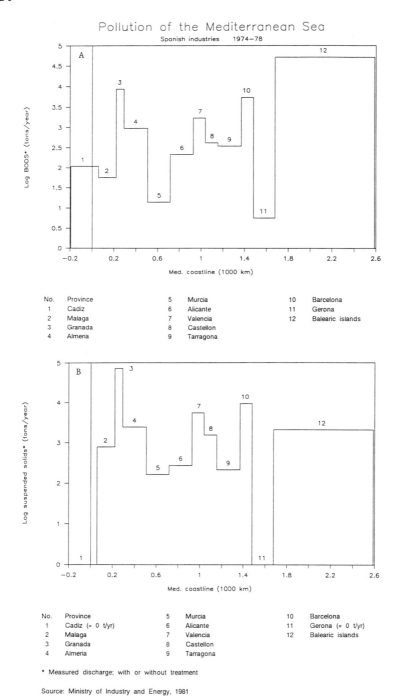

Figure 2.14 Industrial discharges along the Mediterranean coast: (A) BOD discharges and (B) suspended solids

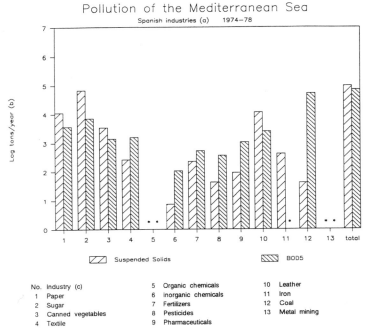

Figure 2.15 Distribution of industrial discharges (per sector) along the Spanish Mediterranean coast

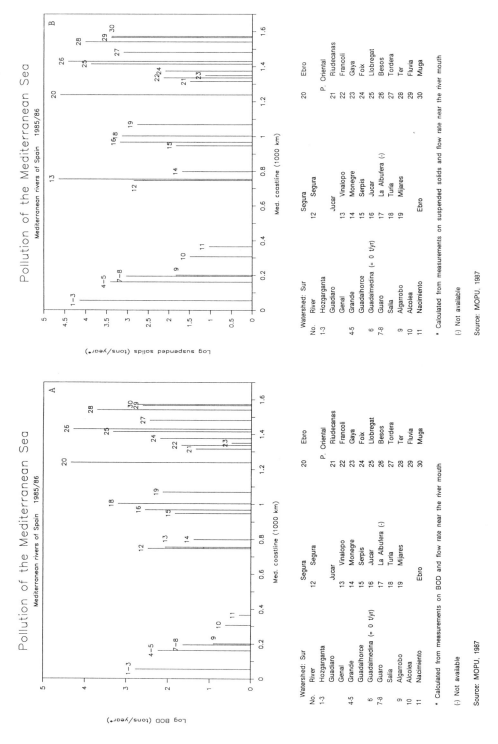

Figure 2.16 Annual (Oct. 1985 – Sept. 1986) discharges of the 30 major rivers into the Mediterranean A: BOD; B: Suspended Solids

227

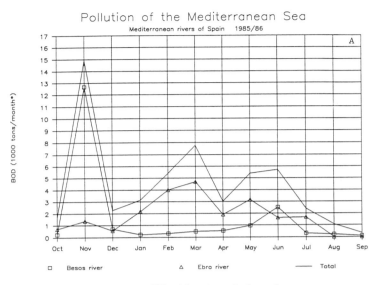

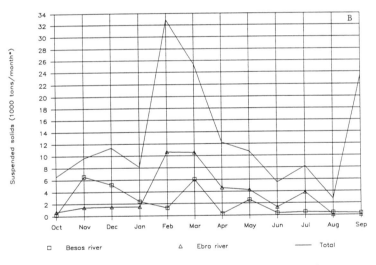

Figure 2.17 Monthly (from Oct. 1985 to Sept. 1986) river dicharges into the Mediterranean (same rivers as previously): (A) BOD, (B) Suspended Solids

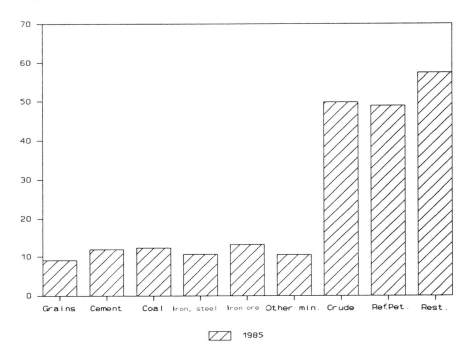

Figure 2.18 Traffic evolution of different types of goods (million tons/year) [MOPU, Direccion General de Puertos y Costas, 1985]

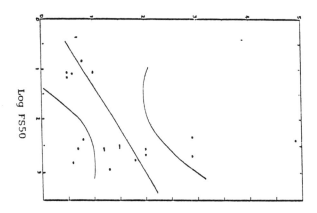

Regression line and 95% confidence limits for the expected morbidity rate of ear ailments associated with a given value of the concentration of faecal streptococci reached in 50% of the samples

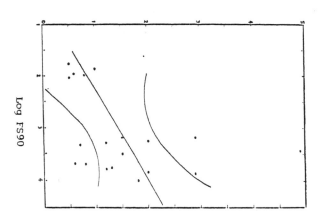

Regression line and 95% confidence limits for the expected morbidity rate of ear ailments associated with a given value of the concentration of faecal streptococci reached in 90% of the samples

Figure 2.19 Relation between ear ailments and faecal streptococci: densities [Mujeriego R., 1986]

Table 2.1   Cadmium concentrations in sediments ($\mu g\ g^{-1}$ DW) and in marine organisms ($\mu g\ kg^{-1}$ FW) along the Spanish coast from Castellon to Cartagena   [De Leon et al., 1985]

Location	sediment range	M. gallopr. m.	M. gallopr. range	D. trunc. m.	D. trunc. range	M. barbatus m.	M. barbatus range
Castellon	0.5 - 0.4	67	38-100	8	5 - 11	5	1 - 13
Sagunta	ND - 0.16	-		12		7	1 - 14
Valencia	0.12- 0.5	86	53-120	22	6 - 41	5	3 - 8
Cullera	0.08- 0.12	61	35- 90	5		2	1 - 4
Alicante	0.17- 0.28	113	41-175			2	1 - 4
Guardamar	0.02- 0.55	100	70-130	6		6	2 - 11
Portman	0.02-10.4	940	930-950				
Cartagena	0.2 - 1.6	400	160-650			7	1 - 23

M = mean

Table 2.2   Heavy metals levels in organisms (Hg) in $\mu g/kg$ FW   [Guerrero, J. & Deya, M.M., 1988]

METAL	SPECIES	SAMPLING SITE	n	Mean	Range
Hg	MULLUS BARBATUS	ALICANTE	138	98	3-260
		GUARDAMAR	138	110	20-220
		CARTAGENA	126	87	20-170
		MAZARRON	144	134	40-290
		AGUILAS	168	52	20-140
		VILLARICOS	48	67	60-80
		CARBONERAS	90	108	20-200
		GARRUCHA	144	71	10-200
		ALMERIA	126	47	10-150
		PALMA MALLORCA	59	218	70-1110
		PORTMAN	6	50	---
	MYTILUS GALLOPRO- VINCIALIS	ALICANTE	310	34	10-100
		GUARDAMAR	300	60	10-120
		CARTAGENA	315	40	20-60
		ALMERIA	1060	37	10-60
		MALAGA	62	38	10-80
		ALGECIRAS	131	58	20-130
		PORTMAN	330	23	10-50
	ARISTEUS ANTENNATUS	ALICANTE	156	303	30-740
		GUARDAMAR	126	507	120-1140
		CARTAGENA	132	309	70-830
		MAZARRON	210	276	20-570
		AGUILAS	162	298	90-710
		GARRUCHA	150	327	90-690
		ALGECIRAS	10	460	---
		PALMA MALLORCA	73	353	70-1230

Table 2.3   Heavy metals levels in organisms (Pb) in ug/kg FW [Guerrero, J. & Deya, M.M., 1988]

METAL	SPECIES	SAMPLING SITE	n	Mean	Range
Pb	MULLUS BARBATUS	ALICANTE	138	60	0-200
		GUARDAMAR	138	79	0-190
		CARTAGENA	126	1744	0-19000
		MAZARRON	144	117	0-540
		AGUILAS	168	70	0-190
		VILLARICOS	48	105	---
		CARBONERAS	90	58	0-130
		GARRUCHA	144	121	0-460
		ALMERIA	126	99	0-480
		PALMA MALLORCA	59	52	0-210
		PORTMAN	6	0	---
	MYTILUS GALLOPROVINCIALIS	ALICANTE	310	730	120-1430
		GUARDAMAR	300	450	100-1290
		CARTAGENA	315	9460	2690-26680
		ALMERIA	1060	1270	230-2800
		MALAGA	62	710	230-1570
		ALGECIRAS	131	1290	230-3610
		PORTMAN	330	72590	37650-125000
	ARISTEUS ANTENNATUS	ALICANTE	156	60	0-160
		GUARDAMAR	126	90	0-300
		CARTAGENA	132	60	0-180
		MAZARRON	210	130	0-390
		AGUILAS	162	110	0-760
		GARRUCHA	150	60	0-150
		ALGECIRAS	10	0	---
		PALMA MALLORCA	73	30	0-90

Table 2.4   Heavy metals levels in organisms (Cd) in ug/kg FW [Guerrero, J. & Deya, M.M., 1988]

METAL	SPECIES	SAMPLING SITE	n	Mean	Range
Cd	MULLUS BARBATUS	ALICANTE	138	5	1-19
		GUARDAMAR	138	3	1-12
		CARTAGENA	126	4	1-10
		MAZARRON	144	6	1-24
		AGUILAS	168	64	1-1000
		VILLARICOS	48	2	---
		CARBONERAS	90	3	1-8
		GARRUCHA	144	3	1-6
		ALMERIA	126	4	1-14
		PALMA MALLORCA	59	6	2-23
		PORTMAN	6	4	---
	MYTILUS GALLOPRO-VINCIALIS	ALICANTE	310	74	30-150
		GUARDAMAR	300	56	12-79
		CARTAGENA	315	278	48-490
		ALMERIA	1060	182	36-1500
		MALAGA	62	173	99-250
		ALGECIRAS	131	165	62-510
		PORTMAN	330	930	190-1450
	ARISTEUS ANTENNATUS	ALICANTE	156	62	11-330
		GUARDAMAR	126	18	1-30
		CARTAGENA	132	53	8-173
		MAZARRON	210	37	9-170
		AGUILAS	162	32	8-100
		GARRUCHA	150	62	1-390
		ALGECIRAS	10	37	---
		PALMA MALLORCA	73	60	1-62

Table 2.5   Concentrations of DDT's and PCB's (in ng/g) of sediments in the Catalonian region [Sánchez-Pardo, and Rovira, 1985]

Sample	Type	pp'-DDE	pp'-DDD	pp'-DDT	Σ DDT's	PCB's(A1254)
1	B	2,82	2,88	2,97	8,67	35,64
2	C1	0,66	0,37	0,74	1,77	9,40
3	a: C3	0,04	n.d.	0,94	0,98	1,43
4	b: B	1,08	0,61	3,36	5,05	66,0
5	a: C2	0,08	0,04	0,08	0,20	4,10
6	a: C2	0,16	0,05	0,20	0,41	0,82
7	a: C1	0,16	0,05	0,24	0,45	5,64
8	a: C2	0,02	0,04	0,08	0,14	2,31
9	b: B	0,64	0,72	8,70	10,06	11,20
10	A	1,55	5,76	6,39	13,70	33,10
11	A	5,45	7,90	4,33	17,68	133,37
12	A	1,01	0,90	2,22	4,13	22,40
13	A	3,11	2,48	4,35	9,94	36,15
14	B	1,75	1,29	1,94	4,98	29,24
15	A	1,48	n.d.	14,52	16,0	10,95
16	A	3,53	0,85	0,45	4,83	2,86
17	A	1,52	0,43	0,13	2,08	1,21
18	A	1,36	0,24	0,10	1,70	0,78
19	A	1,13	0,42	n.d.	1,55	1,23

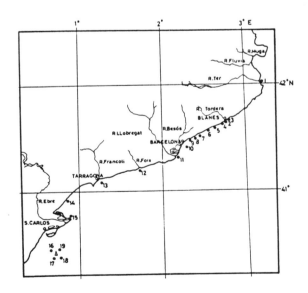

Map of sampling points.

Table 2.6   Concentrations of DDT's and PCB's (in ng/g) in sediments of Balear region [Sánchez-Pardo, and Rovira, 1985]

Sample	Type	pp'-DDD	pp'-DDT	pp'-DDE	ΣDDT's	PCB's(A1254)
1	B	0,20	0,10	1,75	2,05	3,45
2	a:B	0,28	0,36	2,12	2,76	8,95
3	c:B	0,12	0,17	0,55	0,84	8,56
4	C1	0,09	0,20	0,73	1,02	2,36
5	A	0,13	0,22	0,61	0,96	2,93
6	a:B	0,35	0,03	0,87	1,25	4,35
7	a:B	0,09	0,42	1,90	2,41	3,82
8	a:B	0,69	1,60	2,87	5,16	8,82
9	C1	0,15	0,17	0,78	1,10	5,81
10	a:B	0,09	0,08	0,07	0,24	0,36
11	a:C1	0,05	0,02	0,04	0,11	0,37
12	B	0,30	0,15	0,16	0,61	1,36
13	a:C1	0,06	0,02	0,11	0,19	0,96
14	a:B	0,11	0,08	0,21	0,40	1,11
15	c:B	0,14	0,04	0,06	0,20	0,30
16	B	0,04	0,02	0,09	0,15	0,28

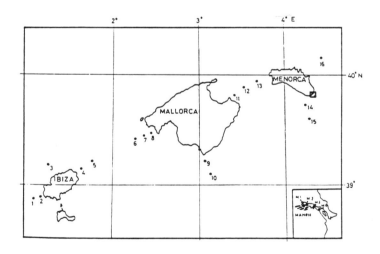

Map of sampling points.

Table 2.7   Dissolved/dispersed petroleum hydrocarbons (ug l^{-1}) [GESAMP Working Group 26, 1987]

Area	Year	Concentrations	Technique	Reference
Western Mediterranean (nearshore)				
Spanish coast				
Castellon	1983	1.36-2.40		De Leon, 1984
Sagunto	"	0.06-3.40		"
Valencia	"	0.63-4.35		"
Cullera	"	0.06-3.10		De Leon, pers. comm.
Benidorm	"	0.60-0.26		"
Alicante	"	0.85-8.26		"
Guardamar	"	1.15-3.15		"
Portman	"	0.26-6.50		"
Cartagena	"	0.26.3.22		"
French coast	1975-78	50-5000 (av. 580)	IR	UNEP, 1980
Banyuls-sur-Mer				
Var Estuary	1981	0.4-1.0	GC-UCM	Burns and Villeneuve, 1982
Gulf of Fos	1983-84	30-200		MEDPOL Phase II
Gulf of Ajaccio	1983-84	0-100		MEDPOL Phase II

Table 2.8   Concentrations of aliphatic hydrocarbons (C_{14} to C_{24}) and ramified alkanes (pristano and fitano) in the deltaic sediments of the Catalonian rivers (ug/kg D.W.) [Sánchez-Pardo and Rovira, 1985]

Station	Zone	$\Sigma C_{14} - C_{24}$	C_{17}	Pristano	C_{18}	Fitano	R_1*	R_2*
1	Ter	132,9	18,5	23,4	10,7	11,9	0,8	0,9
2	Todera	565,0	19,7	12,3	19,7	17,2	1,6	1,1
3	Besós	423,6	28,7	57,4	21,5	14,4	0,5	1,5
4	Besós	77,6	1,0	1,6	13,5	1,3	0,6	10,4
5	Llobregat	347,7	23,8	40,7	15,1	40,7	0,6	0,4
6	Foix	81,3	15,0	29,1	8,0	18,3	0,5	0,4
7	Francolí	70,6	7,4	4,6	2,8	1,8	1,6	1,6
8	Ebro	118,9	3,3	13,6	2,0	9,5	0,2	0,2
9	Ebro	348,3	42,6	31,2	7,6	32,2	1,4	0,2
Average		240,6	17,8	23,8	11,3	16,4	0,9	1,8
Standard deviation		183,4	13,1	18,0	6,7	13,0	0,5	3,2

* $R_1 = C_{17}$/pristano, $R_2 = C_{18}$/fitano.

Table 2.9 Concentrations of DDT's in samples of fishes and other organisms of the Ebro Delta area (ug/kg DW) [Sánchez-Pardo, J. & Rovira, J., 1985]

Studied species	p,p'-DDE	p,p'-DDD	p,p'-DDT	Σ DDT's
Arnoglossus laterna [1]	323,5	279,3	143,7	746,5
Murex brandaris [1]	13,5	4,8	—	18,3
Spicara sp.[1]	292,4	71,8	122,1	486,3
Carcinus maenas [1]	332,5	206,9	24,2	553,7
Astropecten aurantiacus [1]	28,4	12,8	13,2	54,0
Holoturia forskali [1]	15,4	17,1	24,9	57,4
Squilla mantis [1]	404,2	90,1	55,0	549,3
Conger conger [2]	979,1	882,7	—	1861,8
Conger conger [3]	2772,7	1812,1	262,9	4847,8
Dicentrarchus labrax [2]	2569,5	1758,6	73,3	4401,4
Dicentrarchus labrax [3]	5984,8	563,1	363,1	6911,0
Merluccius merluccius [2]	6466,5	4193,8	1035,3	11695,6
Average	1681,9	824,4	176,5	2625,9
Standard deviation	2323,8	1244,6	293,0	3680,4

(1): soft tissue or muscle; (2): liver; (3): gonads; (—): under the limit of detection of the technique used.

Table 2.10 Concentrations of PCB's in samples of fishes and other organisms of the Ebro Delta area (mg/kg D.W.) [Sánchez-Pardo, J., Rovira, J., 1985]

Studied species	Aroclor 1254	Aroclor 1260	Σ PCB's	Σ PCB's/Σ DDT's
Arnoglossus laterna [1]	1,1	2,4	2,1	2,81
Murex brandaris [1]	0,1	0,1	0,1	5,46
Spicara sp.[1]	0,6	1,2	1,1	2,26
Astropecten aurantiacus [1]	—	—	—	—
Carcinus maenas [1]	0,8	1,5	1,4	2,53
Holoturia forskali [1]	0,1	0,1	0,1	1,74
Squilla mantis [1]	1,6	3,5	3,0	5,46
Conger conger [2]	3,4	9,0	7,1	3,81
Conger conger [3]	12,0	29,4	23,0	4,74
Dicentrarchus labrax [2]	7,2	14,6	11,8	2,68
Dicentrarchus labrax [3]	17,8	26,9	12,3	1,78
Merluccius merluccius [2]	15,9	24,4	26,5	2,27
Average	5,0	9,4	7,4	2,96
Standard deviation	6,6	11,4	9,2	1,63

(1): soft tissue or muscle; (2): liver; (3): gonads. (—): under the limit of detection of the technique used.

Table 2.11 Hydrocarbons in biota samples from the Spanish Mediterranean Coast (in ug g^{-1} dry wt) [UNEP/MAP TRS No. 19, 1988]

Species	Area (year) fraction (UCM)	Saturate fraction (crude oil eq.)	Aromatic
Mytilus (10 samples)	Palamós	106- 190	-
	Barcelona	500-3200	-
	Ebro Delta	8- 216	-
Mullus sp. (muscle) (14 samples)	Palamós	12.6	4.4
	Barcelona	22.2	9.3
	Ebro Delta	5.8	11.1
Merluccius sp. (muscle) (14 samples)	Palamós	1.5	1.7
	Barcelona	0.2	3.9
	Ebro Delta	0.2	2.4
Trachurus sp. (muscle) (14 samples)	Palamós	11.2	4.2
	Barcelona	1.4	10.9
	Ebro Delta	5.4	3.7
Engraulis sp. (muscle) (19 samples)	Barcelona	7.7	7.8

Table 2.12 Concentrations of aliphatic hydrocarbons (C_{14} to C_{24}) and ramified alkanes (pristano and fitano) in samples of fishes and other organisms from the Ebro Delta area (mg/kg D.W.) [Sánchez-Pardo and Rovira, 1985]

Studied species	ΣC_{14}-C_{24}	C_{17}	Pristano	C_{18}	Fitano	R_1	R_2
Arnoglossus laterna (1)	9,6	1,4	0,8	1,2	0,2	1,7	6,0
Murex brandaris (1)	11,4	1,2	0,4	1,2	0,4	3,0	3,0
Spicara sp. (1)	5,8	0,4	1,2	0,7	—	0,3	—
Carcinus maenas (1)	9,4	1,1	0,4	1,1	0,2	2,7	5,5
Astropecten aurantiacus (1)	2,3	0,2	0,4	—	0,4	0,5	—
Holoturia forskali	12,0	1,2	0,3	1,7	0,2	4,0	8,5
Squilla mantis (1)	10,7	2,1	2,3	1,9	0,4	0,9	4,7
Conger conger (2)	243,7	1,8	1,4	30,1	—	1,3	*
Conger conger (3)	548,2	—	—	63,8	—	*	*
Dicentrarchus labrax (2)	168,3	4,6	—	16,7	—	*	*
Dicentrarchus labrax (3)	49,2	—	—	4,6	—	*	*
Merluccius merluccius (2)	697,7	4,8	4,8	77,4	—	1,0	*
Medias	147,3	1,6	1,0	16,7	0,1	1,7	5,5
Desviaciones típicas	236,9	1,6	1,4	26,8	0,2	1,2	2,0

(1): soft tissue or muscle; (2): liver; (3): gonads. $R_1 = C_{17}$/pristano; $R_2 = C_{18}$/fitano. (—): under the limit of detection of the technique used; (*): not found.

Table 2.13 Virus isolation from beaches expressed as MPNCU l^{-1}. Samples were taken 4 or 5 m offshore from June 1979   [Lucena et al., 1982]

	Castelldefels beach			Barceloneta beach			Badalona beach		
		95% Confidence limits			95% Confidence limits			95% Confidence limits	
Isolation date	MPNCU l^{-1}	Lower	Upper	MPNCU l^{-1}	Lower	Upper	MPNCU l^{-1}	Lower	Upper
0.7.6.79	—	—	—	0.12	<0.02	0.36	0.12	<0.02	0.36
08.6.79	0.12	<0.02	0.36	0.14	<0.02	0.8	NT	—	—
09.6.79	NT	—	—	0.14	<0.02	0.8	1.60	0.28	5.2
11.6.79	0.12	<0.02	0.36	0.18	<0.02	0.8	NT	—	—
12.6.79	0.92	0.16	4.8	1.12	0.4	6	NT	—	—
13.6.79	NT	—	—	NT	—	—	0.14	<0.02	0.8
15.6.79	0.28	0.04	0.92	NT	—	—	NT	—	—
16.6.79	NT	—	—	NT	—	—	0.12	<0.02	0.36
18.6.79	NT	—	—	NT	—	—	0.12	<0.02	0.36
19.6.79	0.12	<0.02	0.36	NT	—	—	NT	—	—

NT: non tested.

Table 2.14 Virus isolation expressed as MPNCU l^{-1}, of samples taken at different distances offshore from Badalona beach. Samples were taken from September 1979   [Lucena et al., 1982]

	2 m			5 m			20 m			1000 m		
		95% Confidence limits			95% Confidence limits			95% Confidence limits			95% Confidence limits	
Isolation date	MPNCU l^{-1}	Lower	Upper	MPNCU l^{-1}	Lower	Upper	MPNCU l^{-1}	Lower	Upper	MPNCU l^{-1}	Lower	Upper
05.9.79	0.30	0.04	0.92	0.36	0.04	0.92	0.15	<0.02	0.80	0.18	<0.02	0.80
06.9.79	0.15	<0.02	0.80	0.18	<0.02	0.80	0.15	<0.02	0.80	0.30	0.04	0.92
07.9.79	NI	—	—	NI	—	—	0.15	<0.02	0.80	NI	—	—
08.9.79	0.36	0.04	0.92	NI	—	—	NI	—	—	0.15	<0.02	0.80
10.9.79	0.18	<0.02	0.80	0.18	<0.02	0.80	NI	—	—	NI	—	—
11.9.79	0.15	<0.02	0.80	0.15	<0.02	0.80	NI	—	—	NI	—	—
12.9.79	NI	—	—	NI	—	—	0.30	0.04	0.92	NI	—	—
13.9.79	NI	—	—	NI	—	—	0.15	<0.02	0.80	NI	—	—
14.9.79	0.36	0.04	0.92	NI	—	—	0.15	<0.02	0.80	0.15	<0.02	0.80

NI: non isolated (<0.12 MPNUC l^{-1}).

Table 2.15   Hectares affected by different degrees of erosion in the provinces of the Spanish Mediterranean littoral
[MOPU, feb. 1987; B-351, part II]

PROVINCE	TOTAL SURFACE	LOW EROSION Has.	%	MODERATE EROSION Has.	%	STRONG EROSION Has.	%
Gerona	588.600	27.374	4,7	123.412	21,0	5.716	7,2
Barcelona	733.300	37.427	4,3	136.958	17,7	79.607	10,9
Tarragona	628.300	30.313	4,8	192.556	30,6	135.849	21,6
Castellón	667.900	30.363	4,5	215.196	32,2	180.753	27,1
Valencia	1.076.300	77.176	7,2	245.143	22,8	300.702	27,9
Alicante	586.300	16.420	2,8	157.332	26,8	188.647	32,2
Murcia	1.131.700	79.525	7,0	245.783	21,7	540.006	47,7
Almería	877.400	8.954	1,0	77.693	8,3	633.645	72,2
Granada	1.253.100	41.345	3,3	196.116	15,7	643.830	51,4
Málaga	727.600	14.803	2,0	145.299	20,0	336.273	46,2
Cádiz	738.500	158.107	21,4	172.470	23,4	155.441	21,0
Baleares	501.400	88.185	17,6	176.267	35,2	72.177	14,4
TOTAL AREA	9.550.400	610.432	8,2	2.079.225	24,8	3.308.833	34,6
SPAIN	50.475.000	5.444.141	10,8	13.923.342	27,0	13.034.682	25,8
% AREA/SPAIN	18,9		11,2		14,9		25,4

Table 2.16   Number of fires, area affected and cost [DOCTER, 1987]

| Year | N. | Damaged surface (ha.) | | | Loss of primary products (mm. of pesetas) | Environmental costs (mm. of pesetas) | Total cost (mm. of pesetas) |
		Wooded	Cleared	Total			
1974	3.980	58.789	81.422	140.211	1.992	7.709	9.701
1975	4.242	111.091	76.223	187.314	4.121	13.855	17.976
1976	4.596	79.853	82.447	162.300	3.974	12.575	16.549
1977	2.148	26.454	41.086	67.540	981	3.393	4.374
1978	8.324	159.264	275.603	434.867	9.205	17.639	26.844
1979	7.167	119.579	152.139	271.718	6.790	16.819	23.609
1980	7.193	92.503	173.451	265.954	6.774	18.992	25.766
1981	10.882	141.667	156.769	298.436	9.557	29.570	39.127
1982	6.443	63.879	87.765	151.644	4.871	25.945	30.816
1983	4.880	57.832	59.767	117.599	4.197	21.640	25.837
1984	7.224	53.653	110.893	164.546	5.797	26.552	32.349
1985	12.837	180.885	288.541	469.426	15.819	71.587	87.406

Table 2.17 Natural interest areas with special protection in the Mediterranean coastal provinces [Espacios Naturales Protegidos. Boletín Official del Estado. 1986]

NAME	CATEGORY	APPROVAL	INTEREST
ISLAS COLUMBRETES	National Park	In progress	Flora/Fauna. 15 Ha
BAHIA DE ROSAS Rosas (Gerona)	Natur. Int. Area	Law 28-10-83 Parl. Cataluña	Marsh area. 2.025 Ha Flora & Fauna
LAGUNA DE FUENTE DE PIEDRA. (Malaga)	Integral Reserve	Law 9-1-84 Parl. Andalucía	Marsh area Fauna. 1.875 Ha
LA GARROTXA Olot (Gerona)	Natural Park	Law 3-3-82 Parl. Cataluña	Geobotanic. 6.000 Ha
MONASTERIO DE POBLET (Tarragona)	Natur. Int. Area	Law 9-11-84 Parl. Cataluña	Historic & Monum. 24.868 Ha
MACIZO DE PEDRAFORCA Barcelona-Lérida	Natur. Int. Area	Law 6-5-82 Parl. Cataluña	Geobotanic Fauna.
TORCAL DE ANTEQUERA Málaga	Natural Park	Act 3.062 27-10-	Landscape 1.000 Ha
SIERRA DE ESPUÑA Alhama de Murcia	Natural Park	Act 3.157 10-11-78	Landscape 10.400 Ha
MACIZOS DEL CADI Y MOIXERO. Barcelona, Gerona	Natural Park	Act 353 15-7-83 Parl. Cataluña	Landscape Flora & Fauna
MONTE EL VALLE Murcia	Natural Park	Act 2.611 7-9-79	Landscape 2.600 Ha
DELTA DEL EBRO Tarragona	Natural Park	Act 357 4-8-83 Parl. Cataluña	Marsh Area Flora & Fauna 16.000 Ha
ALBUF. VALENCIA Valencia	Natural Park	Act 89 8-7-86 Gen. Valencia	Marsh Area Flora & Fauna 3.200 Ha
SIERRA GRAZALEMA Norte de Cádiz	Natural Park (*)	Act 316 18-12-84 Parl. Andalucía	Flora & Fauna 27.000 Ha

(*) To be declared National Park.

Table 2.18 Natural interest areas without special protection in the Mediterranean coastal provinces
A). Continental Coast

PROVINCE	NAME	MUNICIPALITY	HECTARES	INTEREST
Gerona	Islas Medas	Torroella de Montgrí	24	Fauna
	Cabo de Creus	Port-Bou y otros	18.564	Cliffs
	Costa Brava	S. Feliú de G. y otros	1.430	Cliffs
	Marisma S. Pedro P.	Castello de Ampurias	1.450	Marsh Area
	Marisma de Pals	Torroella de Montgrí	500	Marsh Area
	Marisma costera	Castello de Ampurias	575	Marsh Area
TOTAL HECTARES GERONA			22.543	
Barcelona	Garraf	Begues y otros	5.300	Cliffs
	E. de la Ricarda	Prat de Llobregat	98	Marsh Area
TOTAL HECTARES BARCELONA			5.398	
Castellón	Lagunas de La Tayola	Almenara y La Llosa	1.400	Marsh Area
TOTAL HECTARES CASTELLON			1.400	
Valencia	Marjal de Sueca	Sueca, Sollana y Silla	9.000	Marsh Area
	Marjal del Júcar	Poliñá del Jucar y otros	5.000	Marsh Area
	Albufera de Anna	Anna	5	Marsh Area
TOTAL HECTARES VALENCIA			14.005	
Alicante	Montgó	Denia y Jávea	825	Landscape
	Dunas	Guardamar y Elche	774	Landscape
	Palmeral	Elche	16.800	Flora
TOTAL HECTARES ALICANTE			18.399	
Almería	Albufera	Adra	75	Fauna
	Cabo de Gata	Almeria	200	Fauna
TOTAL HECTARES ALMERIA			275	
Málaga	Acantilados de Maro-La Herradura	Nerja y La Herradura	230	Paisaje Fauna
TOTAL HECTARES MALAGA			230	
TOTAL HECTARES CONTINENTAL COAST			61.530	

Table 2.19 Natural interest areas without special protection in the Mediterranean coastal provinces B). Balearic Islands Coast
[Inventario Abierto de Espacios Naturales de Protección Especial. 1977-1980. ICONA 1983]

ISLAND	NAME	MUNICIPALITY	HECTARES	INTEREST
MALLORCA	Sierra de Alfabia	Calviá, Andraitx,	50.130	Landscape Flora/Fauna
	Albufereta de Pollensa	Alcudia	140	Fauna
	La Victoria	Alcudia	1.525	Flora Landscape
	Albufera de Alcudia	Muro, La Puebla y Alcudia	1.400	Fauna
	Acantilados de Farrutx	Artá	50	Landscape
	Isla Dragonera	Andraitx	300	Fauna
	Islotes de la Bahía de Palma	Palma, Andraitx y Calviá	5	Landscape
	Cabo Blanco	Lluchmayor	130	Cliffs
	Salobrar de Campos	Campos del Puerto	155	Landscape Fauna
	Estany de Ses Gambes	Ses Salines	65	Marsh Area Fauna
	Islas Cabrera, Conejera e Islotes	Palma	1.360	Landscape Fauna
MENORCA	Acantilados N.O.	Menorca	190	Landscape
	La Vall	Ciudadela	823	Scientific
	Santa Ana	Ciudadela	785	Landscape
	Islas e islotes junto a Menorca	Mahón y S. Luis	125	Landscape
	Albufera del Grao	Mahón	320	Landscape Fauna
IBIZA	Islas e islotes junto a Ibiza	S. Antonio Abad y otros	225	Landscape
	Salinas de Ibiza	San José	1.400	Landscape Fauna
FORMENTERA	I. del Espardell y del Espalmador	S. Fco. Javier	175	Landscape
	Estany De's Peix	S. Fco. Javier	112	Marsh Area
	Estany Pudent y Salinas de Formentera	S. Fco. Javier	800	Landscape Fauna
	La Mola	S. Fco. Javier	1.700	Landscape Flora Fauna

TOTAL HECTARES BALEARIC ISLANDS   61.915

TOTAL LITTORAL (Has.)   123.445

Table 2.20 Atmospheric pollutant emissions - SO_2, particles and NO_x - from industrial sources in the Mediterranean coastal region in 1980 [MOPU, feb. 1987; B-351, part II]

PROVINCE	SO_2 Tm/Year x 10^3	PARTICLES Tm/Year x 100	NO_x Tm/Year x 10^3
Gerona	4,29	1,07	0,94
Barcelona	142,99	47,72	23,20
Tarragona	37,69	13,68	7,80
Castellón	84,14	8,77	21,60
Valencia	16,19	30,89	3,03
Alicante	4,24	6,38	1,27
Murcia	44,10	10,49	10,30
Almería	7,14	2,40	1,57
Granada	5,18	0,45	1,16
Málaga	15,80	3,15	5,55
Cádiz	50,80	1,73	9,73
Baleares	23,30	11,37	15,04
TOTAL AREA	439,89	138,10	101,19
SPAIN	1.570,07	999,86	305,97
% AREA/SPAIN	28,0	13,8	33,1

Table 2.21 Sewage treatment in the Mediterranean area. 1986 [MOPU, feb. 1987; B-351 part II]

PROVINCE	No. MUNICIP.	TREATMENT PLANTS	% SEWAGE VOLUME TREATED	TREATMENT
Gerona	22	12	80	BIO (12)
Barcelona	27	5	80	BIO (4); PC (1)
Tarragona	20	2	1	BIO (2)
Castellón	16	13	80	PRS (13)
Valencia	20	12	80	BIO (6); PRS (6)
Alicante	15	12	65	BIO (12)
Murcia	6	4	5	BIO (4)
Málaga	13	58	67	BIO (4); PRS (54)
Almería	13	10	45	PRS (10)
Granada	9	0	0	------
Cádiz	5	8	---	PRS (8)
Baleares	--	11	100	BIO (11)
Total	166	147		BIO (55); PC (1); PRS (91)

(TREATMENT: BIO=Biological, PC=Physical & Chemical, PRS=primary sedimetation)

Table 2.22 Points of sewage discharge into the Spanish Mediterranean sea
[Estudio de Actualización del inventario de zonas litorales del Mediterráneo español más contaminadas. Ambio. July 1984]

PROVINCE	COAST LENGTH (Km.)	DIRECT DISCHARGES	SEA OUTFALLS	RIVERS	TOTAL
Gerona	198	215	23	8	246
Barcelona	101	198	33	3	234
Tarragona	216	119	35	5	159
Castellón	116	20	7	2	29
Valencia	109	37	7	5	49
Alicante	212	17	7	4	28
Murcia	208	7	8	0	15
Almería	217	4	8	3	15
Granada	71	12	3	2	17
Málaga	171	92	60	16	168
Cádiz	60	32	9	6	47
Baleares	910	4	36	1	41
TOTAL	2,579	757	236	55	1,048

Table 2.23   Depth and length of sea outfalls
                  [MOPU, 1984]

PROVINCE	EXISTING OUTFALLS	UNKNOWN DATA	DEPTH LESS THAN 15 m.	LENGTH LESS THAN 1 Km.	PROPOSED DRAINS
Gerona	23	7	5	19	3
Barcelona	33	10	17	28	9
Tarragona	35	5	11	17	8
Castellón	7	3	2	3	2
Valencia	7	1	3	1	0
Alicante	7	5	1	5	0
Murcia	8	2	6	1	0
Almería	8	6	0	4	1
Granada	3	2	0	2	2
Málaga	60	15	30	33	6
Cádiz	9	7	0	2	0
Baleares	36	26	1	24	1
TOTAL	236	89	76	139	32

Table 2.24   Number of industrial facilities discharging
                  pollutants into the Mediterranean (entire
                  coastal provinces)
                  [MOPU, feb 1987; B-351 part II]

PROVINCE	No. OF PLANTS
Gerona	6
Barcelona	1.062
Tarragona	42
Castellón	56
Valencia	46
Alicante	39
Murcia	39
Almería	14
Granada	7
Málaga	47
Cádiz	63

246

Table 2.25 Maritime Traffic of Mediterranean Spanish Ports (1985) (tons)
[MOPU DG Puertos y Costas, 1985]

	Bulk (106)	General cargo (106)	Petroleum (106)	Containers (103)	Supplies (106)	Fish Desembarked (103)	Others	All Products (106)
Almaría	7.33	–	–	–	–	12.7	0.14	7.74
Barcelona	6.18	5.92	2.9 (*)	3.632	0.44	–	–	19.08
Tarragona	4.63	0,75	14.42	1.71	0.15	–	1.11	21.23
Valencia	3.00	4.57	1.20 (*)	2.912	0.19	–	–	11.87
La Coruña	1.57	–	7.12	–	0.37	97.9	0.52	9.68
Cartagena	1.35	–	8.92	2.59	–	–	1.12	11.65
Alicante	1.17	–	–	3.49	–	–	1.13	2.65
Algéciras	0.84	–	11.67	3.178	0.54	31.6	4.72	20.98
La Luz y Las Palmas	–	2.89	4.16 (*)	1.237	1.95	301.1	5.87	10.53
Palma de Mallorca	–	2.02	–	1.231	0.22	–	1.66	3.11
Castellón	–	–	6.45	–	–	–	1.07	7.52
Malaga	–	–	5.87	–	–	–	1.35	7.22
Others	13.21	9.17	6.77 (*)	5.61	1.89	296.9	4.17	31.9
Subtotal	47.69	31.42	84.88	13.53	5.75	754.3	22.12	198.24
	24.05%	15.85%	42.82%	0.02%	2.9%	0.99%	0.44%	(83.5% of all Spanish ports/ offsh.)

(*) Ports without refinery

Table 2.26   Principal diseases in the Spanish beaches
              [Ministerio de Sanidad y Consumo, 1983]

	Morbidity rate per 100 bathers		
Diseases	1979(a)	1981(b)	1982(c)
Eye infection	3,8	2,38	8,5
Nose infection	2,5	2,5	4,9
Ear infection	2,9	1,37	2,3
Throat infection	1,4	0,9	2,7
Skin rash	3,9	3,96	3,3
Skin parasites	3,4	1,55	2,9
Diarrhea	1,6	1,7	2,6
Allergies	1,7	1,13	1,6

(a) Málaga and Tarragona.
(b) Cádiz, Castellón, Huelva, Ibiza and Mallorca.
(c) Cádiz, Cantabria, Castellón, Ibiza and Mallorca.

Table 2.27 Microbiological data and observed human irritations/diseases at the main Spanish beaches (Data from 1982) [Ministerio de Sanidad y Consumo, 1983]

Beach	WHO Criteria E. Coli	Concentration Chlorophyll-a	Eye infection %	Nose infection %	Skin rash %	Throat infection %	Skin parasite %	Ear infection %	Diarrhia %	Allergy %
Covachos	S*	2,476 ppm	10	3	0,8	2,5	2,5	3	0,8	0
Sardinero	NS	2,625 "	8	2,6	1	1	2	2	0	0,6
Pegina	S	1,743 "	4	8	6,5	3	4	5	5	1
Punlilla	NS	4,489 "	8,6	5,3	1	5,3	5,3	4	3	0
Peñiscola	S	34,666 "	15,2	11,5	9	8,9	1,5	3	6	4
Pinar	NS	49,175 "	18,4	7,3	6	4	2,6	4	6,5	6,3
Salinas	S	0,115 "	0,5	3	1,6	0,5	0	0,5	0	0
Den Bossa	NS	0,404 "	1	6	5	2	1	3	0	0,5
Correlation coefficient (r)		0,872	0,49	0,55	0,61	-0,41	-0,07	0,45	0,84

*S: satisfactory (<100/100 ml in 50% of the samples)

Table 2.28 Economic impacts of soil erosion (estimated) [Calderón, E., et al, 1987]

REGION	(A)	(B)	(C)	(D)
Andalucía	40.70	1,881	1,500	67,72
Baleares	14.40	36	1,100	0,95
Cataluña	12.90	250	3,100	18,60
Murcia	47.70	317	1,500	11,41
Valencia	28.60	333	1,500	11.99
Total		2,817		110,67

(A): Percentage of regional surface subject to severe erosion.
(B): Estimated surface rendered unproductive because of erosion (Ha).
(C): Output per Ha. for average crop.
(D): Total money loss (m. pts) at average of 24 pts/kg.

Table 2.29 Economic impacts of forest fires [Calderón, E., et al, 1987]

REGION	(A)	(B)	(C)	(D)
Andalucía	662	22,091	415.00	0.150
Baleares	110	400	11.80	0.190
Cataluña	394	9,412	582.97	0.630
Murcia	24	91	2.14	0.004
Valencia	470	22,500	535.03	0.450
Total		54,494	1,546.94	

(A): Total number of reported fires
(B): Surface affected (Ha)
(C): Estimation of direct money loss (M PTS)
(D): Percentage of Regional Agricultural income

Table 2.30   Relevant regulatory framework relating to the Mediterranean environment   [MOPU, 1987]

Continental water protection

Law 29/85	02/08/85	National	Continental water use regulation and quality protection
Act	18/06/82	National	Sanitary regulation of the drinking water
Act 32/85	05/02/85	Andalucia	Regulation of the drinking water fluorization
Law 5/81	04/06/81	Cataluña	General sewerage regulation in the territory of Cataluña
D.O	10/05/82	Cataluña	Creation of the Cataluña's Land, Water and Climate Data Bank
Act 305/82	13/07/82	Cataluña	Development of the Law 5/81
D.O.	19/10/82	Cataluña	Technical regulation of sewerage operation and financing
Act 150/84	16/03/84	Cataluña	Regulation of sewerage fees perception in Cataluña
Act 42/84	28/05/84	Baleares	Regional subventions for sewerage and water treatsent.

Sea water and littoral protection

Coasts Law	10/03/80	National	General regulation of the marine coasts use, warning and protection
O.M.	26/05/76	National	Technical regulation to prevent the shipping pollution
Law 21/77		National	Regulation of penalities on pollutants discharges from ships and plains
Shellfish Law 1969		National	General regulation of shellfisch catching and culture
Act 2486/84	26/09/84	National	General regulation of pleasure ports related activities
Act 178/84	19/06/84	Andalucia	Regulation and vigilance of sanitary and environmental conditions of beaches
Act 204/84	17/07/84	Andalucia	Starting a new process of land planning in the coastal strip of Andalucia
New Coasts Law 1988?		National	In procedure: limitation of building in a 100 m. width coastal strip.

Nature conservation: flora & fauna protection

Law 15/75		National	General regulation of National Parks and other relevant natural areas
Act 2676		National	Development of the Law 15/75
Law 1/84	14/03/84	Baleares	Planning and protection of the relevant natural areas of the Balearic islands
Law 12/85	13/06/85	Cataluña	Planning and protection of the relevant natural areas of Cataluña
Act 545/83	15/12/83	Cataluña	Organizing the restoring of the burnt forestry areas
Act 53/85	18/02/85	Cataluña	Regulation of the coral extraction ban in the Cataluña's coast
D.O.	13/12/84	Murcia	Regulation of the oysters catching in the Mar Menor
D.O.	10/05/85	Baleares	Regulation of the coral extraction ban in the Balearic islands coast

Table 2.31 Highly polluted areas on the Spanish Mediterranean Littoral [derived from MOPU's and Ambio's data: Estudio de Actualización del inventario de zonas litorales del Mediterráneo español más contaminadas. Ambio. Julio 1984. Acciones para la Protección del Mediterráneo. MOPU 1987]

AREA OR TOWN	CAUSES
-BLANES (Gerona, Cataluña)	Industrial discharges (Textile).
-COASTAL SECTION MATARO-LLOBREGAT (Barcelona, Cataluña)	Industrial discharges (Dyeing)
-TARRAGONA (Tarragona, Cataluña)	Industrial & Petrochemical discharges.
-DELTA DEL EBRO (Tarragona, Cataluña)	Agricultural pollution.
-SAGUNTO (Valencia, País Valenciano)	New chemical complex and super-port.
-ALBUFERA DE VALENCIA (Valencia, País Valenciano) (15 Km. south from the city)	Multiple problems (industrial, urban, agricultural discharges). Starting a general anti-pollution plan.
-PORTMAN (Murcia)	Mining discharges. This is probably the biggest coast pollution problem.
-CARTAGENA (Murcia)	Industrial & Petrochemical discharges. Starting an anti-pollution plan.
-MOTRIL Y SALOBREÑA (Granada, Andalucía)	Urban waste water.
-MALAGA (Málaga, Andalucía)	Urban waste water. Soon solutionned.
-BAHIA DE ALGECIRAS (Cádiz, Andalucía)	Industrial & Petrochemical discharges.

Table 2.32 Forecast of investments on general sanitation in the autonomous region of Cataluña (millions of Pesetas) [MOPU, feb. 1987; B-351, part II]

ZONA No	AFECTED PROVINCE	TOTAL FORECASTS	ACHIEVED UNTIL 1987
5	Barcelona	122.154,77	45.773,0
2-3	Gerona	10.617,71	3.768,7
6-7	Tarragona	6.119,10	1.970,4
4	Gerona-Barcelona	2.910,60	1.119,9
12	Lérida	3.505,00	829,0
TOTAL		145.307,18	53.461,0

Table 2.33 Forecast of investments and actions for the sanitation plans in the coastal municipalities of the autonomous region of Valenciana. 1986-1989 [MOPU, feb. 1987; B-351, part II]

LOCAL PLAN	UNDERWATER DRAINS*	TREATMENT PLANTS*	Pts (Mill.)	POPULATION*	START	CAPACITY M³/DAY*
Vall de Uxó	-	1	350	26.500	1988	6.600
Vila-Real	-	1	350	40.000	1989	10.000
Vinaroz	1	-	120	60.000	1987	
Peniscola	1	-	100	75.000	1987	
TOTAL CASTELLON	2	2	920	201.500		16.600
Valencia P.Colec.	1	1	5.000	300.000	1988	75.000
Quart Benachert	-	1	3.000	255.000	1989	63.750
Camp del Turia	-	2	-----	170.000	1988-89	
Saneam. rio Buñol	-	1	395	11.000	1989	2.750
Saneam. rio Albaida	-	3	2.200	72.000	1988-89	18.000
Sueca	-	1	140	70.000	1986	17.500
Canet de Berenguer	1	-	72	10.000	1987	2.500
Sagunto	-	1	250	60.000	1989	15.000
La Safor	-	2	500	30.000	1988-89	7.500
Colec. O Albufera	-	-	2.200	200.000	1988-89	
TOTAL VALENCIA	2	12	13.757	1.178.000		202.000
Saneam. Vega Baja	-	3	890	91.000	1988-89	22.750
Elda Petrer Sax	-	1	170	25.000	1987	6.250
Santa Pola	-	1	250	110.000	1987	27.500
Novelda Monforte	-	1	200	25.000	1988	6.250
Alicante, Campello, San Juan	2	1	1.050	125.000	1988	31.250
Denia	1	-	151	80.000	1986	
Alcoy	-	1	780	70.000	1989	17.500
Des. río Serpis	-	1	450	16.000	1989	4.000
Altea	-	1	205	30.000	1986	7.500
Jávea	2	1	1.200	50.000	1989	12.500
TOTAL ALICANTE	5	11	5.346	622.000		135.500
TOTAL REGION	9	25	20.023	2.001.500		354.100

* forecast of projects in Sanitary Plan

Table 2.34 Forecast of investments and actions for the sewage plan in the coastal municipalities of the autonomous region of Murcia. 1985
[MOPU, feb. 1987; B-351, part II]

LOCALIZATION	TREATMENT PLANTS	SEA OUTFALLS	SEWERS	PUMP STATIONS	INVESTMENT COST (Mill. Pts.)
Aguazaras (4)	1	-	-	-	12,4
Centi Lorqui (3)	1	-	-	-	72,4
Mazarrón 1ª f. (3)	1	-	-	-	61,3
Alcázares (1)	1	-	-	-	59,9
Zarandorra (2)	1	-	-	-	59,9
San Javier 1ª f. (4)	1	-	-	-	30,7
Beniel (redress) (4)	1	-	-	-	2,1
Jumilla (1)	1	-	-	-	74,4
Mula (5)	1	-	-	-	71,4
Santonera (extension) (5)	1	-	-	-	11,7
Los Alcázares (2)	-	-	1	-	28,6
Mazarrón (2)	-	-	-	1	19,5
S. Pedro del Pinatar (2)	-	1	-	-	42,1
Ceheguin (2)	-	-	1	-	33,4
Other works					21,1
TOTAL	10	1	2	1	600,9

(1) Working
(2) finished
(3) under construction
(4) approved
(5) submitted for approval

Table 2.35 Forecast of investments and actions for the sewage plans in the coastal municipalities of the autonomous region of Andalucia
[MOPU, feb. 1987; B-351, part II]

LOCALIZATION	TREATMENT PLANTS	UNDERWATER DRAINS	INVESTMENTS (Mill. Pts.)
Adra	1	1	36
Aguadulce	1	-	20
Almeria	-	1	30
TOTAL ALMERIA	2	2	86
La Herradura	1	-	110
Castillo de Baños	1	-	10
Almuñécar	1	-	325
Carchuna y Calahonda	1	-	70
Salobreña-La Guardia	1	-	205
Motril Varadero	1	-	325
Torrenueva	1	-	176
Casteldeferro	1	-	115
TOTAL GRANADA	8	-	1.336
Málaga	1	-	700
V. del Trabuco	1	-	40
Alhaurin de la Torre	1	-	40
Ojen	1	-	20
TOTAL MALAGA	4	-	800
San Roque	2	-	110
TOTAL CADIZ	2	-	110
TOTAL ANDALUCIA ORIENT.	16	2	2.332

Table 2.36 Forecast of treatment plants and und. drains construction by the autonomous region of Baleares [MOPU, feb. 1987; B-351, part II]

LOCALIZATION	POPULATION	TREATMENT PLANTS	UND. DRAINS	START	CAPACITY M³/DAY
Mallorca	154.000	9	5	May 86-Dec 88	33.750
Ibiza	70.500	3	2	Jun 86-Dec 87	14.125
Menorca	14.500	4	-	Jun 86-Dec 87	3.625
Formentera	3.000	-	1	Dec. 86	------
TOTAL	242.000	16	8		51.500

Table 2.37 Provinces and municipalities with coastal projects and required investments by INEM [Revista MOPU, Nov. 1986]

Province	number of projects		Investment (milion Pts.)		
	beaches * coastal projects	river mouth structures	beaches * coastal projects	rivermouth structures	TOTAL
Almeria	12	12	56	17	73
Cadiz	9	5	51	29	80
Granada	7	15	30	45	75
Malaga	13	8	60	30	90
Baleares	21	3	67	9	76
Barcelona	28	-	50	-	50
Gerona	20	-	50	-	50
Tarragona	19	-	51	-	51
Murcia	7	-	50	-	50
Alciante	15	9	60	43	103
Castellon	14	8	55	37	92
Valencia	15	14	80	40	120
TOTAL	180	74	660	250	910*

* Mediterranean investments represent 25% of total Spanish investment.

3.   FRENCH MEDITERRANEAN ENVIRONMENT

The French coastal zone occupies 2.5% of the national territory, while 11% of the total French population lives in that zone. The demographic concentration is accompanied by high industrial densities. A major input of pollutants results from the Rhône River discharge, the largest river in the western Mediterranean. Consequently, the problems of preservation of the marine coastal environment are numerous.

In section 3.1 the extent of marine pollution in the French Mediterranean environment is investigated. Sections 3.2 and 3.3 deal with the pressures on the environment and the economic impacts, respectively. In section 3.4 the institutional/legal framework is examined. Section 3.5 deals with the past and future trends in environmental expenditures.

3.1.   State of the environment

The national Network of Observation of the marine environment (RNO), with the scientific and technical support of the French Institute of Marine Research (Ifremer, initially Cnexo and ISTPM), has monitored the quality of the coastal waters since 1974. The French coast is divided into 43 monitoring zones, which are grouped together in 8 regions. Each region has at least one principal zone called "supporting point" where the water quality, living matter and sediments are monitored. In the other zones only the living matter and sediments are monitored (Figure 3.1).
The Mediterranean coast can be divided into three zones, with different departments:
- the region Languedoc-Roussillon to the west of the Rhône Delta is characterized by a flat and sandy coast and by an extensive continental shelf;

- the region Provence-Alpes-Côte d'Azur to the east of the Rhône Delta is characterized by a rocky coast and by a limited continental shelf;
- the island of Corsica is characterized by a rocky coast to the west and by a flat coast to the east; very important <u>Posidonia</u> beds surround the island.

3.1.1. Water quality

A comparison between the average French Atlantic coastal waters (Seine, Brest) and the average French Mediterranean coastal waters (Fos, Villefranche) for temperature, salinity, nitrates and phosphates is shown in Table 3.1. The Mediterranean coastal waters are lower in nutrients (nitrogen and phosphorus). Especially the Bays of Cannes and Villefranche (0.04 µmoles/l phosphates) and those of Beaulieu and Menton have the lowest values. The maximum nitrate concentrations are observed in the Gulf of Fos (3.7 µmoles/l nitrates) and the Berre Lagoon.

The range of the values of nitrates, ammonia, silicates and phosphates at the eight sites monitored between 1975 and 1980 along the French Mediterranean coast is shown in Table 3.2. The Fos-Berre zone (near the mouths of the Rhône and Durance rivers) displays the major pollution from land-based sources with maximum nitrate values of 18 µmoles/l. Cortiou has high phosphate values (0.83 µmoles/l) near the Marseille municipal sewage outfall. The values of the Thau Lagoon are representative for the lagoons in the Languedoc-Roussillon region. The nitrate, ammonia and silicate concentrations are at an intermediate level, but phosphate has the highest levels of all sites (2.62 µmoles/l).

Measurements of the concentrations of cadmium, copper, zinc and lead in the waters of the Gulf of Fos, during two cruises in 1975 (Benon et al., 1978), showed heavy-metal contamination of the whole area, with particularly high values in the northern and eastern parts which are bordered by large industrial and urban complexes. Surface-water samples were collected in spring 1975 from a network of 35 stations (Figure 3.2). In the fall, samples of surface, 5-meter depth and bottom water were collected at 16 stations. The average surface cadmium (Cd) and copper (Cu) concentrations were 0.8 and 5.2 µg/l, respectively (Table 3.3), both defined by the European Community (EC) as heavily polluted. The average surface zinc (Zn) concentration was 46.3 µg/l defined as very heavily polluted, while the average surface lead (Pb) concentration was 4.3 µg/l defined as intermediately polluted.

During monitoring by RNO in 1979 the highest cadmium (0.86 µg/l) value was found near Cortiou (Marseille), while the highest mercury concentration (0.44 µg/l) was detected in the Thau Lagoon (Table 3.4). In the Bay of Cannes the cadmium and mercury values are very low. They are used as reference levels for the coastal Mediterranean waters in France.

The average annual heavy-metal concentrations in seawater (Table 3.5) showed both decreasing and increasing values between 1977-1980. The offshore values were generally lower than the coastal values. The heavy-metal concentrations in the Languedoc-Roussillon/Provence Rhodanienne region were always higher than the corresponding values in the Côte d'Azur/Corsica region.

The overall range of heavy-metal concentrations in filtrated seawater in European coastal waters is given in Table 3.6. The maximum offshore concentrations of mercury (Hg), zinc (Zn) and lead (Pb) in the French Mediterranean coastal waters are higher (up to 2.6 times for Hg, 1.7 for Zn and 5.4 for Pb) than the corresponding concentrations of the European coastal waters. The offshore concentrations in the Mediterranean are also higher than the general European values.

In a July 1987 survey by the EC, the concentrations of five heavy metals (mercury, cadmium, lead, copper and zinc) along the French Mediterranean coast were monitored (CEC, 1988). Figure 3.3 shows the geographical distribution of heavy-metal concentrations.

The maximum mercury concentration of 65 ng/l was recorded at Sète (La Corniche). The most elevated cadmium and lead concentrations of 1.1 µg/l and 10.2 µg/l, respectively, were observed near the Petit Rhône Delta (les Saintes-Maries-de-la-Mer). The maximum copper and zinc concentrations of 15.7 µg/l and 22.8 µg/l respectively were recorded in Ajaccio (Tahiti Beach).

The values of organic pollutants (hydrocarbons, detergents, organochlorines) have been decreasing significantly over the last 10 years and certain parameters (pesticides, organochlorines) are reaching detection limits. Total aromatic hydrocarbons, polychlorinated biphenyl (PCB) and the organochlorine biocide lindane were measured in a survey that took place during the period between September 13 and September 27, 1984, in the French Mediterranean coastal waters (RNO, 1987). Total aromatic hydrocarbons (petroleum hydrocarbons) were measured by spectrofluorescence UV, using dichloromethane as solvent for the extraction, expressed as a blend of Arabian Light/Iranian Light equivalents. The lowest concentrations of aromatic hydrocarbons (from 0.5 to 1.1 µg/l) are observed in the coastal waters of Corsica (Figure 3.4). Along the coast of Provence-Côte d'Azur and Languedoc-Roussillon the values generally are less than 2 µg/l, not indicating any significant contamination. The contribution of the Rhône River, ranging from 18 to 23 µg/l at the mouth, is quite significant. The urban sewage outfall in Cortiou Cove, with 104 µg/l at the end of the outfall, constitutes the most important source.

In the offshore zone measurements for hydrocarbons in two stations gave 1.0 and 1.4 µg/l, respectively (RNO, 1987). These two stations corresponded to the sampling stations ET-R_2 and GY-L of the PHYCEMED campaign during April 1981 for which the values of 0.49 and 0.47 µg/l of chrysene equivalent (Hô et al., 1982) were determined, respectively, using hexane as extraction solvent. The higher concentrations measured in 1984 compared to those in 1981 are due to the use of the more effective dichloromethane as extraction solvent.

The highest PCB concentrations have been recorded at the end of the Marseille marine outfall at Cortiou, reaching 57 ng/l (for the undiluted sewage effluent the measured concentration was 100 ng/l), and at the mouth of the Rhône with values from 10 to 11 ng/l. Elsewhere, along the coasts of Languedoc-Roussillon, Provence-Côte d'Azur, in the Gulf of Lions and around Corsica, the measured concentrations are generally below the detection limit (2 ng/l). These low levels confirm the decreasing trend over the past 10 years, during which monitoring has taken place.
The rapid decline in hydrocarbons and PCB in coastal waters with increasing distance to the shore is shown in Figure 3.5. Hydrocarbons and PCB are still detected up to 20 to 25 kilometers from the shore near Fos.

Lindane behaved similar to the hydrocarbons and PCB. Lindane reached 60 ng/l at the end of the Marseille marine outfall at Cortiou. The undiluted sewage effluent had a concentration of 120 ng/l. At the mouth of the Rhône a concentration of 8 ng/l (Figure 3.4) was measured. The decline in lindane concentrations in both coastal and offshore waters of the Gulf of Fos is shown in Figure 3.6. The detection limit, due to the improvement of the analytical methods, was lowered from 50 ng/l to 1 ng/l. The maximum of 115 ng/l reached in 1977 decreased to 13 ng/l in 1985.

3.1.2.   Sediments

A distinction must be made between the geochemical base levels and anthropogenic sources, when analyzing data on metal pollutants in sediments. The naturally occurring and the anthropogenic toxic levels of mercury (Hg), cadmium (Cd), copper (Cu), lead (Pb) and zinc (Zn) are shown in Table 3.7.

The Berre Lagoon was investigated in a survey that took place in July 1976 (Arnoux et al., 1980b). The highest mercury concentrations of up to 2.82 mg/kg were measured in the southwestern part (east of Martigues). The pollution originates from the region of the Caronte Canal (Figure 3.7). An exceptional elevation of the mercury concentration, up to 3.16 mg/kg, was measured in the sediments of the Saint-Chamas Bay, an isolated sector of the Berre Lagoon in the north, where a factory of explosives was in operation during the first half of this century.

Lead concentrations were particularly high, up to 314 mg/kg, in the Vaine Lagoon (an isolated part in the southeast of the Berre Lagoon) due to the large number of industries and the sedimentation in this area having very little renewal (Figure 3.8).

The sediments of the Gulf of Fos were analyzed in three surveys that were conducted in 1976, 1978 and 1979 (Arnoux et al., 1980c). The mercury pollution was distributed into three categories, depending on the degree of contamination, as shown in Figure 3.9. The lowest concentrations, of less than 0.5 mg/kg, were localized around the They de la Gracieuse and in the north of the gulf. The highest values, of more than 1 mg/kg, were identified once in the mouth of the Rhône but more frequently in the zone of Port-de-Bouc-Lavéra where the maximum was 4.4 mg/kg. Between these two extremes, intermediate concentrations characterized the central part of the gulf, Carteau Inlet and the Darse zone. The mercury pollution carried by the Rhône is diluted inside the Gulf of Fos, while that originating from the eastern region creates a pollution zone ranging from the Berre Lagoon-Caronte Canal to the south of the gulf.

The distribution of lead and copper is presented in Figures 3.10 and 3.11, respectively. It establishes both the contribution of the Rhône (either direct or through the Port-Saint-Louis Canal) and in particular that of the industrial activities in the western region.

Around the sewage outfall of Cortiou (Marseille) and in a radius of up to about 1.5 kilometers a zone with toxic concentrations of mercury, copper, lead and zinc was defined (Arnoux et al., 1980a). Figures 3.12, 3.13, 3.14 and 3.15 show the distribution of the above metals in the sediments. The observed concentrations of mercury and copper farther away (from 2 to 4 kilometers from the exit) were comparable to those in the Gulf of Fos, while lead concentrations were similar to those in the Berre and Vaine lagoons.

A comparison between the heavy-metal concentrations in the sediments of the Bays of Cannes and Villefranche reference zone and the polluted Gulf of Fos (Table 3.8) shows that the latter are 2 to 8 times higher (Ministère de l'Environnement, 1987).

The analysis of organic pollutants in the sediments reveals not only their current supply but also their persistence in the marine environment.

The establishment of the petroleum terminal and later of the Shell-CFR petrochemical complexes around the Berre Lagoon has caused a significant hydrocarbon contamination in the southern part, as was shown in a survey that took place in July 1976 (Arnoux et al., 1980b). A considerable accumulation was observed in the sediments of the Vaine Lagoon; an isolated annex in the southeast of the Berre Lagoon. The hydrocarbon content was often higher than 10,000 mg/kg (Figure 3.16).

PCB showed two zones of pollution with concentrations greater than 100 µg/kg: a major area in the Vaine Lagoon, and a smaller in the mouth of the Caronte Canal (Figure 3.17). Although the concentrations were generally lower in the rest of the lagoon, there were accumulations of 50 and 100 µg/kg in a number of sampling sites near small rivers, such as the Arc and the Touloubre, and near the Istres-Saint-Chamas-Miramas industrial region.

Lindane (Figure 3.18) showed maximum concentrations of up to 31.8 µg/kg in the zone directly near the discharge of the EDF Canal.

Two sediment surveys took place in the Gulf of Fos in 1976 and 1979 (Arnoux et al., 1980c). The hydrocarbon distribution showed two zones separated by a line that runs south of the They de la Gracieuse to Port-de-Bouc (Figure 3.19). North of this line, the concentrations were low, despite the tanker traffic and the proximity to the ESSO refineries. In the south much more elevated concentrations, around 1,000 mg/kg, showed the influence of the region of the Berre Lagoon and Port-de-Bouc-Lavéra, extending to the south.
In the central part of the Gulf of Fos, the contribution of the Rhône and the Caronte Canal resulted in PCB concentrations between 50 and 100 µg/kg (Figure 3.20).

Around the sewage outfall of Cortiou and in a radius of 1 kilometer, the concentrations of hydrocarbons were as high as those in the Vaine Lagoon (Arnoux, 1980a), as presented in Figure 3.21.
High PCB concentrations, of more than 1,000 µg/kg, were measured around the sewage outfall of Cortiou up to a radius of 1.5 kilometers (Figure 3.22).
The distribution of the insecticide dichlorodiphenyl trichlorethane (DDT) (pp' DDT) is presented in Figure 3.23, with high values ranging from 100 to 160 µg/kg around the sewage outfall.

A September 13-27, 1984 survey by RNO (RNO, 1987) of organic contaminants in marine sediments along the French Mediterranean coast found significant concentrations in the Rhône Delta (274 mg/kg hydrocarbons), in the Gulf of Fos (420 mg/kg hydrocarbons) and in Cortiou Cove (9,166 mg/kg hydrocarbons) as shown in Figure 3.24A. Other maximum values are observed in Cortiou Cove for PCB (21,615 µg/kg), DDT, DDD and DDE (86 µg/kg) and lindane (less than 10 µg/kg) (Figures 3.24B-D).

The decline in the concentration of hydrocarbons, PCB and DDT, DDD and DDE in sediments with increasing distance away from the sources is shown in Figures 3.25A-C. Their persistence in the marine environment is evident as they are being detected in sediments at distances up to 70 to 100 kilometers offshore.

The annually progressing deterioration of the bottom-sediment quality near the Cortiou outfall between 1967 and 1980 is shown in Figure 3.26. Research is currently on-going to determine the quality improvement after the start of the sewage treatment plant operation in 1987.

3.1.3.   Living matter

The concentrations of heavy metals in oysters can be up to 20 times higher than the concentrations in mussels.
Low mercury and cadmium concentrations are present in mussels and oysters (Figure 3.27). The lowest mercury concentrations are measured in the zones of Thau and East Corsica. The industrialized zone of the Gulf of Fos has lower mercury concentrations in the living matter (mussels) than those measured in the same species exposed to urban waste waters in the Marseille area.
A regulation to reduce mercury wastes in the electrolytic industry was established in 1974. As a result, the concentrations of mercury in mussels (sedentary species) have decreased from 0.3 to 0.03 mg/kg of drained flesh, i.e. a reduction by a factor of 10. A reduction of 70% in the content of mercury in mussels in the more polluted Gulf of Fos between 1972 and 1984 is shown in Figure 3.28.
Cadmium shows a large range among the different sites. The concentrations are similar to those measured in the same species along the French Atlantic coast.
Lead concentrations are higher in Marseille, Toulon and Cannes due to discharge of domestic and industrial waste waters.
The high metal concentrations encountered at Banyuls are of geological origin and not due to anthropogenic sources. The high metal concentrations near Toulon are due to industrial

activities (military arsenals and shipyards). The low concentrations measured near Corsica can be used as reference values for the Mediterranean coast.

The geographical distribution of mercury in the mussel (Mytilus galloprovincialis) shows a maximum of 0.87 mg/kg dry weight near Marseille. This is close to the upper limit of 1 mg/kg dry weight as established by the Oslo-Paris Convention for strongly polluted shellfish (Figure 3.29). The highest cadmium concentration of 12.5 mg/kg dry weight was detected at the mouths of the Orb and Hérault rivers. This is well above the 5 mg/kg dry weight limit established by the Oslo-Paris Convention for strongly polluted mussels.

The three most important organic pollutants (hydrocarbons, PCB and DDT, DDD and DDE) are present in the living matter at all sites, often in low concentrations (Figure 3.27). The levels are higher in the more polluted zones: hydrocarbons in Toulon (25 mg/kg dry weight) and Gulf of Fos (30 mg/kg dry weight); and PCB in Toulon (4,000 µg/kg dry weight). The DDT concentrations, which are higher along the French Mediterranean coast as compared to the French Atlantic coast north of the Loire estuary, have been decreasing (by at least a factor of 10) between 1972 and 1979.

The geographical distribution of concentrations of hydrocarbons and PCB, respectively, in the mussel (Mytilus galloprovincialis) are shown in Figure 3.30. The maximum hydrocarbon concentration of 15 mg/kg dry weight found in the Gulf of Fos is 3 times higher than that found in the Bay of the Seine. The PCB maximum of 1,440 measured in Toulon is twice as high as the limit of 700 µg/kg dry weight established by the Oslo-Paris Convention for strongly polluted shellfish.

3.1.4.   The quality of the bathing waters

Since 1975 water samples from all French coastal bathing areas are analyzed annually for various bacteriological parameters. The bathing waters are classified into four categories based on the levels of total coliforms (TC) and fecal coliform (FC) bacteria:

A: water of very good quality (TC ≤ 500/100 ml and FC ≤ 100/100 ml in 80% of the samples)
B: water of average quality (TC < 10,000/100 ml and FC < 2,000/100 ml in 95% of the samples)
C: water liable to be polluted temporarily (TC < 10,000/100 ml and FC < 2,000/100 ml in less than 67% of the samples)
D: water of poor quality (TC > 10,000/100 ml and FC > 2,000/100 ml in more than 3 consecutive samples).

The percentage of French Mediterranean beaches complying with the European Directive has increased from 83.6% in 1976 to 95.7% in 1986 (Table 3.9). In 1986, 606 Mediterranean beaches were monitored. During the monitoring period (between June 15 and September 15) only three beaches were found to be systematically polluted. Up to 96% of the monitored beaches were found to comply with standards for classes A and B.

The results of an EC July, 1987 survey (CEC, 1988) for 29 French Mediterranean beaches are shown in Figure 3.31. All, except 6, Argeles (P. de Pins), les Saintes-Maries-de-la-Mer, Cavalaire (P. du Centre Ville), Bastia (Thalassa Beach), Porto-Vecchio (Le Port) and Saint-Florent, complied with the EC guide number (recommended criteria limit). The above-mentioned 6 beaches exceeded the EC guide number, but complied with the required standard.

3.1.5. Air pollution

An aerosol sampling campaign for copper, cadmium, nickel, cobalt, lead, zinc, manganese and titanium was conducted in the Fos-Berre industrial region in June 1983 (Gomes et al., 1985).

The contribution of the sulfur dioxide (SO_2) in the Fos-Berre region represents between 4.5 and 9% of the total daily emission in France. The percentage for copper (Cu), nickel (Ni), zinc (Zn) and manganese (Mn) is comparable to that of SO_2. The relative contributions of cadmium (Cd) and cobalt (Co) are larger and for lead (Pb) are smaller than those of SO_2 (Table 3.10).

The high value for cadmium is attributable to the large concentration of refineries in this area. The high value of cobalt is due to an underestimation of this element for the whole of France, as only the fossil combustible contribution was quantified. The principal source of atmospheric lead is from automobile exhaust. The low percentage of this element in the Fos-Berre region is not surprising, as only the industrial contribution was taken into account excluding the automobile contribution.

3.2.   Pressures on the environment

The pressures on the environment are considered for each of the three regions along the coast, i.e. Languedoc-Roussillon, Provence-Alpes-Côte d'Azur and Corsica.

There are many tourist resorts along the coast of Languedoc-Roussillon, while the big cities of the region, i.e. Montpellier, Béziers, Narbonne and Perpignan, are located about 15 kilometers from the coast. There are no large rivers in this region. The Têt, Aude, Orb and Hérault rivers are rather small. There are only a few industries along the coast. The industrial areas of Port-la-Nouvelle and Sète (the most important French Mediterranean fishing port) are the largest.

The Provence-Alpes-Côte d'Azur region is visited by many tourists. The big cities, i.e. Marseille, Toulon, Nice and Cannes, are located on the coast. The Rhône discharges in the western part of the region. The river has a very significant impact on the Mediterranean Sea, due to its large average outflow of 1,700 m^3/sec, which greatly varies throughout the season. The industrial pollutants are primarily discharged to the Gulf of Fos-Berre Lagoon. The

above zone is currently the most important industrial zone of France.
At the island of Corsica the two important cities, i.e. Ajaccio and Bastia, are located on the coast. The outflow of the rivers is, on average, rather small, with important seasonal variations. The land use is predominantly agricultural. Tourism is an important activity.

The administrative regions that correspond to the above three zones are:
- Languedoc-Roussillon (27,376 km^2) It contains the four coastal departments, i.e. Pyrénées-Orientales, Aude, Hérault and Gard (22,209 km^2);
- Provence-Alpes-Côte d'Azur (Paca) (32,397 km^2) It contains the three coastal departments, i.e. Bouches-du-Rhône, Var and Alpes-Maritimes, of 15,355 km^2;
- Corsica (two departments of 8,680 km^2).

The above regions have 1,700 kilometers of coastal length, of which 214 belong to Languedoc-Roussillon, 687 to Paca and 802 to Corsica.

3.2.1. Municipal discharges

The seasonal distribution of the population along the French Mediterranean coast shows its maximum in the Bouches-du-Rhône region (Figures 3.32A,B).
The French Mediterranean coastal regions nearly double their population during the summer. In Corsica the population increases even by a factor of 2.3. Consequently, the measures against domestic pollution must take into consideration both the resident and tourist population.

Figure 3.33 shows the coastal distribution of the domestic sewage discharges in tons per year of organic matter, as biochemical oxygen demand (BOD), and suspended solids for the year 1982 expressed on a logarithmic scale. A total of 76 x 10^3 tons of BOD per year and 120 x 10^3 tons of suspended solids per year was released on the basis of summer

discharges. Most of the domestic discharges are concentrated in the department of Bouches-du-Rhône.
Table 3.11 shows that 28% of the discharged domestic load was treated in 1985. The treatment percentage is expected to increase to 65% in 1990.

A total of 185 treatment plants was installed as of 1982 for treatment of industrial and municipal waste waters in the departments of the French Mediterranean region. The installed capacity (3.05 million population equivalents) and the selected processes are shown in Table 3.12.
The most common treatment processes are:
- activated sludge (37%-88%)
- trickling filters (0-55%)
- physical-chemical (0-42%)
- lagoons (0-50%)

The above four types are being used by 153 of the total 154 municipal treatment plants. In only one station the method is a decanter-digester.
The principal pollutants of domestic origin are:
- suspended solids, which are removed by simple sedimentation enhanced by addition of flocculants;
- organic matter and detergents, which are well eliminated by biological treatment;
- pathogenic microorganisms, which are inactivated by disinfection.

The degree of elimination of the pollution in cities of more than 50,000 inhabitants varied between 0 and 75% in 1984. The number of treatment plants and the installed capacity have increased gradually between 1975 and 1984 (Figure 3.34). The eliminated pollution in population equivalents from 1975 to 1984 is about half of the installed total capacity indicating poor operating conditions. Most of the installed capacity was for cities treating between 50,000 and 100,000 inhabitants.

3.2.2. Industrial discharges

The industrial pollution in the Mediterranean coastal waters comes predominantly from the chemical industries, the refineries, the paper industries, the alcohol distilleries and the wine-liquor industries.
Recent tabulated data from the Agence de Bassin Rhône-Méditerranée-Corse show a net discharge of 22.5×10^3 tons of BOD, 22.7×10^3 tons of suspended solids and 0.43×10^3 tons of equitox toxic substances in 1986. The coastal distribution of BOD and suspended solids discharges is shown in Figure 3.35, using a logarithmic scale, while their decline between 1982 and 1986 (31% for suspended solids and 12% for BOD) is shown in Figure 3.36. The toxics (Figure 3.37) showed a 14% decrease similar to that of the BOD.

In the coastal region of the Provence-Alpes-Côte d'Azur, the industrial discharges are mainly concentrated in Marseille, Toulon and Nice and in particular around the Gulf of Fos-Berre Lagoon. The location of the 109 major industrial establishments is given in Figure 3.38.

A special program was established to combat water pollution in the heavily industrialized zone of Fos-Berre Lagoon. It had the following objectives:
- construction of biological treatment plants, especially in the petrochemical and refinery industry, for older industrial establishments;
- imposition of strict standards for discharge of wastewater effluents as applied to new industrial establishments. Meeting the standards is a prerequisite for a construction permit.

The Berre Lagoon covers an area of 15,500 hectares, with a maximum depth of 9 meters; estimates put the volume of water at 900 million m^3. The lagoon receives urban effluents from 11 communities with 180,000 population equivalents and inflows from three rivers with a catchment area of 155,000 hectares. A diversion canal from the Durance River is designed to feed 5 hydroelectric plants. The lagoon is connected to the sea by a navigation canal 9 meters deep. Each year the Durance diversion canal carries a volume of water into the lagoon four times that of the lagoon itself. The volume of these inflows (up to 800 m^3/sec) and the pronounced disparity in salinity (measured vertically) leads to marked density variations.

The enrichment of the lagoon with discharges of organic matter and mineral salts (of urban, industrial and agricultural origin) has resulted in increased eutrophication. The breakdown of the organic matter discharged into the lagoon or produced "in situ" consumes substantial quantities of dissolved oxygen, exacerbated by the density gradient which hinders reoxygenation from the surface. As a result, the Berre Lagoon often suffers from anoxic conditions, particularly during the summer.

Since 1972 major efforts have been made to cut down the industrial and urban discharges. This resulted in a reduction in the load of organic material of more than 90%. Suspended matter and several pollutants such as total hydrocarbons, phenols and mercury showed a similar reduction. Discharges of silt from the Durance Canal are still very high, although the use of settling tanks has reduced the discharges by 40%.

The results of the reduction in industrial pollution in the zone of Fos-Berre Lagoon, from 1972 until 1983, are shown in Figure 3.39. The chemical oxygen demand (COD) load decreased from 177 tons per day to 10 tons per day.

A similar decrease was observed along the Rhône River due to recently installed industrial and municipal treatment capacity. The COD load decreased from 78 tons per day to 24 tons per day or a 70% reduction, as is shown in Figure 3.40.

The coast of Corsica, being only slightly industrialized, does not suffer from major industrial pollution.
No data are readily available on production and disposal of

industrial hazardous waste.

3.2.3. River discharges

The rivers also discharge a considerable amount of pollutants into the Mediterranean. The 16 rivers surveyed between October 1985 and September 1986 discharged a load of 212.4×10^4 tons of suspended solids per year and 290×10^3 tons of BOD per year, whereby the Rhône River accounted for 74% of the suspended solids and 94% of the organic load. The 16 individual river discharges along the coast and their seasonal pattern with maxima in the high-water period of April through May are shown in Figures 3.41 and 3.42, respectively. Figure 3.43 shows an important decrease in discharge of suspended solids for 11 rivers and no change in BOD discharge during the 6-year data period. Part of the decrease in discharge of suspended solids can be attributed to the 19% decrease in flow rate during that period and to sedimentation of the particles in the riverbed.

The major discharges into the Rhône River are geographically shown in Figure 3.44 with large agglomerates near Lyon and Arles. As a result of the discharges, both ammonia (NH_4^+) and phosphate (PO_4^{3-}) concentrations increase steeply downstream from Lyon (Figure 3.45). The average CNR (Compagnie Nationale du Rhône) data collected between 1979 and 1985 show that the NH_4^+ concentration exceeds the limits for class 1B (it falls into the moderate quality class) and PO_4^{3-} exceeds that for class 2 (it falls into the poor quality class). Table 3.13 shows the water quality classes (in general and for fish breeding) and the corresponding concentrations of the pollutants. A limited decrease in ammonia concentrations downstream from its maximum is attributed to nitrification. Phosphate decrease is likely due to adsorption onto particulates and removal by sedimentation. The daily load from the Rhône River entering the Mediterranean as measured between September 15 and September 22, 1986, is shown in Figure 3.46. The load varies between 100-400 tons of BOD per day, 80-200 tons of total nitrogen (N) per day and 10-25 tons of phosphate (expressed in P) per day. These values correspond to 2.5-10 million population equivalents on the basis of organic matter, 5.3-13.3 million population equivalents on the basis of nitrogen and 2.5-6 million population equivalents on the basis of phosphorus.

3.2.4.   Maritime discharges

The pollution of the sea (pelagic or open sea pollution) can result from discharges from ships or from navigation accidents. Ships that transport hydrocarbons, chemical products and other dangerous substances constitute the greatest potential pollution source.
France has ratified the MARPOL Convention, which determines the rules that apply to reception facilities. Regulation number 12 of the above Convention obliges the Mediterranean countries to equip their ports with reception facilities. The French Mediterranean regions have constructed such facilities. The recently established Fos-sur-Mer station is one of the most modern in Europe. The hydrocarbon concentrations of the treated effluents are a few mg/l. The reception stations at Marseille-Marignane, Lavéra and Sète have only gravity type separators (primary treatment). The latter two deballasting stations discharge effluents of a lower quality, but within the standard of 20 mg/l. This is similar to that for the effluents of refineries.
The estimated cost of additional reception facilities is $1,550,000 (U.S.).

3.3.   Economic impacts of low environmental quality

France's coastal lagoons are of major economic importance. In addition to the role of lagoons as tourist attractions they perform three important functions:
- refuge and feeding of economically valuable species,

which periodically return to the lagoon to complete their biological cycle;
- a well-stocked fishing area;
- a substantial shellfish industry (mussels and oysters).

The most important areas are the Berre and Thau Lagoon.

Fishing remains an important economic activity in the Berre Lagoon, despite its prohibition under a decree issued in August 1967. The estimated catch in 1984 was 3,000 tons, of which eels made up two-thirds.

The significant reduction in discharge of trace pollutants has lowered their levels in edible fish. A recent study showed that levels of lead, cadmium and PCB in fish are generally now well below the maximum permitted limits. Maximum values for total mercury (virtually all in methylated form) correspond to the maximum values tolerated for this element.

Eutrophication of the lagoon waters, however, still takes place and can cause anoxic fish and benthic organism kills. A stricter control must be imposed on the discharges of organic matter and nutrients.

The Thau Lagoon has a surface area of 7,500 hectares and an average depth of 5 meters. One-fifth of the Thau Lagoon is occupied by some 700 shellfish farm areas. Its significance for shellfish farming is considerable. Figure 3.47 shows 3 selected shellfish farms. The official production figures of 7,000 tons of oysters per year and 5,000 tons of mussels per year are greatly underestimated. Annual production is as much as 17,000 to 20,000 tons of oysters and 8,000 tons of mussels.

The environment of the Thau Lagoon is relatively free of major pollution. Massive shellfish mortality occurred three times during the past 15 years (in 1975, 1982 and 1987). In 1987 the losses were estimated at 10,000 tons. The cause was anoxic conditions in the seabed, associated with specific climatic conditions (absence of wind for a long period combined with intense sunshine).

Research aimed at preventive action through treatment measures (removal of oxygen-consuming sludges accumulated under the breeding installations) should be stepped up in the future.

3.4.   The institutional/legal framework for environmental protection

The Ministry of the Environment is the leading administrative body, which coordinates all actions (legal, technical, financial) concerning the environment. Three other administrative bodies play an active role in planning, project implementation and monitoring:
- the Regional Branch of the Ministry of Industry, which is the key authority in the establishment of standards for industrial pollution and is responsible for ensuring that they are implemented;
- the Regional Water Agency ("Agence de bassin"), or, at a national level, the Agency for Water Quality, which finances investments or subsidizes operational costs of treatment plants;
- the municipalities, which decide on the need for investments and either operate the plants directly or delegate their operation to private enterprises.

The most important laws for the protection of the environment since 1960 are the following:
- Law No. 64-1245 of December 16, 1964, containing the measures that must be taken to reduce water pollution;
- Law No. 75-633 of July 15, 1975, concerning the elimination of wastes and stimulation of recycling of materials;
- Law No. 76-629 of July 10, 1976, concerning the protection of nature;
- Law No. 79-5 of January 2, 1979, concerning the pollution of the sea by hydrocarbons;
- Law No. 84-608 of July 16, 1984, concerning the French

Institute of marine research;
- Law No. 85-661 of July 3, 1985, concerning the installations for the protection of the environment;
- Law No. 85-1273 of December 4, 1985, concerning the protection of the forests;
- Law No. 86-2 of January 3, 1986, concerning the proper development, protection and usage of the coastal regions.

Laws 76-599 and 76-600 of July 7, 1976, provide for the prevention of sea pollution by discharges from ships and aircraft and the burning of refuse at sea. They establish the principle of prior authorization and specify penalties for unauthorized dumping and incineration; they also empower the authorities to seize vessels in breach of the regulations.

Law 77-530 of May 26, 1977, refers to the civil liability of shipowners and the owner's obligation to have insurance coverage. It establishes the obligation regarding compensation for pollution damage laid down by the 1969 Brussels Convention.

Law 85-542 of May 22, 1985, states that sea fishing must be carried out in accordance with the appropriate regulations (Regulation 170 of January 15, 1983) within the framework of the Community policy for the conservation of natural resources.

A tabulation of French laws and decrees that are relevant to the Mediterranean pollution are summarized in Table 3.14.

3.5. Past and future trends in environmental expenditures

The Ministry of the Environment has formulated two major environmental goals:
- increasing the current removal efficiency of 40% in 1985 to 60% in 1995. This will be achieved by installing treatment plants in industries that lack these facilities and by increasing the operational efficiency of the existing treatment plants;
- protecting coastal areas and lake shores and in addition unique forests and agricultural and urban landscapes, together with the development of national parks.

The percentage of treated municipal waste water is expected to increase to 65% in 1990, due to construction of new treatment plants (Table 3.15). The start of the operation of the Perpignan treatment plant (summer 1987) will improve the quality of the waters of the Têt River and the seaside bathing waters. The operation of the treatment plant at Marseille (1.5 million population equivalents) in 1988 and the operation of the treatment plant at Nice, with a complete biological treatment, in 1987 will greatly improve the coastal water quality. East Toulon sewage is currently treated and facilities are planned for Toulon West.

Financial aid is assigned by the "Commission des Aides de l'Agence de Bassin Rhône-Méditerranée-Corse" for the local communities of each region of the French Mediterranean. This is for the creation, extension and improvement of sewage systems and for treatment plants (Table 3.16). In 1986 a total 62.3 million francs was given as grants, 45 million francs in advance payments and 4.6 million francs in loans for a total of 112 million francs.

The most important operations have been the following:
- aid of 17.7 million francs for the treatment plant at Marseille;
- 6.5 million francs for the treatment plant at Saint-

Tropez;
- 5.8 million francs for the extension of the treatment plant at Narbonne;
- 5.3 million francs for the treatment plant in Saint-Cyr-sur-Mer.

The financial aid assigned by the Commission des Aides for each region of the French Mediterranean for measures against industrial pollution is shown in Table 3.17. Grants to industrial facilities were 30 million francs, advance payments 17.8 million francs, loans 2.7 million francs for a total of 50.5 million francs. Specific examples of industrial treatment plants are given in Table 3.18. The largest plant was the ARCO plant in Fos-sur-Mer, costing more than 31.3 million francs.

3.6. Conclusion

France has a well-established program to monitor the quality of the marine environment. Construction of new municipal and industrial sewage treatment plants has eliminated large amounts of pollutants. This is reflected by an improving marine water quality and a decline in the concentration of trace contaminants in sediments and living matter. The quality of bathing waters has also improved considerably.

Still, further effort is needed to clean up the rivers and coastal waters, especially with regard to the Rhône basin. Comprehensive river basin management will decrease the levels of pollutants in this area. More investments for plant improvement and expenditures to increase operating efficiencies are still necessary.

References

AGENCE DE BASSIN RHONE-MÉDITERRANÉE-CORSE, Qualité du fleuve Rhône. Synthèse des connaissances. April 1988.

ARNOUX, A., BLANC, A., JORAJURIA, A., MONOD, J. L. and TATOSSIAN, J., État actuel de la pollution sur les fonds du secteur de Cortiou (Marseille). Ves Journ. Étud. Pollut. pp. 459-470. CIESM. Cagliari 1980a.

ARNOUX, A., MONOD, J. L., BOUCHARD, P. and AIRAUDO, C. B., Évolution et bilan de la pollution des sédiments de l'étang de Berre. Ves Journ. Étud. Pollut. pp. 433-446. CIESM. Cagliari 1980b.

ARNOUX, A., MONOD, J. L., TATOSSIAN, J., BLANC, A. and OPPETIT, F., La pollution chimique des fonds du Golfe de Fos, Ves Journ. Étud. Pollut. pp. 447-458. CIESM. Cagliari 1980c.

BENON, P., BLANC, F., BOURGADE, B., DAVID, P., KANTIN, R., LEVEAU, M., ROMANO, J.-C. and SAUTRIOT, D., Distribution of some heavy metals in the Gulf of Fos. Mar. Pollut. Bull., 1978, 9, 71-75.

BULLETIN D'INFORMATION du Comité et de l'Agence de Bassin Rhône-Méditerranée-Corse. No. 25. June 1987.

CEC, 26.7.1987-1.8.1987 plages propres. Résultats de la campagne de sensibilisation. Brussels 1988.

CENTRE OCÉANOLOGIQUE DE BRETAGNE, Synthèse des travaux de surveillance 1975-1979 du réseau national d'observation de la qualité du milieu marin. Brest 1981.

CITY OF MARSEILLE, Station d'épuration de Marseille. Ouvrages de traitement des eaux. April 1985. Direction Général des Services Techniques.

DOCTER, European environmental yearbook 1987. 1988. DocTer International U.K., London.

GOMES, L., BERGAMETTI, G. and DUTOT, A. L., The atmospheric emissions of an industrial area (Fos/sur/Mer, France) and their implications for the heavy metal cycles in the western Mediterranean Sea. 5th International Conference on Heavy Metals in the Environment. Vol. 1, pp. 177-179. Athens 1985.

HÔ, R., MARTY, J.-C. and SALIOT, A., Hydrocarbons in the western Mediterranean Sea, 1981. Intern. J. Environ. Anal. Chem., 1982, 12, 81-98.

MINISTERE DE L'ENVIRONNEMENT, État de l'assainissement en zone littorale. Direction de la Prévention des Pollutions, Service de l'Eau, 1982.

MINISTERE DE L'ENVIRONNEMENT, La protection de l'environnement Méditerranéen. Contribution de la France. 1987.

MINISTERE DE L'ENVIRONNEMENT et Conseil Général des Bouches-du-Rhône, Direction Régionale de l'Industrie et de la Recherche Provence-Alpes-Côte d'Azur, La lutte contre la pollution industrielle dans le département des Bouches-du-Rhône. Bilan et perspectives en 1984.

MINISTERE DE L'ENVIRONNEMENT et Ministère de la Santé et de la Famille, État sanitaire des zones de baignade en mer. 1987.

MINISTERE DES AFFAIRES SOCIALES ET DE L'EMPLOI, Qualité des eaux de baignade. May 1988.

RNO, Dix années de surveillance 1974-1984. IFREMER et Ministère de l'Environnement, May 1985.

RNO, Résultats de la campagne à la mer INTERSITE II (13-27 septembre 1984). 1987. IFREMER et Ministère de l'Environnement. DERO-87.20-EL. 356 pp.

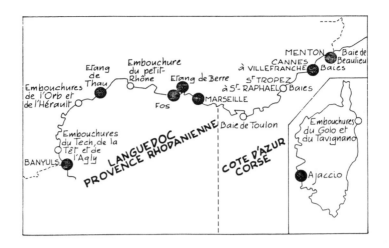

Figure 3.1   RNO monitoring stations:
● principal and ○ complementary

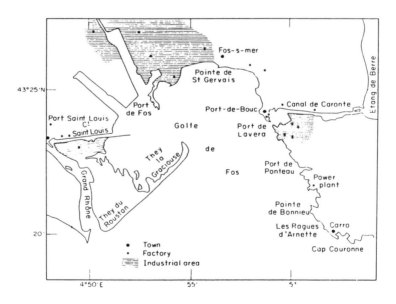

Figure 3.2 Map of the Gulf of Fos showing the two main outputs of fresh water (Rhône and Canal de Caronte) and the surrounding industrial area [Benon, P., et al., 1978]

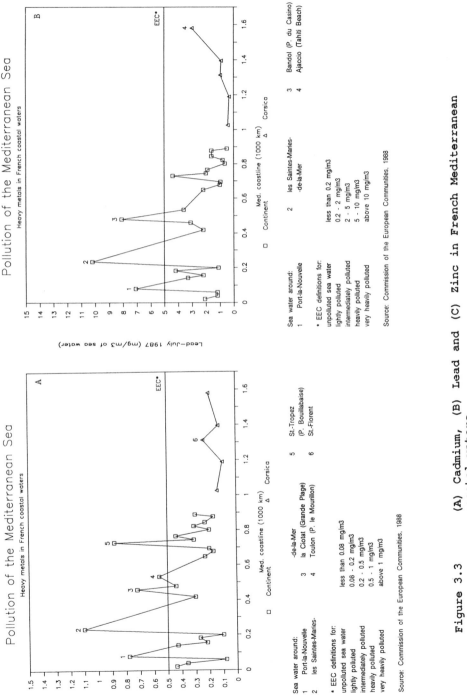

Figure 3.3 (A) Cadmium, (B) Lead and (C) Zinc in French Mediterranean coastal waters

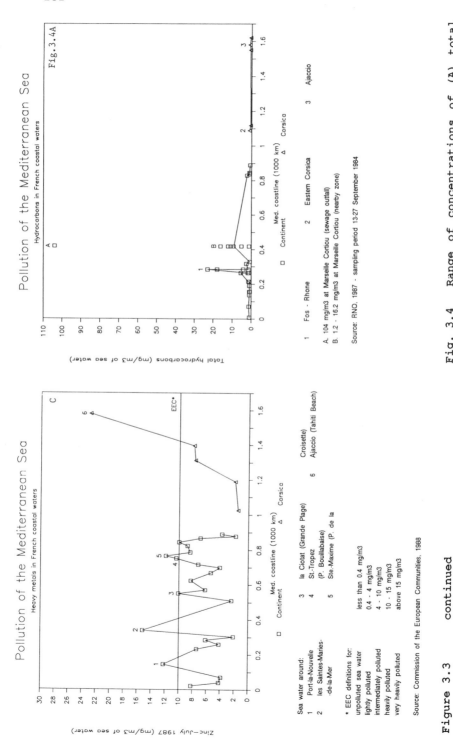

Figure 3.3 continued

Fig. 3.4 Range of concentrations of (A) total aromatic hydrocarbons, (B) PCB and (C) lindane in different sectors of the French Mediterranean coastal waters

283

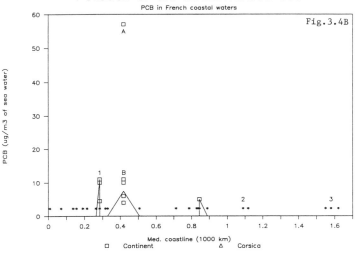

A. 57 ug/m3 at Marseille Cortiou (sewage outfall)
B. 4 - 11 ug/m3 at Marseille Cortiou (nearby zone)

* less than 2 ug/m3

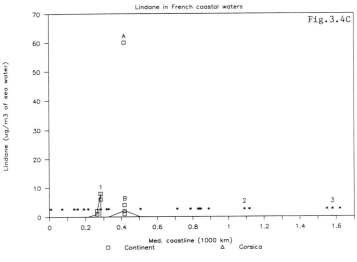

A. 60 ug/m3 at Marseille Cortiou (sewage outfall)
B. < 1 - 4 ug/m3 at Marseille Cortiou (nearby zone)

* less than 1 ug/m3

Source: RNO, 1987 - sampling period 13-27 September 1984

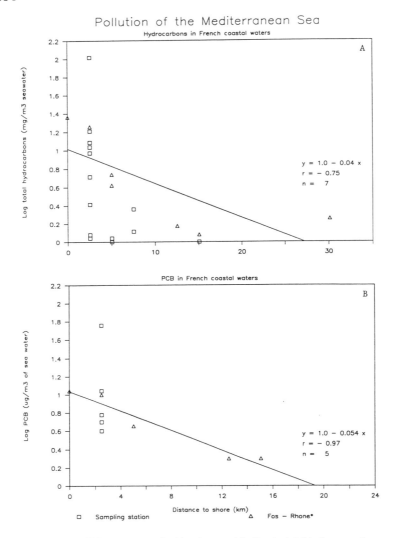

Figure 3.5   Decrease of concentrations of (A) total aromatic hydrocarbons and (B) PCB with distance to the shore

285

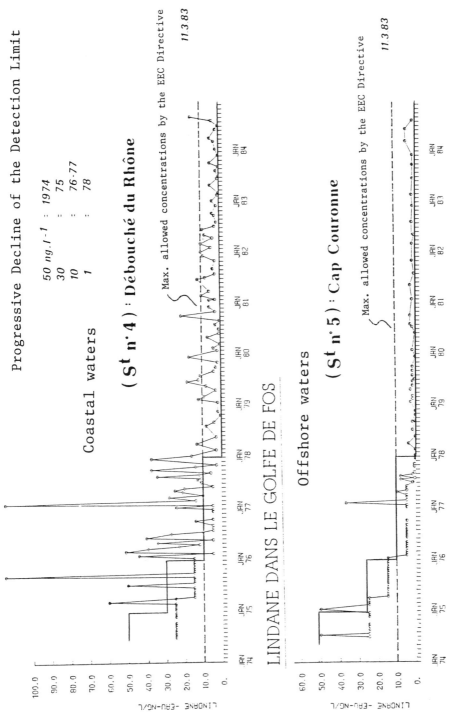

Figure 3.6 Progressive decline of the Detection limits and measured lindane in the Golfe de Fos

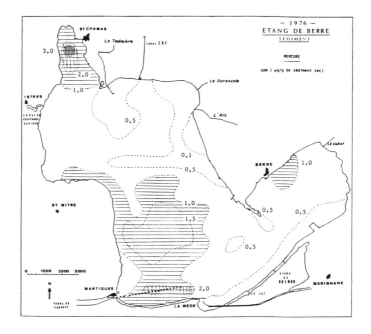

Figure 3.7   Distribution of Hg in the sediments of the
              Etang de Berre [Arnoux, A., et al., 1980b]

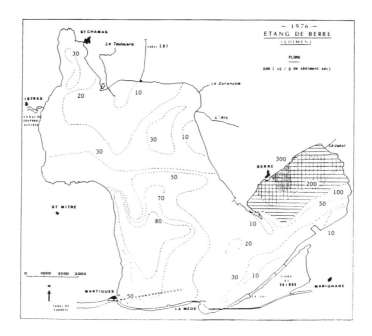

Figure 3.8   Distribution of Pb in the sediments of the
              Etang de Berre [Arnoux, A., et al., 1980b]

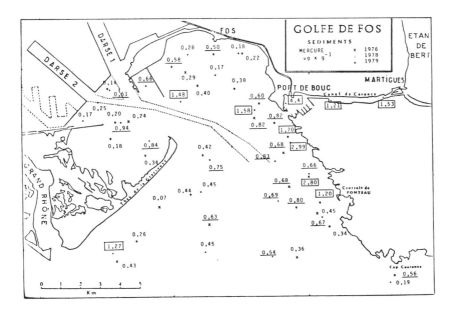

Figure 3.9 Distribution of Hg in the sediments of the Golfe de Fos [Arnoux, A., et al., 1980c]

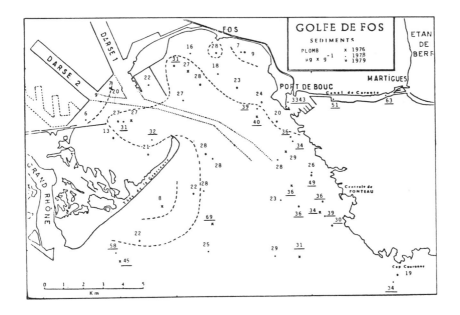

Figure 3.10 Distribution of Pb in the sediments of the Golfe de Fos [Arnoux, A., et al., 1980c]

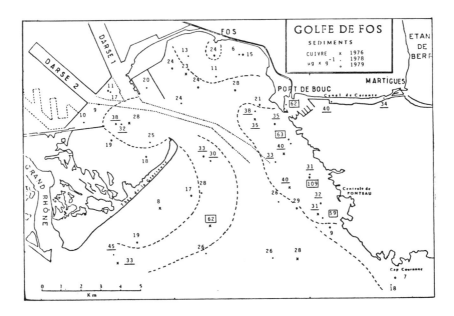

Figure 3.11   Distribution of Cu in the sediments of the Golfe de Fos   [Arnoux, A., et al., 1980c]

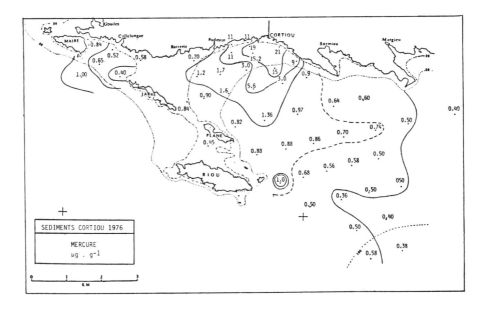

Figure 3.12   Distribution of Hg in the sediments of Cortiou (Marseille)   [Arnoux, A., et al., 1980a]

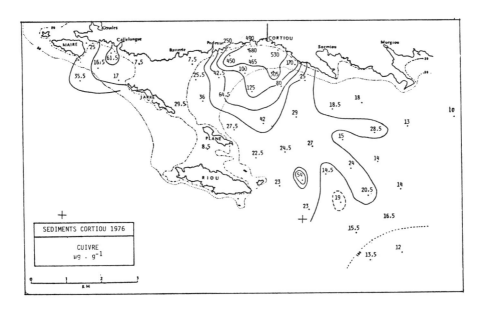

Figure 3.13 Distribution of Cu in the sediments of Cortiou
(Marseille)   [Arnoux, A., et al., 1980a]

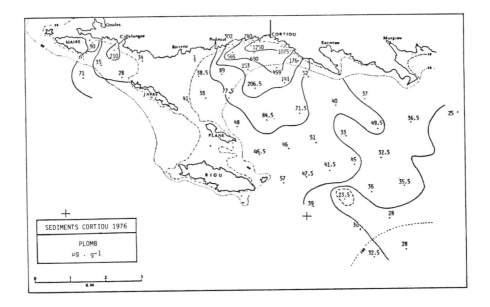

Figure 3.14 Distribution of Pb in the sediments of Cortiou
(Marseille)   [Arnoux, A., et al., 1980a]

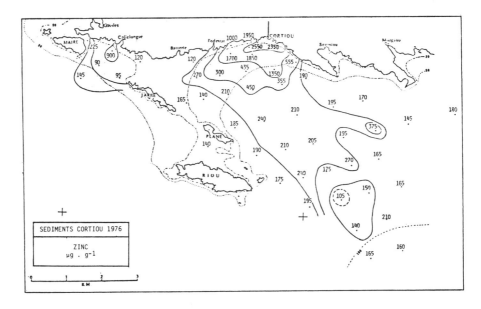

Figure 3.15   Distribution of Zn in the sediments of Cortiou (Marseille)   [Arnoux, A., et al., 1980a]

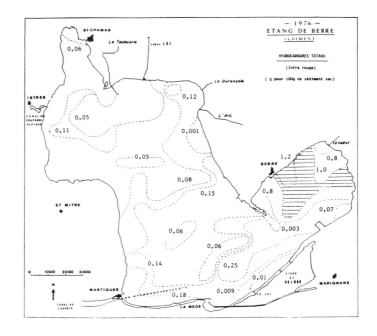

Figure 3.16   Distribution of Total Hydrocarbons in the sediments of the Etang de Berre [Arnoux, A., et al., 1980b]

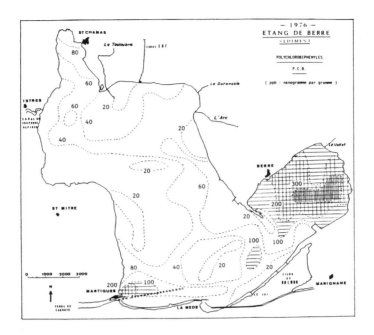

Figure 3.17 Distribution of PCBs in the sediments of the Etang de Berre [Arnoux, A., et al., 1980b]

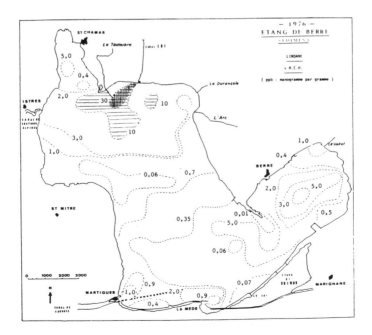

Figure 3.18 Distribution of Lindane in the sediments of the Etang de Berre [Arnoux, A., et al., 1980b]

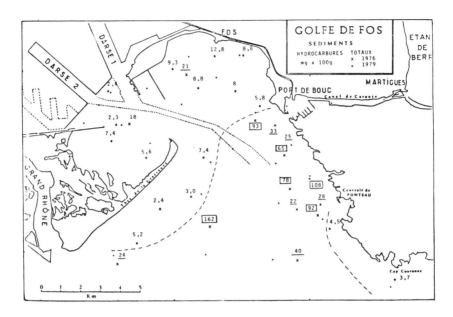

Figure 3.19 Distribution of Total Hydrocarbons in the sediments of the Golfe de Fos
[Arnoux, A., et al., 1980c]

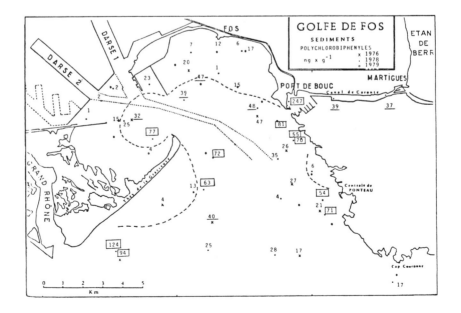

Figure 3.20 Distribution of PCBs in the sediments of the Golfe de Fos   [Arnoux, A., et al., 1980c]

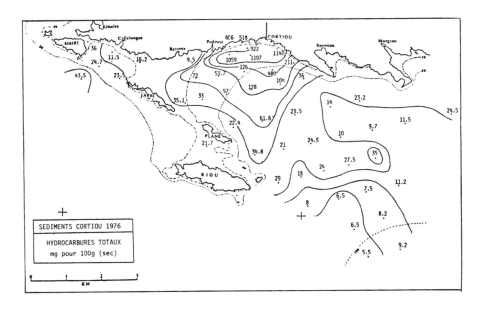

Figure 3.21   Distribution of Total Hydrocarbons in the sediments of Cortiou (Marseille) [Arnoux, A., et al., 1980a]

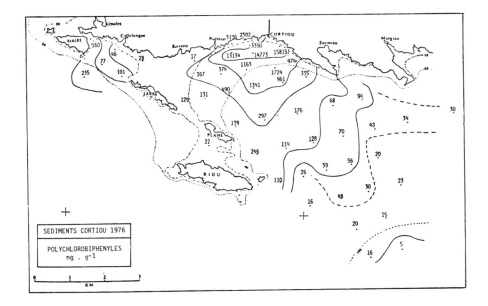

Figure 3.22   Distribution of PCBs in the sediments of Cortiou (Marseille) [Arnoux, A., et al., 1980a]

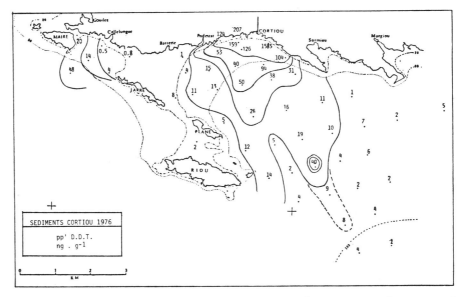

Figure 3.23 Distribution of pp' DDT in the sediments of Cortiou (Marseille) [Arnoux, A., et al., 1980a]

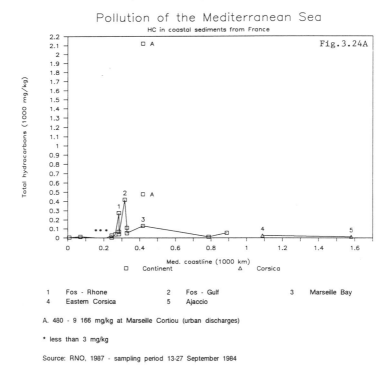

Figure 3.24 Range of concentrations of (A) hydrocarbons, (B) PCB, (C) DDT and (D) lindane in sediments along the French coast

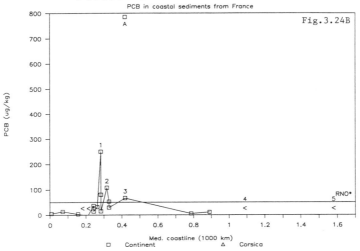

Fig. 3.24B

| 1 | Fos - Rhone | 2 | Fos - Gulf | 3 | Marseille Bay |
| 4 | Eastern Corsica | 5 | Ajaccio | | |

* RNO definitions for:
unpolluted sediment	less than 0.5 ug/kg
lightly to intermediately polluted	5 - 40 ug/kg
heavily polluted	50 - 400 ug/kg
very heavily polluted	above 1 000 ug/kg

A. 785 - 21 615 ug/kg at Marseille Cortiou (urban discharges)

< less than 2 ug/kg

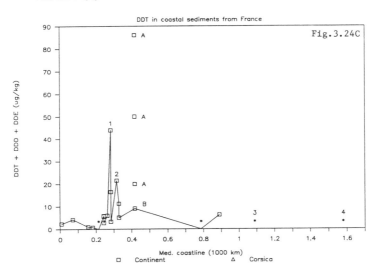

Fig. 3.24C

| 1 | Fos - Rhone | 2 | Fos - Gulf | 3 | Eastern Corsica |
| 4 | Ajaccio | | | | |

A. < 20 - 86 ug/kg at Marseille Cortiou (urban discharges)
B. 9.1 ug/kg at Marseille Bay

* less than 1 ug/kg

Source: RNO, 1987 - sampling period 13-27 September 1984

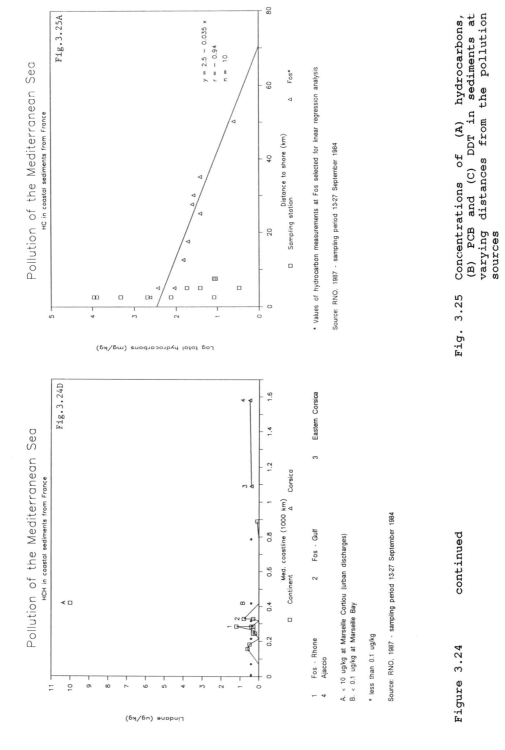

Fig. 3.25 Concentrations of (A) hydrocarbons, (B) PCB and (C) DDT in sediments at varying distances from the pollution sources

Figure 3.24 continued

Pollution of the Mediterranean Sea

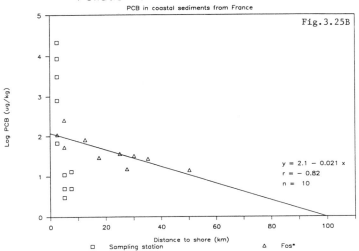

Fig.3.25B

PCB in coastal sediments from France

$y = 2.1 - 0.021 x$
$r = -0.82$
$n = 10$

* Values of PCB measurements at Fos selected for linear regression analysis

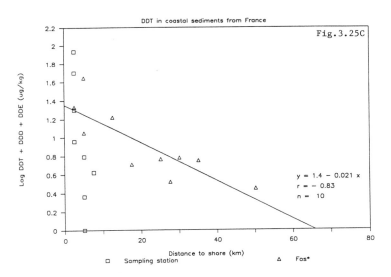

Fig.3.25C

DDT in coastal sediments from France

$y = 1.4 - 0.021 x$
$r = -0.83$
$n = 10$

* Values of DDT measurements at Fos selected for linear regression analysis

Source: RNO, 1987 - sampling period 13-27 September 1984

298

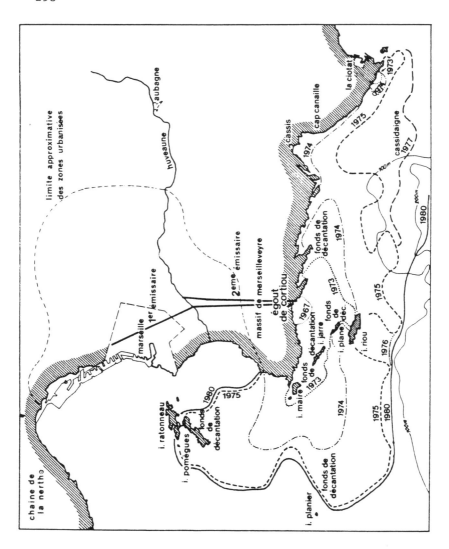

Figure 3.26 Extension of areas disturbed by sewer systems of Marseille and La Ciotat at the sea floor and benthos from 1967 to 1980 [City of Marseille, April 1985]

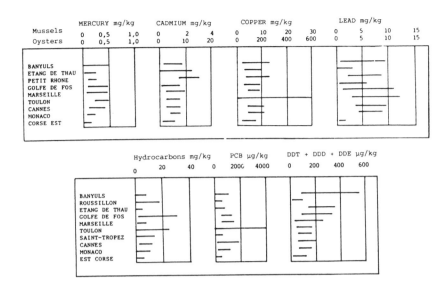

Figure 3.27 Levels of heavy metals (1979-1983) and of organic pollutants (1979-1982) in the living matter, expressed in mg/kg of dry matter, in the French Mediterranean coastal waters (average and standard deviation) [RNO, 1985]

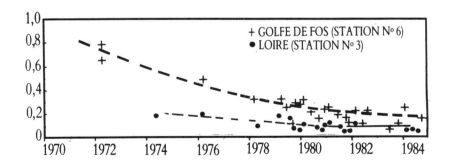

Figure 3.28 Decrease of the concentration of Mercury in the mussels, between 1972 and 1984, expressed in mg/kg of dry matter in the Golfe de Fos
[RNO, "Dix années de surveillance 1974-1984", May 1985]

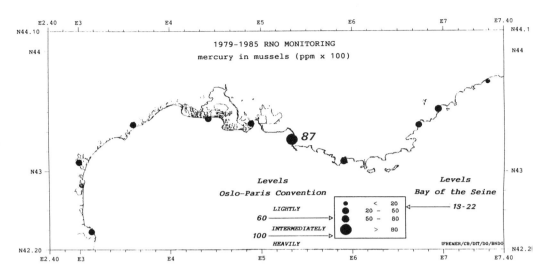
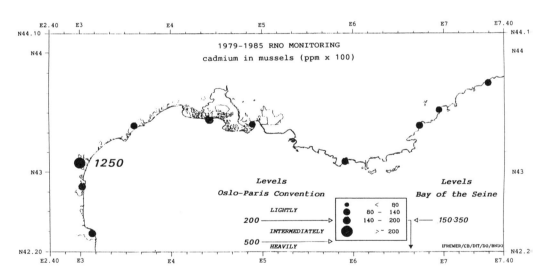

Figure 3.29 Geographical distribution of the concentrations of Mercury and Cadmium in the dry matter of the mussel (*Mytilus galloprovincialis*) along the French Mediterranean coast

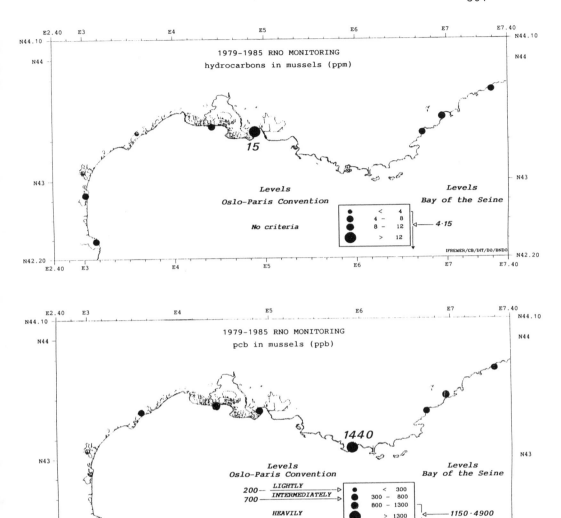

Figure 3.30 Geographical distribution of the concentrations of Hydrocarbons and PCB in the dry matter of the mussel (<u>Mytilus galloprovincialis</u>) along the French Mediterranean coast

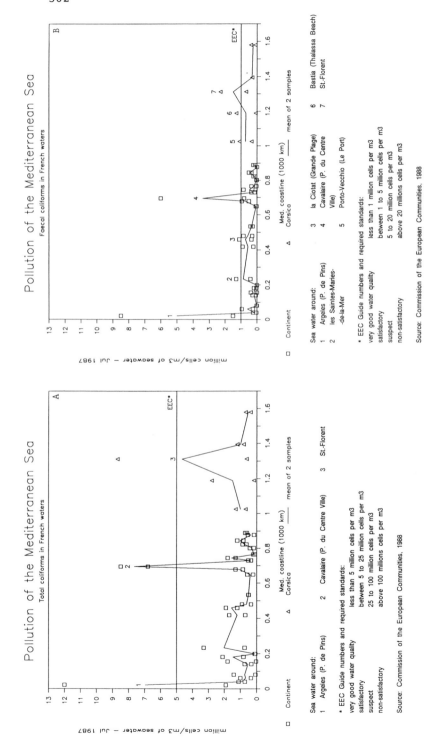

Figure 3.31   Bacterial quality of the bathing waters along the French coast
A: total coliforms; B: faecal coliforms; C: faecal streptococci

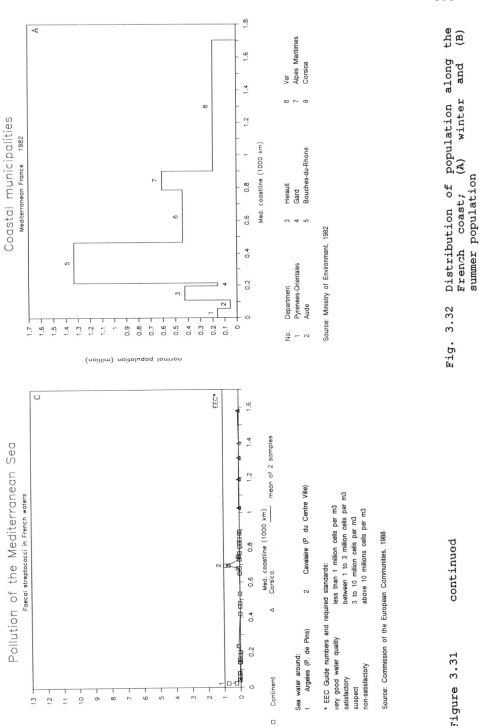

Figure 3.31 continued

Fig. 3.32 Distribution of population along the French coast; (A) winter and (B) summer population

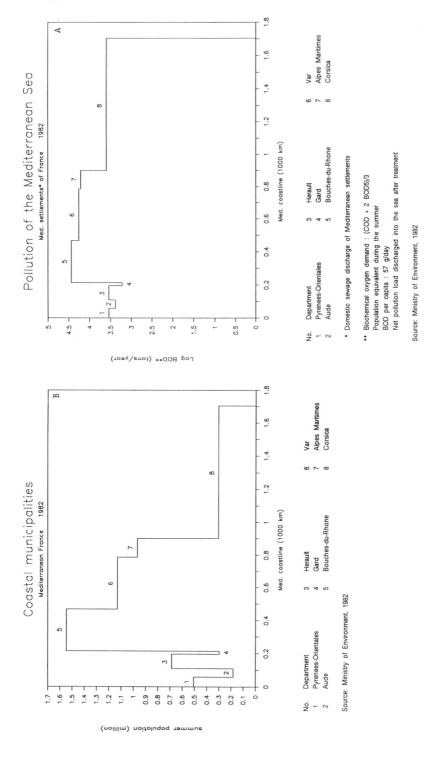

Figure 3.32 continued

Fig. 3.33 (A) BOD and (B) Suspended Solids discharged by coastal municipalities in the different departments

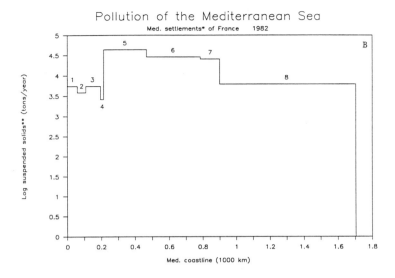

No.	Department				
1	Pyrenees-Orientales	3	Herault	6	Var
2	Aude	4	Gard	7	Alpes Maritimes
		5	Bouches-du-Rhone	8	Corsica

* Domestic sewage discharge of Mediterranean settlements

** Population equivalent during the summer
   Suspended solids per capita : 90 g/day
   Net pollution load discharged into the sea after treatment

Source: Ministry of Environment, 1982

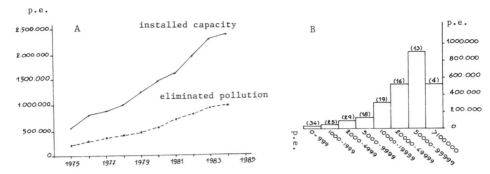

Figure 3.34 A: Evolution of the installed capacity of the treatment stations and of the pollution that is eliminated, in population equivalents (p.e.), from 1975 to 1984 (total numbers)
B: Capacities of the treatment stations in 1984 [Ministere de l'Environnement, "La Protection de l'Environnement Mediterraneen", France 1987]

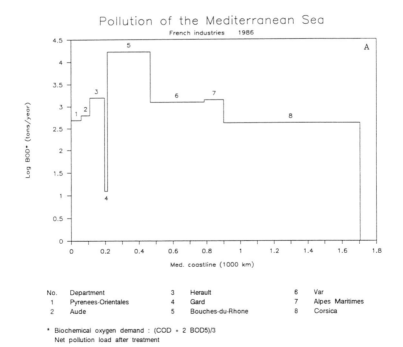

No.	Department					
1	Pyrenees-Orientales	3	Herault	6	Var	
2	Aude	4	Gard	7	Alpes Maritimes	
		5	Bouches-du-Rhone	8	Corsica	

* Biochemical oxygen demand : (COD + 2 BOD5)/3
  Net pollution load after treatment

Source: Agence de bassin Rhone-Mediterranee-Corse, 1988

Fig. 3.35 Industrial discharges along the French coast; (A) BOD and (B) suspended solids

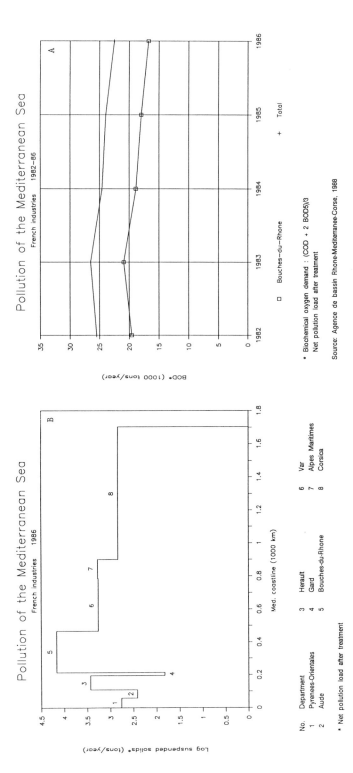

Fig. 3.36 Reduction of industrial discharges along the French coast between 1982 and 1986; (A) BOD and (B) suspended solids

Figure 3.35 continued

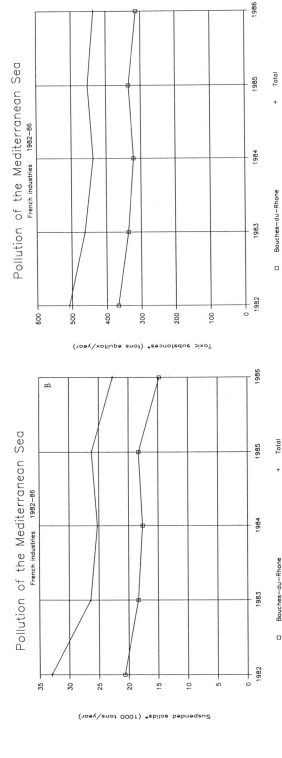

Figure 3.36 continued

Fig. 3.37 Reduction of discharges of toxics along the French coast between 1982 and 1986

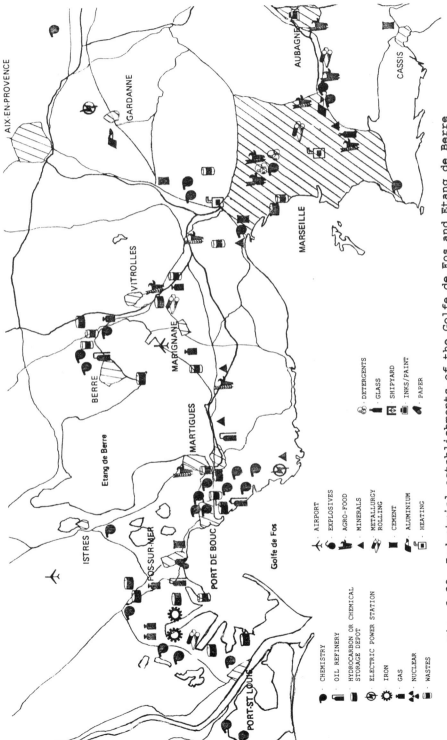

Fig. 3.38 Industrial establishments of the Golfe de Fos and Etang de Berre [Ministere de l'Environnement, 1984]

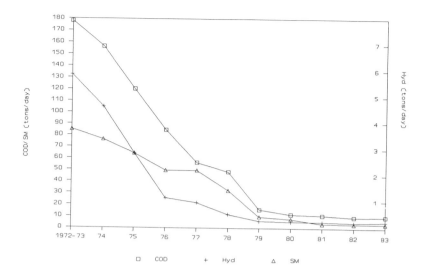

Figure 3.39   Reduction of discharge in the existing plants of the Zone de Fos-l'Etang de Berre, between 1972 and 1983 for COD, Hydrocarbons and Suspended Matter [Ministere de l'Environnement, Conseil General des Bouches-du-Rhône, 1984]

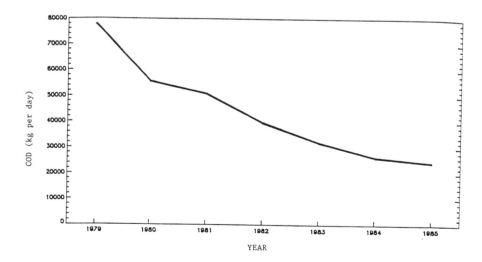

Figure 3.40   Reduction of the discharges of chemical compounds in the river Rhône, south of Lyon, between 1979 and 1985
[Agence de Bassin Rhône-Méditerranée-Corse, "Qualité du Fleuve Rhône", April 1988]

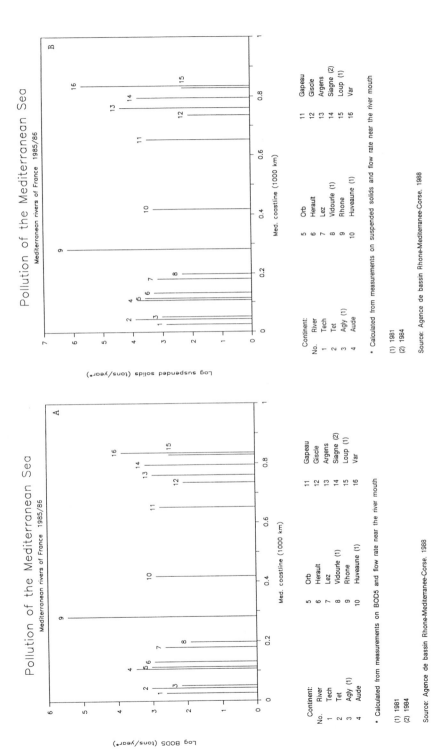

Figure 3.41 Annual (Oct. 1985 - Sept. 1986) discharges of the 16 major rivers into the Mediterranean: (A) BOD5, (B) Suspended Solids

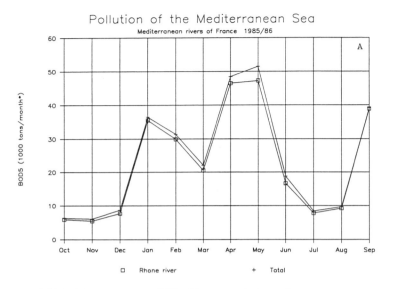
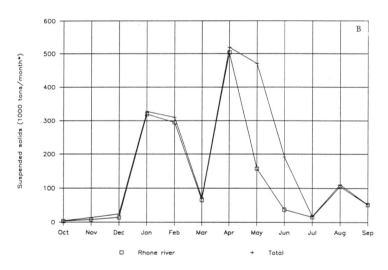

Figure 3.42   Monthly (from Oct. 1985 to Sept. 1986) river discharges into the Mediterranean (same rivers as previously)
A: BOD5; B: Suspended Solids

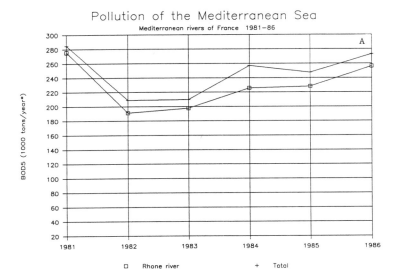

Figure 3.43 Change of discharge of Rhône and of the sum of 11 French Mediterranean rivers between 1981 and 1986 (same rivers as previously except: Agly, Vidourle, Huveaune, Siagne and Loup, which were not included due to lack of data)
A: BOD5; B: Suspended Solids

314

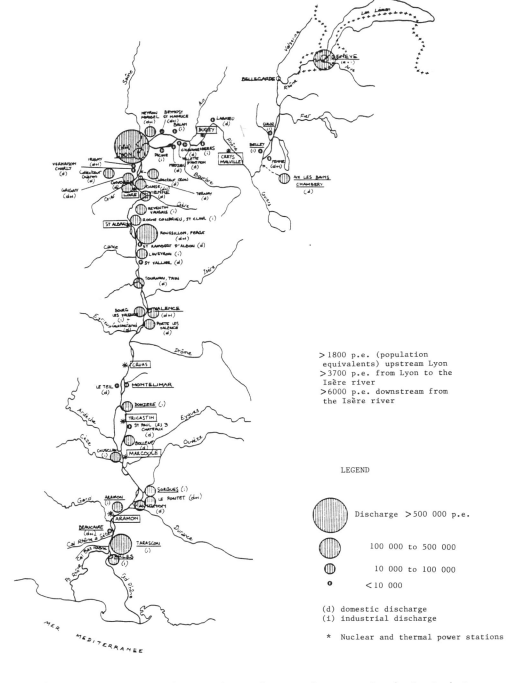

Figure 3.44 Location of major urban and industrial discharges into the Rhône river in population equivalents (net pollution)
[Agence de Bassin Rhône-Méditerranée-Corse, "Qualité du Fleuve Rhône", April 1988]

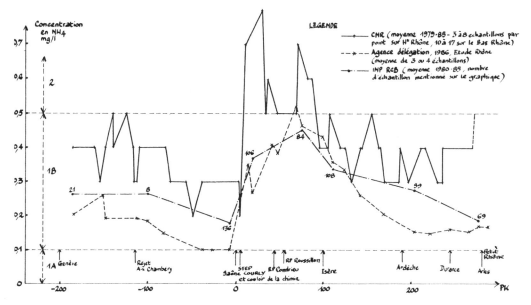

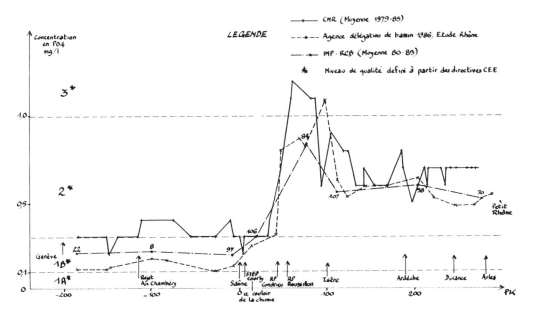

Figure 3.45 Longitudinal profile of NH_4 and PO_4 concentrations along the Rhône river and water quality classes (comparison of CNR, INP-RCB and Agence de Bassin Data)
[Agence de Bassin Rhône-Méditerranée-Corse, "Qualité du Fleuve Rhône", April 1988]

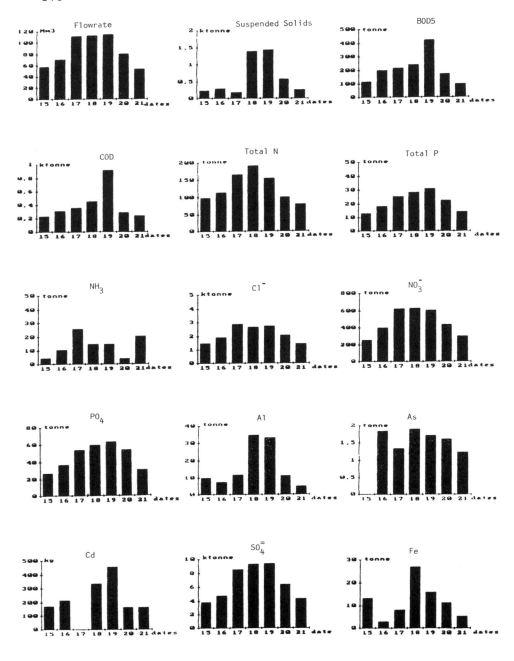

Figure 3.46　　Daily load of the Rhône river discharging into the Mediterranean as measured between September 15-22, 1986 [Agence de Bassin Rhône-Méditerranée-Corse, "Qualité du Fleuve Rhône", April 1988]

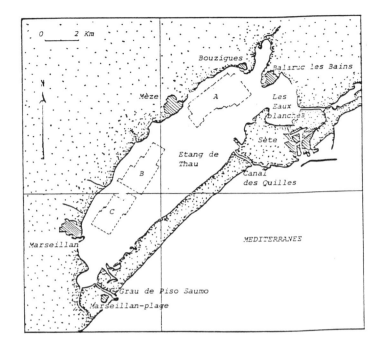

Figure 3.47   Etang de Thau: shellfish farms

Table 3.1 Comparison between the average French Atlantic coastal waters (Seine, Brest) and the average Mediterranean French coastal waters (Fos, Villefranche) for temperature, salinity, nitrates and phosphates [RNO, "Dix annees de surveillance 1974-1984", May 1985]

	Temperature (gr.C)	Salinity (g/kg)	Nitrates (μmol/l)	Phosphates (μmol/l)
Seine	12.0	30.5	31.6	3.45
Brest	12.4	34.4	7.0	0.34
Fos	15.2	33.6	3.7	0.28
Villefranche	17.8	37.7	0.38	0.04

Table 3.2 Range in the values of nitrates, ammonium, silicates and phosphates in seawater collected at eight sites along the French Mediterranean coast [RNO, 1975-1980]

		BANYULS	THAU	FOS	BERRE	CORTIOU	VILLEFRANCHE	MONACO	AJACCIO
Nitrate μmol/l	coastal	0,05-1,19	0,11-2,21	0,06-7,31	0,82-18,0	0,05-4,53	0,05-0,72	0,44-2,09	0,05-0,37
	offshore	0,10-1,19	0,05-3,55	0,16-2,50	0,26-19,0	0,05-5,07	0,05-0,58	0,43-2,34	0,05-0,18
Ammonium μmol/l	coastal	0,05-0,08	0,05-3,63	0,57-2,99	0,85-4,41	0,05-5,02	0,03-0,29	0,16-1,19	0,30-1,66
	offshore	0,05-0,06	0,05-2,17	0,46-1,77	1,17-4,70	0,05-1,03	0,03-0,24	0,22-2,19	0,37-1,72
Silicate μmol/l	coastal	0,32-2,64	1,6-13,3	0,1-23,9	13,9-51,9	0,12-6,19	0,25-1,78	0,64-1,77	0,85-4,09
	offshore	0,36-1,82	1,3-4,39	0,11-4,64	10,5-48,6	0,39-4,81	0,11-1,14	0,63-1,59	0,44-1,48
Phosphate μmol/l	coastal	0,03-0,15	1,06-2,62	0,08-0,48	0,15-0,83	0,03-0,83	0,01-0,07	0,02-0,07	0,01-0,23
	offshore	0,03-0,15	0,08-0,71	0,05-0,28	0,19-1,02	0,06-0,40	0,01-0,11	0,02-0,07	0,01-0,15

Table 3.3 Average levels and range of concentration (ug/l) of cadmium, copper, zinc and lead in the Gulf of Fos   [Benon, P., et al., 1978]

		Cd	Cu	Zn	Pb
Gulf of Fos 23-24 apr. 75					
Surface	average (35)	0.51		63.0	4.45
	lowest	0.07		6.0	0.55
	highest	5.5		120	12.5
Gulf of Fos 25-26 nov. 75					
Surface	average (16)	1.02	5.2	29.7	4.1
	lowest	0.4	2.5	6.2	1.1
	highest	2.75	17.0	65.7	11.1
5 m	average (16)	2.8	6.2	34.2	6.3
	lowest	0.05	2.7	10.8	0.4
	highest	9.9	11.4	65.2	13.7
bottom between 7 and 44 m	average (16)	1.6	5.7	48.5	4.7
	lowest	0.5	2.2	3.2	0.4
	highest	11.5	10.4	400	12.5

Table 3.4 Concentrations in the water of heavy metals in the principal zones ("supporting points") of RNO of the French Mediterranean coast, in 1979 [RNO, 1984, Ministere de l'Environnement, "La Protection de l'Environnement Mediterraneen", France 1987]

	Cadmium (µg/l) safe limit: 0.8 maximum allowable: 10.0	Mercury (µg/l) safe limit: 0.3 maximum allowable: 3.0
Etang de Thau	0.27	0.44
Etang de Berre	0.11	0.04
Fos	0.13	0.06
Cortiou (Marseille)	0.86	0.24
Cannes	0.062	0.00
Menton	0.40	0.02
Ajaccio (Corse)	0.26	0.17
Biguglia (Corse)	0.05*	non available

*Etang de Biguglia: maximum concentration of Cadmium was 3 µg/l, in May 1986

Table 3.5 Average annual concentrations in filtrated seawater for heavy metals in both regions of the French Mediterranean coast, near the coast and in the open sea
[Centre Oceanologique de Bretagne, "Synthese des traveaux de Surveillance 1975-1979 du Reseau National d'Observation de la Qualite du Milieu Marin", Brest 1981]

Hg (µg/l)

	Languedoc-Roussillon/ Provence rhodanienne		Cote d'Azure/Corse	
Year	coastal	off-shore	coastal	off-shore
1977	0.04	0.04	0.01	0.01
1978	0.06	0.06	0.02	0.04
1979	0.06	0.05	0.02	0.01
1st sem. 1980	0.11	0.13	0.01	0.01

Cd (µg/l)

	Languedoc-Roussillon/ Provence rhodanienne		Cote d'Azure/Corse	
Year	coastal	off-shore	coastal	off-shore
1977	0.2	0.1	0.03	0.02
1978	0.4	0.1	0.04	0.03
1979	0.1	0.1	non available	0.03
1st sem. 1980	0.1	0.1	0.03	0.02

Cu (µg/l)

	Languedoc-Roussillon/ Provence rhodanienne		Cote d'Azure/Corse	
Year	coastal	off-shore	coastal	off-shore
1977	1.7	0.9	0.2	0.2
1978	1.5	0.7	0.2	0.3
1979	1.2	0.6	non available	0.2
1st sem. 1980	0.8	0.1	0.3	0.1

Pb (µg/l)

	Languedoc-Roussillon/ Provence rhodanienne		Cote d'Azure/Corse	
Year	coastal	off-shore	coastal	off-shore
1977	2.2	1.7	0.4	0.4
1978	1.5	4.3	0.4	0.5
1979	2.6	1.7	0.2	0.3
1st sem. 1980	2.9	1.8	0.4	0.6

Zn (µg/l)

	Languedoc-Roussillon/ Provence rhodanienne		Cote d'Azure/Corse	
Year	coastal	off-shore	coastal	off-shore
1977	6.0	3.5	3.0	2.4
1978	14.0	12.0	3.0	2.1
1979	3.7	3.5	non available	2.1
1st sem. 1980	25.0	3.6	0.72	0.60

Table 3.6 Range of heavy metal concentrations in filtrated seawater in the European coastal waters [Centre Oceanolonique de Bretagne, "Synthèse des travaux, de surveillance 1975-1979 du Reseau National d'Observation de la Qualité du Milieu Marin, Brest 1981]

	Coastal	Off-shore	Open sea
Hg (µg/l)	0.03-0.3	0.001-0.05	non-available
Zn (µg/l)	5-10	1.5-7.0	0.005-0.300
Pb (µg/l)	0.6-10	0.04-0.8	0.005-1.0
Cu (µg/l)	1.0-15	0.1-1.5	0.005-1.0
Cd (µg/l)	0.1-3	0.02-0.5	0.001-0.015

Table 3.7 Natural and toxic concentrations of heavy metals in sediments [Centre Oceaneg. de Bretagne, "Synthese des travaux de surveillance 1975-79 du RNO", Brest 1981]

Metal	Hg	Cd	Cu	Pb	Zn
Naturally occuring concentration (mg/kg)	0.02-0.35	0.1-2.0	5-30	10-70	20-150
Toxic considered concentration (mg/kg)	>2.5	>10	>300	>500	>600

Table 3.8 Comparison of the concentrations of heavy metals in sediments in Fos and in the Baies de Cannes and de Villefranche, expressed in mg per kg of sediments (dry weight), for grain fractions less than 63 microns
[RNO, 1984, Ministere de l'Environnement, "La Protection de l'Environnement Mediterraneen, France 1987]

	Mercury (mg/kg)	Cadmium (mg/kg)	Copper (mg/kg)	Lead (mg/kg)
Golfe de Fos	1.2	5.8	112	68
Baies de Cannes et de Villefranche	0.2	0.7	19	46

Table 3.9 Bathing water quality in the French Mediterranean coast (1976-1986)
[Ministere de l'Environnement et Ministre de la Famille, "Etat Sanitaire des Zones de Baignade en Mer"]

Year	No. of municipalities	No. of points	No. of samples	Number (N) and percentage (%) of beaches classified into categories												% of beaches according to the European Directives
				A		AB		B		C		CD		D		
				N	%	N	%	N	%	N	%	N	%	N	%	
1976	123	372	3 825	157	42.2	66	17.7	88	23.7	52	14.0	2	0.5	7	1.9	83.6
1977	94	394	4 739	128	32.5	70	17.8	118	29.9	54	13.7	7	1.8	17	4.3	80.2
1978	119	458	5 257	42	9.2	142	31.0	170	37.1	74	16.2	22	4.8	8	1.7	77.3
1979	135	488	5 612	72	14.8	170	34.8	156	32.0	67	13.7	18	3.7	5	1.0	81.6
1980	153	569	6 858	134	23.6	177	31.1	146	25.7	84	14.8	25	4.4	3	0.5	80.3
1981	155	568	6 884	154	27.1	192	33.8	155	27.3	65	11.4	0	0.0	2	0.4	88.2
1982	163	587	6 773	235	40.0	196	33.4	87	14.8	61	10.4	3	0.5	5	0.9	88.2
1983	164	583	6 999	139	23.8	213	36.5	147	25.2	83	14.2	0	0.0	1	0.2	85.6
1984	166	598	7 027	179	29.9	207	34.6	147	24.6	62	10.4	2	0.3	1	0.2	89.1
1985	165	599	7 553	196	32.7	189	31.6	190	31.7	22	3.7	2	0.3	0	0.0	96.0
1986	166	606	7 305	200	33.0	160	26.4	220	36.3	13	2.1	1	0.2	3	0.5	95.7

Table 3.10 Comparison between daily emissions of the Fos-Berre industrial complex with daily emissions for the whole of France [Gomes, L., et al., 1985]

	SO_2 (tons/day)	Cu (tons/day)	Cd (tons/day)	Ni (tons/day)
France (whole)	6000-9000 5000	1.23	0.47	2.47
Fos-Berre	400-450	0.08	0.10	0.16
(Fos-Berre/France) %	4.5-9	6.5	21.3	6.5

Co (tons/day)	Pb (tons/day)	Zn (tons/day)	Mn (tons/day)	Ti (tons/day)
0.28	28.9	16.8	3.27	–
0.05	0.76	2.31	0.15	0.55
17.8	2.6	13.7	4.6	–

Table 3.11 Produced and Treated Domestic Pollution by Sewage Treatment Plants on the Mediterranean Coast
[Ministere de l'Environnement, "La Protection de l'Environnement Mediterraneen", France, 1987]

Coastal Mediterranean Communities (popul. equiv.)	1980	1985	1990 (expected)
Pollution produced by the autocht.	2,506,798	2,588,130	2,652,433
Pollution produced by the tourists	1,830,898	2,324,274	2,381,580
Total pollution produced	4,337,696	4,912,404	5,034,013
Total pollution treated	974,795	1,362,633	3,277,050
Pollution discharged	3,362,901	3,549,771	1,756,963
Pollution treated/pollution produced	22.5%	27.7%	65.1%

Table 3.12 Number of sewage treatment plants per each department of the French Mediterranean regions together with installed capacity in popul. equivalents, and processes used
[Ministere de l'Environnement, Direction de la Prevention des Pollutions, "Etat de l'Assainissement en zone Littorale", 1982]

Department	Sewage treatm. plant	Installed capacity	Selected processes (%)						
			Act. Sl.	Tricl. filt.	Lagoons	Phys. chem.	biol. d.ses	sedi- ment	chlo- rinat
Pyrenees Orientales	26	475,000	73	23	19	15	0	7	7
Aude	14	106,000	50	0	50	0	0	7	7
Herault	37	825,490	64	21	16	5	0	2	5
Gard	9	251,500	88	0	11	11	0	0	0
Bouches du Rhone	26	479,100	73	7	7	7	0	0	26
Var	14	276,600	57	14	7	42	0	7	7
Alpes Maritime	11	469,300	n.a.	n.a.	n.a.	n.a.	n.a.	n.a.	n.a.
Corse du Sud	19	114,520	n.a.	n.a.	n.a.	n.a.	n.a.	n.a.	n.a.
Haute Corse	29	55,700	37	55	0	0	0	0	0
Total	185	3,053,210							

Table 3.13 Water quality classes and their limits
[Agence de Bassin Rhône-Méditerranée-Corse, "Qualité du Fleuve Rhône", April 1988]

Classes
1A: excellent quality   1B: good quality   2 : moderate quality
3 : poor quality   4 : very polluted

| | | GENERAL WATER QUALITY | | | | | WATER FOR FISH LIFE | | | |
| | | | | | | | Salmon water | | Water for carp | |
		1A	1B	2	3	4	Guide-lines	Imper-atives	G	I
Total. colif.	N/100 ml									
Fecal colif.	N/100 ml									
Fecal streptoc.	N/100 ml									
Salmonella, water										
T water	°C	<20	20 - 22	22 - 25	25 - 30	>30		<21,5		<28
pH		6,5 - 8,5			5,5 - 9,5	<5,5 or >9,5		6 - 9		6 - 9
Conductivity	µS/cm									
Suspended matter	mg/l	<30			31 - 70	>70	<25		<25	
BOD 5	mg/l	< 3	3 - 5	5 - 10	10 - 25	>25	< 3		< 6	
COD	mg/l	<20	20 - 25	25 - 40	40 - 80	>80				
Oxydability	mg/l	< 3	3 - 5	5 - 8	>8					
Nitrogen (Kjeldhal)	mg/l									
Dissolved oxygen	mg/l	> 7	5 - 7	3 - 5	<3		50% >9 or 100% >7	50% > 9	50% >8 or 100% >5	50% >7
Saturation degree	%	>90	70 - 90	50 - 70	<50					
Hardness	°F									
NH 4 +	mg/l	<0,1	0,1 - 0,5	0,5 - 2	2 - 8	>8	<0,04	< 1	< 0,2	< 1
Cl -	mg/l									
SO 4 --	mg/l									
NO 2 -	mg/l						<0,01		<0,03	
NO 3 -	mg/l	<44			44 - 100	>100				
PO 4 ---	mg/l									
NH 3	mg/l						<0,0005	<0,025	<0,005	<0,025
Fluorine	mg/l	<1,5	>1,5							
Total iron	mg/l	<0,5	0,5 - 1	1 - 1,5	1,5					
Manganese	mg/l	<0,10	0,1 - 0,25	0,25 - 0,5	>0,5					
Arsenic	mg/l	<0,05		0,05 - 0,1	>0,1					
Cadmium	mg/l	<0,005			>0,005					
Total chromium	mg/l	<0,05			>0,05					
Copper	mg/l	<0,04		>0,04			<0,04		<0,04	
Mercury	mg/l	<0,001			>0,001					
Lead	mg/l	<0,05			>0,05					
Selenium	mg/l	<0,01			>0,01					
Zinc	mg/l	<0,3	0,3 - 1	>1					<0,3	< 1
Cyanide	mg/l	<0,05			>0,05					
Deterg. anionic	mg/l									
Phenols	mg/l	<0,001		0,001 - 0,05	0,05 - 0,5	>0,5				
Subs. extr. chlor	mg/l	<0,2	0,2 - 0,5	0,5 - 1	>1					
Total pesticides	ng/l									
Ind. Reference - Indice Biotic		< 1	1 - 3	3 - 5	5 - 7	>7				

Table 3.14 Overview of French laws and decrees relevant to
the Mediterranean pollution [DOCTER, 1987]

Year	Reference	Topics
1963	Loi n. 63-1178 du 28 novembre 1963 relative au domaine public maritime (modifiée par Décret n. 72-879 du 19 septembre 1972)	coasts, sea protection
1964	Loi' du 26 décembre 1964 n. 64/1331 relative à la pollution des eaux de mer par les hydrocarbures	pollution: water, sea protection
1965	Circulaire du 28 avril 1965 relative aux recherches archéologiques sous-marines	cultural heritage
1966	Décret du 14 septembre 1966 n. 66/699 relatif aux Comités de bassins	pollution: water
1966	Décret du 14 septembre 1966 n. 66/700 relatif aux agences financières de bassins, modifié par décret du 8 avril 1974 n. 74/284	pollution: water
1967	Décret du 15 décembre 1967 n. 67/1093 relatif aux perimètres de protection des eaux	pollution: water
1967	Décret du 15 décembre 1967 n. 67/1094 sanctionnant les infractions à la loi du 16 décembre 1964 n. 64/1265 relative au régime et à la répartition des eaux et à la lutte contre leur pollution	water supply and river management, pollution: water
1968	Décret du 5 avril 1968 n. 68/355 relatif à la coordination interministerielle dans le domaine de l'eau	pollution: water
1975	Décret du 12 mars 1975 n. 75/177 portant application de l'article 6/3 de la loi du 16 décembre 1964 n. 64/1245 relative au régime et à la répartition des eaux et à la lutte contre leur pollution	water supply pollution: water
1975	Décret du 23 avril 1975 n. 75/310 relatif à la coordination interministerielle dans le domaine de l'élimination des déchets	pollution: waste
1975	Arrêté du 8 juillet 1975 relatif aux conditions d'emploi des polychlorobiphényles	pollution: waste
1975	Loi du 10 juillet 1975 n. 75/602 portant la création du Conservatoire de l'espace littoral et des rivages lacustres	coasts, sea protection
1975	Décret 75-1136 du 11 décembre 1975 portant application de la loi du 10 juillet 1975	coasts, sea protection
1976	Loi du 7 juillet 1976 n. 76/599 relative à la prévention et à la répression de la pollution marine par les opérations d'immersion effectuées par les navires et aéronefs, et à la lutte contre la pollution marine accidentelle	pollution: waste, sea protection
1976	Loi du 10 juillet 1976 n. 76/629 relative à la protection de la nature	information, energy, fauna and flora, environmental impact assessment, parks and nature reserves
1976	Loi du 19 juillet 1976 n. 76/663 relative aux installations classées pour la protection de l'environnement	water supply and river management, cultural heritage, energy, pollution: air, waste
1977	Décret du 8 mars 1977 n. 77/254 relatif à la réglementation du déversement des huiles et lubrifiants dans les eaux superficielles, souterraines et de la mer	pollution: waste
1977	Décret du 12 octobre 1977 n. 77/1141 pris pour l'application de l'article 2 de la loi du 10 juillet 1976 n. 76/629 relatif à la protection de la nature	organizational structure, energy, environmental impact assessment, land reclamation
1979	Loi du 2 janvier 1979 n. 79/5 relative à la pollution des mers par les hydrocarbures	pollution: water, sea protection
1979	Décret 70-716 du 25 août 1979 approuvant la directive d'aménagement national relative à la protection et l'aménagement du littoral (Art. R III.27 code de l'urbanisme)	coasts
1982	Décret du 29 septembre 1982 n. 82/842 pris pour l'application de la loi du 7 juillet 1976 n. 76/599 relative à la prévention et à la répression de la pollution marine par les opérations d'immersion effectuées par les navires et aéronefs et à la lutte contre la pollution marine accidentelle	pollution: waste, sea protection
1983	Loi n. 83-8 du 7 janvier 1983 relative à la répartition des compétences entre les communes, les départements, les régions et l'État	coasts, organizational structure, parks and nature reserves
1983	Loi du 12 juillet 1983 n. 83/630 relative à la démocratisation des enquêtes publiques et à la protection de l'environnement	environmental impact assessment, information
1984	Décret n. 84-185 du 14 mars 1984 modifiant le décret n. 80-470 du 16 juin 1980 portant application de la loi n. 76-646 du 16 juillet 1976 relative à la prospection, à la recherche et à l'exploitation des substances minérales	mines and quarries, sea protection
1986	Loi du 3 janvier 1986 relative à la protection, l'amenagement et la mise en valeur du littoral	coasts

Table 3.15 Existing treatment plants, planned and under construction for the coastal cities of the French Mediterranean regions
[Ministere des Affaires Sociales et de l'Emploi, "Qualite des Eaux de Baignade", mai 1988]

City	Existing treatm.plant	Planned treatm.plant	Under constr. treatm.plant
Perpignan	+		
Narbonne Plage-Gruissan		+	
Marseille	+		
Nice	+		
Antibes			+
La Ciotat			+
Toulon Ouest		+	
Saint-Cyr-sur-Mer			+
Bormes les Mimosas			+
Sainte-Maxime			+
Villefranche-sur-Mer		+	
Hyeres		+	
Saint-Tropez			+
Cavalaire			+
Menton		+	
Frejus-Saint Raphael		+	
Bastia			+

Table 3.16 Distribution of Financial Aid, in Million Francs, per region of the French Mediterranean, for the creation extension and improvement of network sewage systems and treatment plants, in the local communities, in 1986
[Bulletin d'information du Comité et de l'Agence de Bassin, Rhône-Méditerranée-Corse, No. 25, Juin '87]

Region	Grant	Advance Payments	Loans	Total
Provence-Alpes-Côte d'Azur	39.4	27.0	___	66.4
Languedoc-Roussillon	18.3	15.0	4.6	37.9
Corse	4.6	3.0	---	7.6
Total	62.3	45.0	4.6	111.9

Table 3.17 Distribution of Financial Aid, in Million Francs, per region of the French Mediterranean, for measures against industrial pollution, in 1986
[Bulletin d'information du Comité et de l'Agence de Bassin, Rhône-Méditerranée-Corse, No. 25, Juin '87]

Region	Grant	Advance Payments	Loans	Total
Provence Côte d'Azur	26.6	9.9	2.1	38.6
Languedoc-Roussillon	3.4	7.9	0.6	11.9
Total	30.0	17.8	2.7	50.5

Table 3.18 Financial Aid given to Industrial Establishments of the French Mediterranean regions, in 1986
[Bulletin d'information du Comité et de l'Agence de Bassin, Rhône-Méditerranée-Corse, No. 25, Juin '87]

Establishment	Capacity 10^3 pop. equiv.	Eliminated pollution 10^3 pop. equiv.	Total plant investment 10^3 F	Financial Aid 10^3 F
RHONE-POULENC Roussillon				
- reduction of pollution of workshop APAP	45	12.1	9,620	2,650
- prevention of accidental pollution	p.m.	p.m.	1,800	504
ATOCHEM Marseille				
- improvement of the RILSAN process	10.6	9.9	5,990	3,894
SHELL-CHIMIE Berre				
- biological treatment plant	19.4	17.1	15,880	3,337
ARCO Fos-sur-Mer				
- treatment of effluents	142	136.7	31,330	7,675
- proper technology	p.m.	142.3	p.m.	7,325

4. ITALIAN MEDITERRANEAN ENVIRONMENT

4.1. State of the environment

4.1.1. Heavy metals

4.1.1.1. Heavy metals in coastal seawaters

A July 1987 European Community (EC) survey (CEC, 1988) measured copper, lead and zinc pollution in Italian coastal waters (Figure 4.1). At Lido di Ostia (near Rome) and Pesaro high concentrations of copper, lead and zinc were detected in the water. Significant mercury (Hg) pollution has been detected in the same survey, near the island of Ischia (province of Naples). The measured concentration of 97 ng Hg/l is above the 70 ng/l limit defined by the EC for heavily polluted seawater. It should be noticed that only a limited number of Mediterranean beaches were investigated.

The Adriatic Sea is the receptor of large quantities of heavy metals of industrial origin. The currents tend to accumulate pollutants towards the Adriatic coast.
Heavy-metal concentrations in seawater along the coast of the Ancona Province are shown in Table 4.1. The sampling zone started 200 meters north of the mouth of the Cesano River and ended 200 meters south of the Musone River. Samples were taken for a 24-month period. Ancona (sampling stations 15, 16 and 17) has the highest concentrations of zinc (Zn). This is due to the discharges of many metal plating and metallurgical industries. The minimum concentrations along the coast are between 14-15 µg/l, above the EC limit of 10 µg/l for heavily polluted waters. It must be noted that elevated zinc concentrations are expected due to the high salinity of the Adriatic Sea.
The concentrations of copper (Cu) in seawater are also highest near Ancona. The values of 9.92 µg/l to 17.16 µg/l

correspond to heavily and very heavily polluted coastal water. The lowest reported value of 5.11 µg/l is still above the EC limit of 5 µg/l for heavily polluted waters.
An elevated concentration of 3.55 µg/l of lead (Pb) was also found in the area of Ancona (sampling stations 14 and 15). Its presence is associated with naval transport and activities in the nearby port.
All cadmium (Cd) concentrations were above the EC limit of 0.5 µg/l for heavily polluted waters. The maximum concentration of 1.41 µg/l indicates very heavily polluted marine waters.

Samples of seawater collected between Porto Santo Stefano and the Albegna estuary also show high concentrations of heavy metals (Breder, 1987). Dissolved lead ranged from 50 to 80 ng/l and cadmium from 5 to 40 ng/l. The total mercury concentrations varied from 13 ng/l to 64 ng/l, while the reactive mercury concentration was between 0.8 ng/l and 1.5 ng/l. Particulate matter is of primary importance to the amount and distribution of dissolved metals in water as it removes metals from the water by adsorption.
The coast from Punta Ala to Piombino (Tyrrhenian Sea) has dissolved lead ($Pb_{diss.}$) concentrations of 60-710 ng/l (Table 4.2). The broad range reflects the influence of the ports of Punta Ala and Piombino, the city of Follonica and a factory on the coast.

Metal pollution in the Gulf of La Spezia (Ligurian Sea) is mainly caused by the shipping industry, in which cadmium, lead and copper are frequently used (Breder, 1987). The maximum concentrations of dissolved cadmium, lead and copper are 77 ng/l, 386 ng/l and 854 ng/l, respectively. The above concentrations were measured inside the port, where the proportion of metals incorporated in or adsorbed onto particulate matter plays a significant role. The ratio of the particulate to the dissolved metal was 1.1 for copper and 3.5 for lead.

4.1.1.2. Heavy metals in marine sediments

Marine sediments of the upper Tyrrhenian are seriously polluted by mercury, which is mainly derived from two sources: the cinnabar (HgS) deposits of southern Tuscany; and a chlor-alkali plant near Livorno. The weathering and leaching of the cinnabar results in mercury (Hg) accumulation in the superficial marine sediments at the mouths of the rivers, in concentrations greater than 1 mg/kg (Figures 4.2A and 4.2B) (Baldi 1982, 1986), and a maximum of 5.3 mg/kg.

In the Adriatic Sea the highest pollutant accumulations have been detected in the prodelta zones of the rivers (off the shore of the Po Delta at depths of more than 20 meters) and in areas of low hydrodynamic energy, i.e. the northern part of the Emilia-Romagna coast, the Gulf of Trieste and the lagoons (Frascari et al., 1988).
The northern area is highly contaminated with Hg (five to ten times natural background levels) due to a chlor-alkali plant located in Marano and the mining district of Idria, as shown in Figure 4.3A. The Ravenna Port Canal (Porto Corsini) has high levels of Hg due to direct dumping of petrochemical waste in the Ravenna salt marches.
High zinc (Zn) concentrations at four times background levels are observed in front of the Lagoon of Venice, shown in Figure 4.3B.
The Brenta-Po di Levante area has anomalous levels (two to four times background levels) of chromium (Cr) (Figure 4.3C), from leather industries centered along the Brenta and Adige rivers.
The highest lead (Pb) content is found in front of the Po Delta and at some locations to the north (Gulf of Trieste) (Figure 4.3D).
Copper values in the Ancona Province are between 74 and 102 mg/kg dry weight which are high in comparison with background data of 19-44 mg/kg (Benetti et al., 1979).

In the Apulia region, the mercury content in the sediments is high in the industrial areas of Brindisi and Taranto (Figures 4.4A and 4.4B).

Total mercury concentrations in the top 3 cm sediment show high levels (0.50-1.75 mg/kg dry weight) in Naples Harbor and at the metallurgical industries in Bagnoli (Figure 4.5). The mercury concentration decreases at distances farther away from these two areas, reaching background levels (0.1-0.26 µg/g) near the bay entrance.

Samples of superficial sediments collected in the Bay of Follonica showed elevated concentrations of trace metals in the Piombino Channel and near Follonica: iron (Fe): 9.9%; manganese (Mn): 2,990 mg/kg; zinc (Zn): 215 mg/kg; lead (Pb): 100 mg/kg; copper (Cu): 77 mg/kg; nickel (Ni): 120 mg/kg; and cobalt (Co): 29 mg/kg (Table 4.3).
In the first area, the high levels of metals are most likely due to the metallurgical industries and the power plant, and to the weathering and the mining of pyrite along the eastern coast of Elba. The high levels of metals in the Piombino Channel are most likely a result of coprecipitation with Fe-Mn hydrous oxides. These two elements are released in large quantities from the Elba mineral deposits.
The pollution near Follonica is also due to the dumping of the chemical plant, producing H_2SO_4 and TiO_2. The trace metals held as impurities in pyrite and ilmenite ores are subsequently released into the environment.

The superficial sediments in the Ligurian Sea near Genoa are contaminated by the high input of industrial and municipal wastes. High concentrations of copper (85.9 mg/kg) and lead (228 mg/kg) were found.

4.1.1.3. Heavy metals in marine organisms

The Mediterranean basin, representing only 1% of the

world's surface, contains as much as 65% of the total world reserve of mercury. One area very rich in mercury is Mount Amiata, which is located in southern Tuscany.

It is well known (Bernhard and Brondi, 1986) that fish and shellfish from the western Mediterranean have significantly higher mercury contents than the corresponding oceanic species. The higher mercury levels are especially found in marine mammals and in tuna and swordfish.

The upper Tyrrhenian receives many anthropogenic pollutants, and run-off from a large area of cinnabar-rich ores. The intens mining over the years and the weathering of mineralized sediments have increased the concentration of mercury in the terrestrial and marine environment.

Research carried out near the Argentario promontory showed mean mercury concentrations in mussel (<u>Mytilus galloprovincialis</u>) and lobster (<u>Nephrops norvegicus</u>) up to 1.83 and 1.92 mg/kg wet weight, respectively, and in striped mullet (<u>Mullus barbatus</u>) up to 2.81 mg/kg wet weight (Leonzio et al., 1981).

The concentrations of lead, iron and nickel in molluscs in the Bay of Follonica were highest at sampling stations in the Piombino Channel, close to the metallurgical industries and the power plant of 640 megawatts, and near Follonica around the TiO_2 producing chemical factory. The most elevated concentrations of lead, iron and nickel in the mussel <u>Mytilus galloprovincialis</u> were 16.4, 270 and 6.9 mg/kg dry weight, respectively. For the limpet <u>Patella caerulea</u> the highest concentrations were 13.3, 331 and 4.7 mg/kg dry weight, respectively (Bargagli et al., 1985).

Levels of heavy metals measured in fish caught in different regions of the Italian Mediterranean Sea are shown in Table 4.4. The highest average concentration of mercury (0.54 mg/kg wet weight) and cadmium (0.111 mg/kg wet weight) are found in fish from the Adriatic Sea near the Slavic coasts. The average concentration of copper (1.17 mg/kg wet weight) is the highest in fish from the Ionian Sea and that of zinc (7.01 mg/kg wet weight) in fish from the Tyrrhenian Sea near the Sicilian coasts.

The heavy-metal concentrations in several fish and crustaceans from the Gulf of Trieste are shown in Table 4.5. The shrimp Squilla mantis has the highest cadmium (Cd) (0.46-0.65 mg/kg wet weight) and copper (Cu) concentrations (15.8-29.8 mg/kg). The highest lead (Pb) concentrations (1.40-1.44 mg/kg wet weight) are found in the anchovies Engraulis encrasicolus, while the garfish Belone belone has the highest zinc (Zn) concentrations (23.6-51.3 mg/kg wet weight).
Bivalve molluscs Chamelea gallina from the Gulf of Trieste show mercury (Hg) concentrations between 0.11 and 0.47 mg/kg wet weight (Table 4.6), while arsenic (As) varies between 1.1 and 2.4 mg/kg wet weight in the mussel Mytilus galloprovincialis.

Several marine organisms in the lower Tyrrhenian Sea and the upper Ionian Sea have high heavy-metal contents (Mojo et al., 1979). The mercury concentrations in the lobster Nephrops norvegicus (maximum: 335 µg/kg wet weight) and the striped mullet Mullus barbatus (maximum: 300 µg/kg wet weight) are high due to their exposure to contaminated sediments. Lead is accumulated in Mytilus galloprovincialis (maximum: 880 µg/kg wet weight) and Nephrops norvegicus (maximum: 865 µg/kg wet weight). The contents of mercury and lead seem to be less in the Tyrrhenian Sea than those in the same species in the Adriatic Sea.

High heavy-metal concentrations have been measured in marine organisms in the west-southwest part of the Bay of Augusta in Sicily. The high levels are due to the presence of oil refinery piers and the discharge of the Augusta sewage. The mussel Mytilus galloprovincialis shows concentrations: 0.64-0.93 mg/kg dry weight for cadmium; 13.1-6.25 mg/kg for copper; 149.4-151 mg/kg for zinc; 5.00-1.36 mg/kg for lead; and 0.99-1.32 mg/kg for mercury (Castagna et al., 1985).

4.1.2. Organic contaminants

In the coastal area the presence of chlorinated compounds is mainly resulting from local sources, especially river outflows.

4.1.2.1. Organics in marine waters and sediments

High polychlorinated biphenyl (PCB) and insecticide dichlorodiphenyl trichlorethane (DDT) (pp' DDT and its analogues) concentrations are detected in the Po Delta and in the area between Po di Levante and the Sile River (Figures 4.6A,B).
The concentrations in the Gulf of Venice are strongly influenced by canals and rivers and the behavior of the lagoon sediment. In the northern zone, where sediments are mainly from the Piave and Sile rivers, a relatively low ratio of PCB to total DDT concentrations of 0.3 points to major agricultural contamination. In the southern part of this area the industrial pollution prevails as the ratio of PCB to total DDT increases to 3-2.4.
The area influenced by the Po River tributaries is moderately contaminated by total DDT concentrations (< 12 µg/kg). Industrial pollution here with PCB concentrations of 20-80 or > 80 µg/kg results in a ratio of 6-7.

The concentration of PCBs in the sediment top 3 cm in the Bay of Naples and adjacent marine areas is shown in Figure 4.7. Inside the Bay of Naples, PCBs attain the highest concentrations (maximum: 3,200 µg/kg dry weight). Outside the bay, PCB levels never fall below 10 µg/kg.
Concentrations of total DDT (mainly as DDE) are lower than 20 µg/kg dry weight in all samples, except in two locations just outside Naples Harbor (52.8 µg/kg dry weight) and three samples from near the Sarno River (43.6 µg/kg). Pesticides commmonly used in agriculture are released in the northern part of this region (Baldi et al., 1983).

In the Gulf of Pozzuoli (west of Naples) significant pollution by PCBs and polycyclic aromatic hydrocarbons (PAH) in surface sediments has been detected (De Simone et al., 1982). PCBs are found in concentrations above 2 mg/kg, with a maximum of 60 mg/kg near Bagnoli. Concentrations of PAH (phenanthrene, fluranthrene and crysene) are generally above 40 mg/kg and reach a maximum of 170.1 mg/kg near Bacoli.

4.1.2.2. Organics in marine organisms

Chlorinated hydrocarbons were measured in mussel (Mytilus galloprovincialis), lobster (Nephrops norvegicus), anchovy (Engraulis encrasicolus), striped mullet (Mullus barbatus) and tuna (Thunnus thynnus) collected in six different fishing areas located between Trieste and Ancona in the Adriatic Sea. The concentrations indicate a uniform distribution at intermediate degrees of contamination. A notable exception were organisms collected inside the lagoons and near the mouths of the Adige and the Po (UNEP/FAO, 1986). PCB residues dominate in all species at all stations regardless of season. In all samples, measurable amounts of DDT, DDE, benzene hexachloride (BHC) and dieldrin were determined, while aldrin was present in minor quantities or in traces.
The concentrations of total PCB, total DDT, total BHC and dieldrin in marine organisms collected off Venice, Ancona and La Spezia far from direct pollution sources are shown in Table 4.7. Residue levels determined in coastal specimens from La Spezia were higher than those from Venice and Ancona.

PCB residues in the striped mullet Mullus barbatus from the Tyrrhenian Sea ranged from 50 to 3,950 µg/kg wet weight, while the mean value was 477 µg/kg wet weight (Geyer et al., 1984). It should be noted that the U.S. Food and Drug Administration (FDA) maximum permissible concentration for PCBs in the edible portion of fish and shellfish is 2 mg/kg wet weight. The above standard was established on October 5, 1979.

Contamination of mussels by polycyclic petroleum hydrocarbons occurred in the Lagoon of Venice (Table 4.8). Mean concentrations are in the range of 12.0-135.1 µg/kg dry weight for 3,4-benzopyrene (BaP) and 1.5-16.9 µg/kg dry weight for perylene (Pe), but values as high as 327 µg/kg dry weight for BaP and 71 µg/kg dry weight for Pe have been measured. The most elevated levels were found in mussels taken from waters adjacent to the Venice industrial port area, within the city of Venice and in an area near the fishing port of Chioggia. The lowest concentrations were encountered in mussels from the Adriatic Sea and in the central basin of the Lagoon of Venice, farthest from industrial and urban areas.

4.1.3. Bathing waters

The total length of the Italian marine coast is 7,162.4 kilometers (not including lagoon zones), of which 4,068.9 is continental shore, 1,244.5 Sicilian and 1,849 Sardinian coast.

An extensive monitoring of the quality of Italian marine coastal waters in 1987 (Ministry of Health, 1988) showed that 86% of the tested samples complied with acceptability limits for bathing prescribed by the Presidential Decree 470 of 1982, which is the law through which EC Directive 76/160 has been adopted in Italy. This percentage is higher than any other value recorded since 1984 (Figure 4.8).

The % noncompliance of marine bathing-water quality of the Mediterranean regions with regard to existing quality requirements are shown in Figures 4.9A-K for the following parameters: total and fecal coliforms; fecal streptococci; Salmonella; pH; color; mineral oils; detergents; phenols; transparency; and oxygen. The analysis of the noncomplying samples clearly shows that the factor limiting the quality of the bathing water is microbiological in nature, especially in Campania and in Calabria. Additionally, Calabria shows noncompliance with acceptability limits for detergents and color. Transparency is very low along the Adriatic Sea (Abruzzi, Marches, Emilia-Romagna and the Veneto). An overview of the marine bathing-water quality for all the coastal regions and provinces in Italy shows a range of acceptability of 40 to 100% (Table 4.9). Caltanisetta has a 40% acceptable quality, while all the coastal provinces have an average of 86%.

An independent investigation of the bacteriological bathing-water quality was conducted by the EC during the period between July 26 and August 1, 1987 (CEC, 1988). The results are shown in Figure 4.10. It should be noticed that only 14 Italian Mediterranean beaches were surveyed. All complied with the EC guide number (recommended criteria limit), except Sorrento (Peters Beach), Maiori and Letoianni which exceeded the EC guide number, but complied with the required standard.

4.1.4.   Eutrophication

Up to the late-1960s, eutrophication was not a significant problem in Italian marine waters. However, in 1978 a number of areas were identified that suffered from eutrophication (Figure 4.11).
The most serious effects of eutrophication are:
1. Change of seawater color and reduction in transparency.
2. Depletion of oxygen in deeper waters and development of toxic algal species leading to death of fish and aquatic animals.
3. Disruption of the equilibrium of the ecosystem, with the most sensitive species being wiped out.

The upper Adriatic Sea has the most severe eutrophication problem. The Lombardy region is the main contributor to this problem (Figure 4.12). The top-four regions (Lombardy,

Emilia-Romagna, Veneto, Piedmont) discharge 43% (25,031 tons per year) of phosphates into the Adriatic Sea. The major sources are domestic discharge and detergents.

Adriatic Sea
The area most acutely affected by eutrophication is the coast of the Emilia-Romagna region where the Po River flows into the Adriatic Sea. The area classified as "hypertrophic" has a length along the coast of 200 kilometers (Figure 4.13). Algal blooms and associated massive deaths of marine organisms are chronic in this area. The dates of the most important "red tides" along the coast of Emilia-Romagna, downstream from the Po outlet, are listed in Table 4.10.
The most serious algal blooms in the Gulf of Trieste were recorded in 1976 and 1977. When red tides are observed, fishing and bathing are generally prohibited for a few weeks in the communities around the affected area. Data collected on the concentration of nutrients show that nitrogen has doubled and phosphorus has quadrupled between 1968 and 1984.

Ionian Sea and Strait of Sicily
"Red tides" are well known in the Gulf of Taranto where they occur with high frequency during the months of June and July in the section named "Mar piccolo di Taranto".

Tyrrhenian Sea
The Tyrrhenian Sea is much less affected by eutrophication as compared to the Adriatic Sea. Algal blooms and fish deaths in the Tyrrhenian Sea have been more episodic, are of shorter duration and are generally restricted to smaller areas. However, the trend shows a rapid deterioration in the near future.
Areas in the Tyrrhenian Sea which are more susceptible to eutrophication include the mouths of the Arno and Tiber rivers, the coast of Pisa, the Bay of Naples and of Palermo and the northwest coast of Sicily.

The waters at the mouth of the Tiber are clearly eutrophic. The discharge of organics and nutrients by the Tiber resulted in an increase in the concentration of various substances in the coastal waters adjacent to the Tiber outfall. However, it has not yet seriously affected the marine ecosystem.

4.1.5. The coastal strip

4.1.5.1. Erosion

The National Research Council studies on coastal erosion found that about 48% of the monitored coasts have regressing shorelines and that only about 6% have advancing shorelines due to soil deposition (Table 4.11). The coastal erosion in the Basilicata region has become dangerous since the entire coastline is subject to regression. Tuscany has the highest percentage of advancing coasts (16%). Even there 43% of its area is regressing. Friuli-Venezia Giulia, Campania and Sicily are the regions with the more stable coasts (97%, 80% and 75% respectively).

Erosion is observed particularly along Adriatic beaches. It also exists along volcanic coasts like Ponza. Intensive erosion has been observed at beaches located near the mouths of the rivers, due to a diminished supply of fluvial sediments to the sea. Many suspended solids are retained behind river dams and accumulate as sediments. Erosion also occurs along coastal strips altered by rapid urbanization and creation of resort areas.

The loss of woodland on sloping coastlines, mainly caused by forest fires, enhances erosion, since rain-water run-off encounters fewer obstacles and is not absorbed by the soil. This results in frequent floods and run-off of contaminants to the seawater.

Land subsidence (geological or induced by excessive groundwater pumping) also damages the coastal strip, particularly in the area of the Lagoon of Venice and nearby Ravenna and Pisa.

The erosive processes are irreversible and will continue. An increased steepness of the bottom slope near the shoreline makes storms more destructive.

4.1.5.2. Destruction and deterioration of forests

Italy has a forested area of 6,384,155 hectares: 3,823,954 hectares (60%) in mountain areas; 2,228,460 hectares (35%) in hilly areas; and 331,155 hectares (5%) in lowland areas (DocTer, 1988).
The continuing existence and the characteristics of national forests are threatened by several factors such as fires, acid rain, pasture expansion, tourism and land development (DocTer, 1988).

Fire
There has been an alarming increase in fires over the past few years. In 1981 more than 14,000 fires devastated some 70,000 hectares of forest. Between 1978 and 1980 43,000 hectares of forest were destroyed by fire per year. Climate and other traditional factors, such as sheep farming, play an important role in this respect. Forests are nowadays also frequented by outsiders who are sometimes unaware of the rules of good behavior towards the environment. The exodus from rural areas has reduced the number of locals who have a direct interest in protecting the forests. The forest destruction by fire due to different causes in 1982 shows the importance of agriculture and other unknown causes (Table 4.12).
Sardinia is the region with the largest amount of damaged area due to forest fires (13,017 hectares), followed by Calabria and Campania which have 6,337 hectares and 5,688 hectares of damaged forest, respectively (Table 4.13). The total amount required to restore the damaged forest areas

is estimated at 44.6 billion lire (Table 4.13).

Deterioration due to acid rain

The first indications of acid rain damage were observed in 1983. A nationwide survey launched by the State Forestry Agency in 1983 showed that approximately 4% (i.e. 300,000 hectares) of woodland area was unhealthy and 1% (i.e. 60,000 hectares) was seriously endangered. The survey also revealed that forest damage of the less industrialized regions, such as Calabria and Abruzzi, was as bad as in the northern regions with the largest concentrations of industry.

Expansion of pasture land

The trend towards intensive livestock farming in many mountain farms raises the problem of compatibility between pasture land and forests. This gave rise to considerable concern at one time. It is still a major problem in southern Italy and the islands, where traditional breeding methods are still in use.

Tourism

The relationship between forests and tourism can become incompatible for a number of reasons. For example, the construction of ski slopes in Alpine and Appennine resorts are on the increase. These slopes are becoming wider and follow less the natural morphology of the land, disfiguring the landscape and leading to the degradation of vegetation, the risk of avalanches and erosion.

Other causes of deterioration

Coastal forest formations (for example, the Bosco della Mesola) have declined due to the reclamation or drainage of marshy areas. The combined action of the influx of tourists, new buildings, roads, sewers and the discharge of pollutants into the sea followed by aerosol formation is damaging the coastal forests (Versilia, Veneto). A related factor is the shrinking of coastlines caused by riverbed

excavation.
Other forest damage has been caused by natural disasters such as infestation by traditional animal and vegetable pests and by very dry summers.

4.1.6. Environmental quality of selected areas

4.1.6.1. The Po basin

The Po and Adige rivers account for nearly one-third of the total freshwater flowing into the Adriatic Sea.
The Po, Italy's major river, is located in northern Italy, which is the most industrialized part of the country. Many towns and industrial plants use the river as a water source and for dumping their waste waters. The river is also important for agriculture in the Po Valley, one of the most productive areas. It has significant influence on the eutrophication of coastal waters of the upper Adriatic Sea, damaging touristic values.

Heavy metals near the Po mouth were detected each year between 1975 and 1980 due to industrial development, as shown in Table 4.14. In particular, the lead and copper concentrations increased significantly during the period 1979-1980; threefold for lead from 0.15 µg/l to 0.55 µg/l and a doubling for copper from 0.45 µg/l to 0.85 µg/l. A high concentration of organic matter as biochemical oxygen demand (BOD) (7.3 mg/l, 1975-1978), indicating organic pollution, was also measured at the Po mouth. Nutrient concentrations at the Po mouth were found to be 1.35 mg N/l for nitrate and 0.23 mg P/l for total phosphorus, respectively, in 1978.

The quality of freshwater for fish life is rather poor throughout the whole basin. The quality at only 28 out of 154 municipalities could support life and reproduction of salmon and carp. The contamination levels in fish, when expressed on the basis of fat content, appear higher in river fish than in fish from the Adriatic Sea, both with regard to total DDT (2.9 mg/kg E.O.M. [extractable organic matter]) and total PCB (21.1 mg/kg E.O.M.). The differences in PCB are even more considerable (Galassi et al., 1981). Accumulation of pesticides and organochlorines in fatty tissues of living organisms was found all along the river course. This reflects the constant use of toxic compounds. Especially below the confluence with the Adda and Lambro River, PCB concentrations exceed the U.S. FDA standard limit of 2 mg/kg wet weight.

Fish collected along the Po contained an average total mercury content of 0.144 mg/kg wet weight and an average methyl mercury content of 0.104 mg/kg wet weight (Figure 4.14).

A similar low quality exists for irrigation and bathing water in the basin. The quality was adequate at only 50 out of 117 municipalities for irrigating the more susceptible crops. Bathing quality was satisfactory at only 6 out of 47 locations.

The bathing quality at the 21 stations, shown in Figure 4.15, is low due to high fecal coliform concentrations; the values frequently exceed the limit of 1,000 fecal coliforms per 100 ml.

Surface waters were below the minimal quality at 96 out of 131 municipalities to produce drinking water. Figure 4.16 shows that only the Crissolo station near the source has quality parameters suitable for drinking-water use. The rest of the stations do not have a sufficient quality.

Groundwater problems also occur, due to infiltration of agricultural chemicals in the river basin. The Ministry of Health had to raise the maximum permissible level of atrazine, molinate and bentazone above 0.1 µg/l for many municipalities.

4.1.6.2.   Lambro, Olona and Seveso basin

In the hydrographic basin of the Lambro, Olona and Seveso rivers, the quality of the environment has deteriorated. These rivers receive waste waters that are only partially treated. The quality of the surface water in much of the basin is such that it is unsuitable for drinking and for irrigation. It is also neither suitable for bathing nor for the survival of fish species such as carp which normally tolerates partially polluted environments.

Drinking-water supplies were contaminated with organo-chlorinated solvents, hexavalent chromium, nitrates or herbicides in 40% of the total number of samples examined.

4.1.6.3.   Adige, Arno and Tiber basins

The Adige River shows an increase in the concentration of heavy metals like lead, cadmium and copper (Table 4.14). Especially cadmium and lead concentrations increased 40-55% per year between 1978 and 1980.
The river has a relatively low level of organic pollution with a mean BOD content of 4.0 mg/l as measured in 1978. Nutrient concentrations for nitrate and total phosphorus were 0.89 mg N/l and 0.19 mg P/l, respectively.

The quality of the Arno River is rather satisfactory in the upper reaches. In the lower part, especially downstream from the city of Florence, the river-water quality deteriorates abruptly. Supersaturated with oxygen near its source, the Arno has an oxygen deficiency near its mouth. In the summer, an oxygen saturation of only 30% is common, due to a limited dilution and slow reaeration process (Pettine et al., 1985).
Nutrient levels at the mouth of the Arno River are 0.99 ± 0.51 mg/l for nitrate and 0.86 ± 0.75 mg/l for total phosphorus (Blundo et al., 1984).
Discharge of approximately 900 tons of surfactants per year

(Blundo et al., 1984) into the river degrades the coastal ecosystem.

The Tiber River has a catchment area of 17,000 km^2. It receives three major tributaries: the Paglia, Nera and Aniene rivers. The Tiber basin has a high industrial and agricultural density.
Upstream from the city of Rome, the dilution and self-purification capacity of the river keeps levels of organic pollution low. Downstream from Rome the inputs of urban discharges and highly polluted waters of the Aniene tributary result in a sharp decrease in the quality of the Tiber with a strong influence on the dissolved oxygen budget.
The Tiber has a high level of organic pollution with a BOD of 9.5 ± 1.0 mg/l at the mouth (Pettine et al., 1985).
Heavy-metal concentrations decreased between 1978 and 1979 as seen in Table 4.14. The area around the Tiber mouth had a cadmium concentration of 0.09 µg/l, which approaches a "minimum risk concentration". The increased population in the Rome area necessitates better control on the discharges of waste water.
Nitrate concentrations increased between 1978 and 1979 from 0.9 to 1.19 mg N/l, while phosphorus orthophosphate remained relatively stable.

PCB concentrations are detected in bottom sediments collected in the coastal area adjacent to the mouths of the Tiber River (Figure 4.17). Most values are in the order of 10-30 µg/kg. The highest values, ranging from 50 to 70 µg/kg, are associated with Tiber River discharges over the past few decades. The contamination appears to be confined to within approximately 10 kilometers of the shoreline and to extend up to about 20 kilometers to the north of Fiumicino. The concentrations are comparable to other Mediterranean urban coastal zones.

4.1.6.4. Province of Naples

Large amounts of untreated pollutants are seriously degrading the environment in the province of Naples. In particular:
- the quality of river water such as the Sarno is closer to that of partially treated waste water than to surface water, with BOD concentrations near the outfalls ranging from 30 to 400 mg/l;
- coastal waters are largely closed to swimming;
- there is widespread fecal pollution of marine sediments in the Bay of Naples. The most contaminated zone of the southern stretch of the bay is located between Torre Annunziata and Castellammare di Stabia (Figures 4.18A and B);
- groundwater is increasingly polluted. Excessive withdrawal causes salt intrusion, while in many areas there are signs of urban pollution;
- the city of Naples is suffering from a chronic shortage of water; drinking water sometimes has unsatisfactory organoleptic characteristics and is colored by the presence of manganese oxide sediment;
- the city of Naples has high levels of atmospheric pollution.
- in the city of Naples and many municipalities along the coast, noise levels are considerably higher than the limits recommended by the Organization for Economic Cooperation and Development and other international organizations.

4.1.7. Damage to fauna and flora

Table 4.15 lists the number of endangered plant and animal species in different Italian regions. Significant extinction of rare species has been noticed especially in the Veneto and in Friuli-Venezia Giulia.

4.2.   Pressures on the environment

4.2.1.   Municipal discharges

The distribution of population along the Italian Mediterranean coast is shown in Figure 4.19. Campania and Latium are the more densely populated regions.
The distribution of the domestic BOD discharges per coastal region is shown in Figure 4.20. The discharge equals 816 x 10^3 tons of BOD per year, with Campania, Latium and Sicily the most contributing regions.
Environmental pressure caused by a growing population increased in all the coastal regions between 1971 and 1981 with the exception of Liguria. The coastal growth is also the result of migration from rural and mountainous areas. An approximately 8% growth was observed in the regions of Campania, Apulia and Sardinia, as shown in Figure 4.21.

4.2.2.   Industrial discharges

Figure 4.22A shows specific coefficients in 29 main industrial sectors (Barbiero et al., 1987) representing the organic load produced per worker. It can be seen that the sugar-beverages-tobacco sector as well as the coal industry have the highest coefficients.
Based on the above indices, industrial BOD discharges were quantified as shown in Figure 4.22B. Food products and chemicals are the largest contributors to the pollution. Not all the load is discharged, as treatment units remove major portions.

The distribution of the overall industrial BOD discharges along the Italian coast is presented in Figure 4.23. The maximum potential load released into the Mediterranean was calculated at 158.5 x 10^4 tons in 1981. Tuscany, Latium, Emilia-Romagna and Veneto regions situated in the north in the most industrialized part of the country contribute the most.

Figure 4.24 shows the change in organic pollution due to industrial activity between 1971 and 1981. In Molise the pollution doubled while in the Veneto a 17% reduction was implemented, although it is still one of the most polluting coastal regions.

4.2.3.   Maritime discharges

There are two main sources of oil release: industrial trade activities and offshore drilling. A large number of harbors are used as oil terminals, and significant amounts of hydrocarbons are discharged accidentally, affecting the coasts. Genoa, Augusta, Trieste and Venice are the main oil terminals in Italy. On the basis of transported amounts of oil products a vulnerability map of the marine coasts was drawn (Figure 4.25). Accidental oil releases registered between 1977 and 1985 amount to approximately 620 tons (390 tons crude, 95 tons hydrocarbons, 38 tons gasoline and 97 tons organic carbon).
Drilling activities in the Mediterranean Sea are largely carried out on the Italian continental platform (Figure 4.26).

4.2.4.   River discharges

Among the 16 Italian rivers discharging a load of 557×10^3 tons of BOD per year into the Mediterranean, the Po (258×10^3 tons of BOD per year), the Tiber (109×10^3 tons of BOD per year) and the Adige (38×10^3 tons of BOD per year) are the major polluting ones, as derived from a literature survey by the United Nations Environment Programme (UNEP) in 1984 (Figure 4.27). Similar results were found by recent measurements of the Arno, Tiber, Metauro, Po and Adige rivers by the EC in 1987 (Figure 4.28A). The load from these 5 rivers was equal to 223×10^3 tons of BOD per year as compared to 557×10^3 tons of BOD per year in the UNEP study. The monthly river BOD discharges are shown in Figure 4.28B. The Po discharge reaches its maximum during the high-water period in May. The BOD discharge of the 5 rivers shows a 13% increase between 1981 and 1986 (Figure 4.29).

The nutrient load discharged into the Mediterranean Sea by the four major Italian rivers (the Po, the Tiber, the Adige and the Arno) is estimated at 114,300 tons of inorganic nitrogen per year and 17,200 tons of total phosphorus per year (Pettine et al., 1985).

Figure 4.30 shows the release of phosphorus among the 7 Italian Mediterranean basins. It appears that the upper Adriatic basin receives by far the largest amount of phosphorus (about 48% of the total). The largest pollution loads are carried by the Po and Adige (Adriatic Sea) and by the Arno and Tiber rivers (Tyrrhenian Sea). The Po and Adige rivers account for nearly one-third of the total freshwater flowing into the Adriatic Sea, while the Tiber accounts for about 22% of the total freshwater input into the Tyrrhenian Sea.

Metals discharged by the Po River into the Adriatic Sea have been estimated in tons per year as follows: arsenic: 243; copper: 1,554; chromium(total): 944; lead(total): 1,312; zinc: 2,646; nickel: 89; and mercury: 65 (Marchetti, 1986). It was estimated that about 30% of the fluxes originates from natural sources.

4.2.5. Pressures in selected areas

4.2.5.1. The Po basin

Many sources contribute to the organic waste discharged into the Po basin. Approximately 13% is of domestic origin, 43% comes from the manufacturing sector and 44% has a zootechnical source. Some 22% of this organic load is discharged into the surface waters of the basin totally untreated (Table 4.16).

Phosphorus and nitrogen quantities (in tons per year) generated and used in the Po basin are as follows:

- nitrogen: 113,410 (domestic and manufacturing sectors), 363,770 (zootechnical) and 280,380 (fertilizers);
- phosphorus: 16,200 (domestic and manufacturing sectors), 60,690 (zootechnical) and 84,470 (fertilizers) (Table 4.17).

The corresponding nutrient-derived waste loads (in tons per year) released into the surface waters of the Po basin are:
- nitrogen: 243,630 (32% of generated load)
- phosphorus: 23,050 (14% of generated load)

The above values are attributed to untreated proportions of domestic and manufacturing waste, zootechnical waste, fertilizers leached from the soil, and run-off from uncultivated land (Table 4.17).

The use of herbicides in the Po basin has grown steadily over recent years. The application of fungicides and insecticides is concentrated in fruit and wine growing areas. Phytoregulating substances are employed mostly for fruit and market garden crops. The latter crops, in particular, are grown on permeable soil causing groundwater contamination by these substances.

The Po Valley area accounts for more than 50% of all industrial waste produced in Italy, or about 25 million tons per year. According to the latest studies (Ministry of the Environment), the aggregate waste processing capacity only suffices to cope with some 25% of the above waste (1.25 million tons per year), indicating that the lack of disposal of this waste from the Po Valley is reaching a crisis point. The 25% is currently treated by a multipurpose public installation operated by the AMIU at Modena and by several private disposal plants for industrial wastes in Mantova, Ferrara, Ravenna, Pavia and Turin.

4.2.5.2. Basin of Lambro, Olona and Seveso rivers

This area is part of the Po basin and includes 381 municipalities. Although it covers 5% of the total Po basin area, its population is 30% of the total basin population. The area's population density is six times the average for the Po basin. It also accounts for about one quarter of the total industrial activity within the entire Po basin.

At present 14% of the population in the area is without sewerage or is served by inefficient sewage disposal systems. Sewage sludge is commonly disposed of on the soil and in the subsoil, with the result that critical situations have been created in the southern regions of the province of Varese and in certain parts of the province of Milan.
Each year the area produces 1,770,000 tons of urban solid waste and other waste products, more than 910,000 tons of sewage sludge, and about 1,790,000 tons of special wastes of which about 800,000 tons can be classified as toxic and noxious. There is a serious shortage of adequate facilities and/or sites for the controlled disposal of refuse.
Annual emissions into the atmosphere amount to around 220,000 tons of sulfur dioxide, 42,000 tons of nitrogen oxides and 850,000 tons of dust.

4.2.5.3. Province of Naples

Only 38% of the organic waste produced in the territory of the province of Naples is adequately treated, while the remainder (5.3 million population equivalents) is discharged into the territory's surface waters untreated. A number of factors (regulatory, administrative and financial) have delayed pollution abatement according to the Special Project No. 3 "Reduction of pollution in the Bay of Naples" of the Cassa per il Mezzogiorno, which now has been disbanded. The above project involves the installation of main collector drains, wastewater treatment installations and systems for disposing of the treated waste water. However, it does not cover sewerage networks on which much work remains to be done, especially in the city of Naples.
Each year, the province of Naples produces more than

1,700,000 tons of degradable urban solid waste and sludge, and more than 910,000 tons of special waste, including around 440,000 tons which can be classified as toxic or hazardous; at the same time, there is a serious lack of facilities to treat and dispose of the waste. About 40 uncontrolled dumps for urban solid waste and/or industrial waste have been identified for which rehabilitation programs have not yet been initiated or financed.

The province of Naples is also subject to heavy emissions of pollutants into its air. Annual inputs into the air amount to around 29,000 tons of sulfur dioxide, 26,000 tons of nitrogen oxides, 6,200 tons of dust and 152,000 tons of carbon monoxide.

4.2.5.4. Lagoon of Venice

The basin permanently discharging into the lagoon covers 185,000 hectares comprising part of the provinces of Venice, Treviso and Padua.

High loads of nutrients are being discharged into the Lagoon of Venice from the surrounding basin (Table 4.18). They are estimated at up to 10,000 tons or more per year of nitrogen and more than 2,000 tons per year of phosphorus, most of it from agriculture and livestock farming (about 50% of the nitrogen and 25% of the phosphorus) and domestic sources (about 40% of nitrogen and 65% of phosphorus).

The continuing uncontrolled industrial discharge and urban and agricultural pollution have converted the sediment of the lagoon into a highly polluting oxygen-deficient sponge like layer. This sediment is able to release phosphorus and nitrogen. Experimental studies suggest that this is responsible for 50% of the concentration of nutrients in the above water layer.

Large nutrient loads contribute to eutrophication in the lagoon promoting proliferation of algae which, in turn, extends the anaerobic zones in which the release of nutrients from sediment is accelerated. This starts a vicious circle leading to uninterrupted production of algae of close to a million tons per year. The present removal capability of about 500 tons of algae per day is insufficient to harvest the production.

The amount and type of solid wastes produced around the Lagoon of Venice are summarized in Table 4.19. At the present, only a part of these wastes is properly disposed of.

4.3.   Economic impacts

4.3.1.   Health impacts

A low environmental quality can have a negative influence on the health of the population. This can result in increasing expenditures for health services and economic losses due to lower productivity.
A study published in ISVET Document No. 28 ("Inquinamento e salute umana," Rome, 1970) has calculated the main economic damages due to the harmful effects of air and water pollution on human health at 84.2 billion lire in 1966. The economic costs calculated for each damage category were as follows (data in million lire):

Costs for workers' premature deaths	19,258
Costs for housewives' premature deaths	2,980
Loss of wages through illness	20,100
Loss of housewives' working time through illness	8,330
Treatment expenses	33,563
Total cost	84,231

The estimated damage (84.2 billion lire) though impressive must certainly be considered lower than the real damage, due to the difficulty of putting a reasonable monetary value on some damages as those resulting from physical and psychological distress, diseases not definitely linked with pollution, and ailments as well as prevention and treatment expenses paid privately. An estimate of the possible future damages was made. The approximate extent of the damage was 102 billion lire in 1968, 130 in 1970, 200 in 1975, 309 in 1980 and 490 in 1985.

4.3.2. Tourism

Tourism is a substantial part of the Italian national economy. A lot of coastal regions are very dependent upon the income from tourism. Environmental problems can lead to severe financial problems and unemployment, as international tourists tend to stay away from areas that receive a negative publicity. Especially health problems due to contamination can become "worldnews" very quickly, with serious consequences for the region.
For example, it has been reported that the explosion at the Farmoplant stationary plant in Massa Carrara (July 17, 1988) has already cost 20 billion lire due to cancelled reservations and tourists leaving the area.
A study published in ISVET Document No. 31 ("Inquinamento, turismo e tempo libero," Rome, 1970) has estimated the economic damages to seaside tourism, lakeside tourism and to sport fishing in 1986. The above damage estimates have been as follows (in 1986 billion lire):

Year	1968	1970	1975	1980	1985
Seaside tourism (internal tourists):	46.6	50.5	60.3	71.8	86.8
Lakeside tourism (internal tourists):	12.5	13.7	17.3	21.9	27.6
Sport fishing (in inland waters):	2.9	3.1	3.8	4.8	6.0

Negative effects on tourism can be more pronounced in cases where coastal water pollution and air pollution damage historic sites.

4.3.3. Historic sites and natural landscapes

International tourists come to Italy because of its mild climate and the beauty of its sea, monuments, historic sites and natural landscapes. A degradation of Italy's rich cultural heritage as well as its natural environment can lead to less income from tourism.
The damage caused by air pollution to historic sites has also direct economic consequences. Measures have to be taken and funds have to be raised for the protection and restoration of historic sites, palaces and monuments damaged by smog and acid rain.

A study published in ISVET Document No. 29 ("Inquinamento e patrimonio dei beni culturali," Rome, 1970) has calculated the economic damage done by air pollutants to cultural assets in Italy. As it was impossible to assess the value of the assets damaged by pollution (and consequently to quantify the relative economic damage), an indirect monetary evaluation was used.
The benefits that could have resulted in this sector from pollution elimination were given by the difference between the expenditure theoretically necessary for the restoration and preventive conservation of cultural assets in the present pollution situation, and the (smaller) expenditure necessary in a hypothetical situation of nonpollution.
A minimum estimate of the economic damage done by air pollution in 1968 was 36 billion lire, including 27 billion related to archeological remains, monuments and medieval and modern works of art and 9 to museums, art galleries, libraries, archives and churches. An estimate of subsequent damage was 42-43 billion lire in 1970, 62-67 in 1975, 91-103 in 1980 and 134-158 in 1985 (still in 1968 lire).

4.3.4. Marine production

The fishing productivity in the different Italian coastal regions (including Sicily and Sardinia) was estimated at

430,630 quintals in 1985. Of the above catch 305,130 quintals were fish, 93,400 were molluscs and 32,100 were crustaceans. Five regions (i.e. Veneto, Emilia-Romagna, Marches, Apulia and Sicily) account for about 72% of the total production.

The enrichment of the sea with nutrients has increased productivity, resulting in more catches. An overenrichment, however, can give rise to algal blooms and depletion of oxygen after algal die-offs, resulting in the deaths of fish and benthic organisms.

In Emilia-Romagna it was estimated that about 20 billion lire were lost due to algal blooms and consequent deaths of marine organisms in the Lagoon of Goro in 1987. The total costs attributed to this factor are 20.2 billion lire (Table 4.20). A new algal bloom occurred at the beginning of August 1988 causing an estimated 1.5 billion lire damage.

The municipality of Venice has estimated that 3.7 billion lire were needed in 1988 to remove algae from the area of the lagoon most heavily affected, and to dispose of them on agricultural soil.

4.3.5.   Agriculture and zootechnics

A study published in ISVET Document No. 30 ("Inquinamento e agricultura," Rome, 1970) made a minimum estimate of the economic damage caused by pollution in 1968 of 12-13 billion lire in terms of net product. The analysis was restricted to the effects of water pollution in the most productive irrigation areas of the plains. The territory considered, represents about 12% of the total agrarian area (2,620,000 hectares) and 50% of the gross national sellable produce (2,850 billion lire). Nearly all the territory affected by pollution is in northern Italy (95% of the territorial area), especially in the northwest (62%). In central Italy there is pollution damage in 3% of the total irrigated plain area.

Later economic damages to farming were estimated at 13.0-

13.5 billion lire in 1970, 17.0-18.0 in 1975, 25.0-30.0 in 1980 and 35.0-46.0 in 1985 (in 1968 lire).

4.4.   The institutional and legal framework

The institutional and legal framework through which environmental issues are dealt with is highly complex. Environmental planning and legislation are duties of the central and regional Governments. There are more than 600 national laws and decrees, and 300 regional ones.
Environmental matters are largely delegated to Regions. Environment-related regional laws generally amount to about 2% of the total number of regional laws. Most of the regional laws deal with control and reduction of environmental pollution (Table 4.21).

The administration of environmental matters is carried out in Italy at 4 levels (i.e. central, regional, provincial and local). At all of the above-mentioned levels there are a number of agencies that are responsible for some environmental aspects. At the central level there are 11 ministries active in the environmental sector. A similar dispersion of competence applies to regional Governments where, in some cases, no specific responsibility for the environment exists.

The situation is more structured and unified at the provincial level and in big cities. Departments ("assessorati") dealing with environmental matters exist in more than half of the Italian provinces and almost in all Italian cities with more than 1 million inhabitants.

4.4.1.   Environmental authorities

A major institutional change, recently enacted in Italy, is the establishment of the Ministry of the Environment with portfolio (Law 8 July 1986, No. 349, and Law 3 March 1987,

No. 59). The Ministry is charged with environmental protection and planning. It plays a central role in the areas of pollution prevention, environmental reclamation, environmental impact assessment and nature conservation.
Due to the interdisciplinary and intersectorial character of environmental issues, the Ministry operates in most cases in cooperation with other ministries.

The Ministry of the Environment has a leading role to oversee the implementation of existing laws for surface-water protection, waste disposal and atmospheric pollution. The Ministry is also responsible for issuing authorizations for sea dumping. Furthermore, it collaborates with the Ministry of the Mercantile Marine as far as marine nature reserves and other aspects of coastal protection are concerned.

Law 319 of 1976 (known as the "Merli Law") provides the State, the Regions and other local bodies with a combination of planning instruments and activities for the control and treatment of waste water. Article 2 (d) of this Law allows the State to set general criteria for a correct and rational use of water for industrial and domestic uses. Within the "Merli Law" problems may arise due to the lack of integration of the regional water-treatment planning and urban planning. Very often the agencies that manage aqueducts are different from those which manage sewerage and/or purification systems. Moreover, there are usually a number of different bodies in charge of aqueducts or sewers or sewage treatment in different parts of the same basin.

The water-use planning is distinct from and not effectively coordinated with plans for protecting waters from pollution both at central, regional, provincial or municipality level.

4.4.2. The adoption of EC directives

Since Italy joined the EC, the Government has been adopting EC directives according to Article 76 of the Constitution:

a) **Quality of bathing waters**
The EC Directive 76/160, which applies both to inland and marine coastal waters, was "partially" adopted in Italy. There are still significant differences between Community regulations and the Italian regulations on bathing water.

b) **Quality of surface water intended for treatment in order to produce drinking water**
The EC Directives 440/1975 and 869/1979 have been adopted in Italy, but only partially implemented.

c) **Quality of water intended for human consumption**
The EC Directive 778/1980 has been adopted in Italy.

d) **Quality of freshwater requiring protection or improvement to be suitable for fish life**
The EC Directive 659/1978 has not yet been adopted by Italy.

e) **Quality of shellfish waters**
The EC Directive 923/1979 has been partially introduced. The Italian legislation, unlike the EC Directive, has a strict hygienic and health-oriented nature and does not contain provisions for water-reclamation programs.

The approach to control water pollution adopted by the "Merli Law" and subsequent modifications is quite different from the one on which the relevant EC directives are based. There are 10 EC directives on pollution caused by the discharge of dangerous substances into the aquatic environment. With the exception of the Directive 280/86, recently adopted in Italy, the adoption of all other directives, including 76/464, 82/176, 83/513, 84/156 and 84/491, is still pending. Similar problems exist also with the EC Directive 68/1980 on the protection of groundwater against pollution caused by certain dangerous substances.

This unfavorable situation is likely to continue until the "Merli Law" is changed. Law 319/1976 establishes maximum admissible concentrations for substances discharged into water, but not (as provided for by the EC directives) maximum quantities.

Law 319/1976 regulates discharges into the soil and subsoil, and states that these are permitted when there is no damage to the groundwater and water-bearing strata.

4.4.3. Marine waters

a) The provisions contained in the Navigational Code prohibit acts (rather than pollution in the technical sense) concerning fouling and deposition of objects into the coastal waters.

b) The legislation on fishing is also useful for the protection of the coastal environment. The protection of the marine biological heritage is carried out through fishing restrictions and prohibitions. The capture of cetaceans, turtles and sturgeons has been prohibited in Italy.

c) Provisions exist for controlling discharges into the sea or coastal activities that pollute marine waters.

d) Specific limitations on human activities can be imposed inside marine reserves (Table 4.22).

e) Article 2 of Law 613 of 1967 governs the State right to explore the continental shelf and to exploit its resources. It establishes that such activities must be carried out in such a way as to avoid unjustified restrictions to navigational freedom, fishing, conservation of the biological resources of the sea or other uses of the open sea according to international law as well as the conservation of the coast, beaches and ports.

The international conventions (on environmental matters) ratified by Italy are shown in Table 4.23.

4.5. Past and present trends in environmental expenditures

Between 1982 and 1987 global environmental expenditures in Italy have been rather stable, corresponding to about 1% of the Gross National Product. This is similar to percentages reported for other European countries. About half of these expenditures comes from the central (State) and regional governments and the other half from the local governments and private sectors. Average per capita environmental expenditure, through State and regions, has increased from 40,000 in 1981 to about 80,000 Italian lire for year 1983 (Figure 4.31).

4.5.1. Environmental expenditures by the central government

Environmental expenditures by the central government reached 2,500 billion lire in 1985 as shown in Figure 4.32. About half has been devoted to mitigate natural disasters (earthquakes, floods and erosion). The other half has been channelled through the Investment and Employment Fund (FIO), which is partly supported by the European Investment Bank.

The areas financed by the FIO include water protection and treatment (i.e. sewerage, purification and reclamation), solid-waste disposal and other items, such as reforestation, hydraulic works, soil protection and environmental protection. A total of 2,957 billion lire was provided by FIO to the regions in 1986-1988 (Table 4.24). About half of the FIO resources were allocated for water protection and treatment in 1986-1988 (Table 4.25).

The Law 441 of October 1987 is the first Italian Law providing loans for improvement of municipal solid-waste disposal systems. Article 1 of this Law provides 275 billion lire for already approved but not financed projects; 625 billion lire for improving existing waste disposal plants; 425 billion lire for building new plants.

The Law 119 of March 1987 has allocated 130 billion lire in 1988, 130 in 1989 and 100 in 1990 for improving the treatment systems of discharges from oil presses.

Article 18 of the Law 67 of March 1988 has allocated 830 billion lire for environmental expenditures. Additional resources (i.e. 200 billion lire) are provided by Article 17 for drinking-water supply and protection and treatment of surface waters in areas in the Po basin threatened by environmental crisis.

The environmental expenditures by the central government in 1988 total 4,166 billion lire as allocated and authorized through FIO from 1986 to 1988, by the Law 441 of 1987 and the Law 67 of 1988 (Table 4.26).

4.5.2. Expenditures by regional governments

Due to lack of uniformity and clarity, the analysis of regional budgets is more difficult than that of ministerial budgets. Since 1982 regional expenditures for the environment are rather stable. They amount to about 1,500 billion Italian lire per year, excluding FIO contributions. Environmental expenditures account, on average, for about

4% of the total regional expenditures and are primarily used for routine interventions and maintenance. Total regional environmental expenditures reached 2,232 billion lire in 1984 (Table 4.27). Gerelli et al. (1986) concluded that regional expenditures are higher in tourist regions and in regions where federal expenditures are higher. No clearcut relationship could be shown between environmental expenditures and level of industrialization or population density.

4.6. State of the environment: projections

Projections concerning a number of parameters (i.e. demography, agro-food industry, manufacturing, energy, tourism, transportation and pollution loads) to the year 2000 and 2025 have been estimated in the framework of the UNEP "Blue Plan", phase II, according to the three following main scenarios:

a) Aggravated continuing scenario, based on the continuation of current trends with increased difficulties due to economic recession in the Mediterranean basin;

b) Moderate continuing scenario, characterized by significant economic development and political cooperation in the Mediterranean basin;

c) Aggregated alternative scenario, characterized by a fully satisfactory economic and institutional cooperation among the countries of the Mediterranean basin.

4.6.1. Demography

The estimated population in Italy in the year 2000 and 2025 will not be very different from the present. The foreseen increase in population, living in cities and in coastal provinces, is rather limited. Average life expectancy will continue to increase and birth rates will decrease. By the year 2000, about one-third of the Italian population is likely to be over 60 years of age.

4.6.2. Agro-food and manufacturing industry

The more pessimistic scenario (i.e. the aggravated continuing one) foresees a significant increase in nonproductive surface due to urbanization and related factors (about 20% of the total for the intermediate future compared with 12.4% in 1985). The above scenario foresees a reduction in food production.

In the case of the aggregated alternative scenario, nonproductive surface will not exceed 20% in the year 2025. Food production is expected to increase remarkably, but the agro-alimentary deficit will remain stable.

As indicated in Table 4.28, projections concerning the manufacturing industry vary remarkably according to the scenario considered. In the case of the aggravated continuing scenario, a considerable reduction is foreseen of both number of production units and of workers. In the case of the aggregated alternative scenario, the number of production units increases significantly, whereas that of workers tends to slightly decrease in the medium and long term.

4.6.3. Tourism and transportation

Only the aggravated continuing scenario foresees a decrease in total and coastal receptive structures (including hotels, vacation homes and guest accommodations) by the years 2000 and 2025. The two other scenarios indicate a significant increase (Table 4.29). All three scenarios expect arrivals to increase and duration of average stay to decrease, depending on the scenario (Table 4.30).

According to the aggravated continuing scenario transportation by road is likely to increase, whereas all other transportation systems should remain stable. The aggregated alternative scenario predicts the further development of transportation by railway, sea and air with an increase in transported goods with a high added value (Table 4.31).

4.6.4. Liquid and solid pollution loads

Italy produced 2,549 kilotons of BOD in 1985. According to the moderate continuing scenario this will increase to 3,246 kilotons in 2000 and 6,731 kilotons in 2025. The treatment efficiency for domestic waste waters will increase from the present 31% to 76% in 2000 and 100% in 2025. The production of domestic solid waste and industrial wastes will increase from 121.3 million tons in 1985 to 201.6 million in 2000 and 546.9 million in 2025.
The aggravated continuing scenario expects a lower production of BOD and chemical oxygen demand, but sewage treatment will only cover about 62% of the need by the year 2000 and about 84% by 2025.
According to the aggregated alternative scenario the treatment efficiency is likely to fulfill the demand much earlier than 2025.

4.7. Estimate of investment requirements and further operational costs

4.7.1. The Lambro, Olona and Seveso basin

On October 1, 1987, governmental authorities declared the basin of the Lambro, Olona and Seveso rivers an area of high environmental risk. As a result, the Minister for the Environment had to prepare a plan for pollution abatement to rehabilitate the basin.

Section I of the Plan provides for the precise identification of the aims of the actions and regulatory measures for eliminating pollution. Section II of the Plan defines the proposed measures and the financing required for their implementation. It provides for measures costing 4,800 billion lire (Table 4.32). The cost of the measures will be spread over the five-year period 1988-1992 (Table 4.33).

The operating costs proposed under the Plan amount to 245 billion lire (column 4/1), while the principal and interest payments amount to 310 billion lire per year (Table 4.34).

4.7.2.   The Po basin

The regions of Emilia-Romagna, Lombardy, Piedmont and Veneto developed proposals within the framework of the Interregional Conference for the reclamation of the Po basin. So far, a total of around 400 proposals have been examined by the Committee requiring a financing for a total of 5,000 billion lire.

These proposals comprise:
- 182 projects for installation of wastewater treatment plants for a total of more than 2,100 billion lire;
- 85 projects for aqueduct works (1,869 billion);
- 18 projects for the purification of aquifers (118 billion);
- 35 projects for cleanup of soils (53 billion);
- 9 projects for studies, experiments and measures to remedy environmental damage (100 billion);
- 33 projects concerning monitoring (43 billion);
- 15 projects concerning extending sewers.

4.7.3.   Province of Naples

On February 26, 1987, the Territory of the Province of Naples was declared an area of high environmental risk. The Minister for the Environment had to prepare a plan for pollution abatement to rehabilitate the province's territory.

Measures and financing requirements are defined by the Plan for a total amount of 1,027 billion lire, to be borne entirely by the State (Table 4.35).

4.7.4. The Lagoon of Venice

The overall reclamation cost of the lagoon is 3,043 billion lire. The distribution of the costs according to specific objectives and years is shown in Table 4.36.

4.7.5. Water treatment in the rest of Italy

The financial needs for water treatment in the other regions of Italy not discussed in the previous sections are estimated at 13,040 billion lire (Table 4.37).
The figures provided in Table 4.37 should be increased by 30% in order to take into account, among others, the inflation rate that has been very significant in recent years in Italy.

In conclusion, it is estimated that approximately 15,000 billion lire are needed to cope with sewerage and purification systems in the rest of Italy. Financial resources are needed over the next 5 years at a rate of about 3,000 billion per year.

4.7.6. Solid-waste disposal

The estimates shown in Table 4.38 have been produced by the Ministry of the Environment in July 1987. A major expenditure in this sector is for the reclamation of abandoned dumps. As many as 6,000 of such dumps have been identified.

It is well known that Italy cannot dispose of about one-fifth of the total industrial wastes it is producing. The cost of building 8 multifunctional disposal centers for industrial waste and wastewater treatment plant sludge has been estimated at 1,100 billion lire for the entire Mezzogiorno area. The same amount is necessary for constructing similar multifunctional centers in the rest of Italy.

Problems could arise in finding and training the estimated

2,000 and more workers for operating these platforms, especially managers and technical staff.

4.7.7. Soil conservation

According to the analysis of the De Marchi report (1969), an investment of 74,000 billion lire is needed over 30 years to improve the entire country's hydrogeological situation.
The urgency is illustrated by the about 40 regional applications that cost FIO approximately 1,400 billion lire to finance water control projects.

4.7.8. Conclusion

Table 4.39 provides an overview of funds needed for the environmental sectors in 1989-1992. More than 34,000 billion additional lire need to be added to the environmental budget normally transferred to the regions. The State is expected to provide 50% of the required amount, whereas the rest could be paid by the local authorities through the imposition of tariffs. Table 4.39 does not include the interventions necessary to improve the seismic resistance of buildings. Although it is not possible to offer reliable figures for these interventions, estimates range from 100,000 to 200,000 billion Italian lire.

References

BALDI, F., The biogeochemical cycle of mercury in the Tyrrhenian Sea. In: Papers presented at the FAO/UNEP/WHO/IOC/IAEA Meeting on the biogeochemical cycle of mercury in the Mediterranean. Siena, Italy, 27-31 August 1984. FAO Fisheries Report No. 325 Supplement, pp. 29-43. FAO, Rome 1986.
BALDI, F. and BARGAGLI, R., Chemical leaching and specific surface area measurements of marine sediments in the evaluation of mercury contamination near cinnabar deposits. Mar. Environ. Res., 1982, 6, 69-82.
BALDI, F., BARGAGLI, R., FOCARDI, S. and FOSSI, C., Mercury and chlorinated hydrocarbons in sediments from the Bay of Naples and adjacent marine areas. Mar. Pollut. Bull., 1983, 14, 108-111.

BARBIERO, G., CICIONI, Gb. and SPAZIANI, F. M., Un sistema informativo per la valutazione dell'inquinamento potenziale. Un'applicazione alle regioni ed alle provincie italiane. CNR-Istituto di Ricerca sulle Acque, Rome 1987.

BARGAGLI, R., BALDI, F. and LEONZIO, C., Trace metal assessment in sediment, molluscs and reed leaves in the Bay of Follonica (Italy). Marine Environ. Res., 1985, 16, 281-300.

BENETTI, E., BERNARDINI, A., PAOLONI, G., MAZZARINI, L., SAVINI, S. and COMUNIAN, E., Contenuto in metalli pesanti delle acque di mare e dei sedimenti delle coste della provincia di Ancona. Nota I. Ricerca del rame, zinco, piombo, nichel, cadmio e mercurio. Inquinamento, 1979, 21, 31-38.

BLUNDO, C., LORETI, L., PAGNOTTA, R., PETTINE, M. and PUDDU, A., Physico-chemical investigation in the coastal area between the Arno and Serchio rivers (Tuscany, Italy). VIIes Journ. Étud. Pollut. CIESM. Lucerne 1984.

BREDER, R., Distribution of heavy metals in Ligurian and Tyrrhenian coastal waters. Sci. Total Environ., 1987, 60, 197-212.

CASTAGNA, A., SINATRA, F., CASTAGNA, G., STOLI, A. and ZAFARANA, S., Trace element evaluations in marine organisms. Mar. Pollut. Bull., 1985, 16, 416-419.

CEC, 26.7.1987-1.8.1987 plages propres. Résultats de la campagne de sensibilisation. Brussels 1988.

DE SIMONE, R., GALASSO, M. and PROTA, G., PCBs, pesticidi clorurati ed altri microinquinanti organici nei sedimenti del Golfo di Pozzuoli. Convegno delle unità operative afferenti ai sottoprogetti. Risorse biologiche e inquinamento marino. P.F. Oceanografia e Fondi Marini (Roma, 10-11 Novembre 1981). Rome 1982.

DOCTER, European environmental yearbook 1987. 1988. DocTer International U.K., London.

FOSSATO, V. U., NASCI, C. and DOLCI, F., 3,4-Benzopyrene and perylene in mussels, Mytilus sp., from the Laguna Veneta, north-east Italy. 1979. Marine Environ. Res. (2), 47-53.

FRASCARI, F., FRIGNANI, M., GUERZONI, S. and RAVAIOLI, M., Sediments and pollution in the northern Adriatic Sea. 1988. Ann. N. Y. Acad. Sciences, pp. 1000-1020.

FRIGNANI, M., RAVAIOLI, M., BOPP, R. F. and SIMPSON, H. J., Contribution to the study of pollution and recent marine sedimentation in front of the Tiber river. Atti 6° Congresso A.I.O.L., pp. 329-336. Livorno, 12-14 April 1984.

GALASSI, S., GANDOLFI, G. and PACCHETTI, G., Chlorinated hydrocarbons in fish from the River Po (Italy). Sci. Total Environ., 1981, 20, 231-240.

GERELLI, E., CELLERINO, R. and GHESSI, G., Quanto spendiamo per l'ambiente. Ricerca, 1986, 69-82.

GESAMP Working Group 26, Review of the state of the Mediterranean environment. UNEP, Athens 1987.

GEYER, H., FREITAG, D. and KORTE, F., Polychlorinated biphenyls (PCBs) in the marine environment, particularly in the Mediterranean. Ecotoxicol. Environ. Saf., 1984, 8, 129-151.

ISTAT, Statistiche ambientali. Volume I. Rome 1984.

ISVET Document No. 28, Inquinamento e salute umana. Rome 1970

ISVET Document No. 29, Inquinamento e patrimonio dei beni culturali. Rome 1970.

ISVET Document no.30, Inquinamento e agricultura. Rome 1970.

ISVET Document no.31, Inquinamento, turismo e tempo libero. Rome 1970.

LEONZIO, C., BACCI, E., FOCARDI, S. and RENZONI, A., Heavy metals in organisms from the northern Tyrrhenian Sea. Sci. Total Environ., 1981, 20, 131-146.

MARCHETTI, R., L'inquinamento del Po e dei suoi principali affluenti: effetti locali e conseguenze per l'Adriatico. Ingegneria Ambientale, 1986, 15, 37-40.

MARCHETTI R. and PASSINO, R., Strategie per interventi integrati nel settore della eutrofizzazione. Convegno sulla eutrofizzazione in Italia. P.F. "Promozione della Qualità dell'Ambiente" AC/2/45-70. Roma, 3-4 ottobre 1978. Rome 1979.

MARCHETTI, R., SAROGLIA, M., TIBALDI, E. and GRAY, R. H., Overview of water quality problems in Italy. In: Toxic substances in the aquatic environment: an international aspect (eds. Mehrle Jr., P. M., Gray, R. H. and Kendall, R. L.), pp. 48-57. Water Quality Section of the American Fisheries Society: Bethesda, Maryland, U.S.A. 1985.

MINISTRY OF HEALTH, Rapporto sulla qualità delle acque di balneazione (D.P.R. 8 giugno 1982, n. 470). Anno 1987. Rome, March 1988.

MOJO, L., MARTELLA, S. and MARTINO, G., Controllo stagionale dei metalli pesanti (Hg, Cd, Pb) in alcuni organismi marini del Mediterraneo centrale. Primi Risultati. Convegno scientifico nazionale. P.F. Oceanografia e Fondi Marini (Roma 5-7 Marzo 1979). Volume II. 1979, 913-924.

PETTINE, M., LA NOCE, T., PAGNOTTA, R. and PUDDU, A., Organic and throphic load of major Italian rivers. Mitt. Geol.-Paläont. Inst. Univ. Hamburg. 1985. 58, 417-429.

UNEP/FAO, Baseline studies and monitoring of DDT, PCBs and other chlorinated hydrocarbons in marine organisms (MED POL III). MAP Technical Reports Series No. 3. UNEP, Athens 1986.

UNEP/UNESCO/FAO, Eutrophication in the Mediterranean Sea: receiving capacity and monitoring of long term effects. MAP Technical Reports Series No. 21. UNEP, Athens 1988.

VIEL, M., DE ROSA, S., FERRETTI, O., DAMIANI, V. and ZURLINI, G., Coastal marine environment research on continental shelf of Apulian region (South Italy): a preliminary study. 5th International Conference on Heavy metals in the Environment. Vol. 2, pp. 411-414. Athens 1985.

VOLTERRA, L., TOSTI, E., VERO, A. and IZZO, G., Microbiological pollution of marine sediments in the southern stretch of the Gulf of Naples. Water, Air, and Soil Pollut., 1985, 26, 175-184.

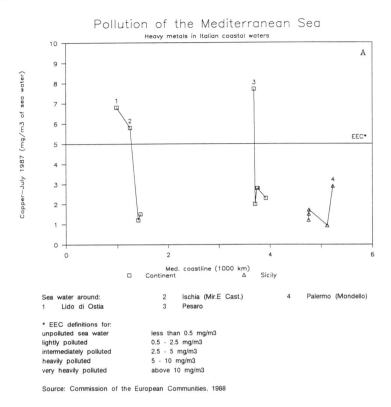

Fig. 4.1   Metal pollution along the Italian coast: (A) Copper, (B) Lead, (C) Zinc

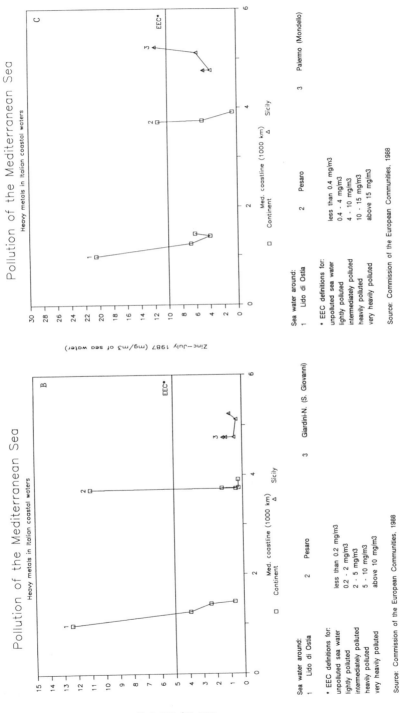

Fig.4.2A

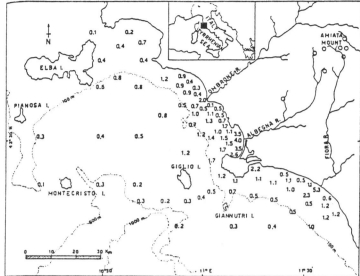

Fig.4.2B

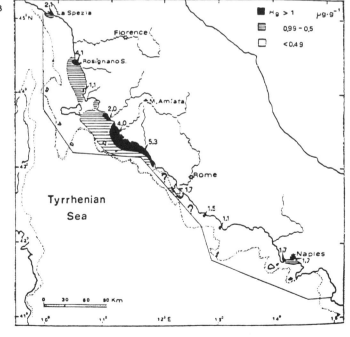

Figure 4.2   Distribution of mercury (mg Hg-T kg^{-1} DW) in surficial sediments from (A) the western Italian coastline from Elba to the river Fiora - note the locations of the cinnabar mines (o) - and (B) along the Tyrrhenian Sea [Baldi and Bargagli, 1982]

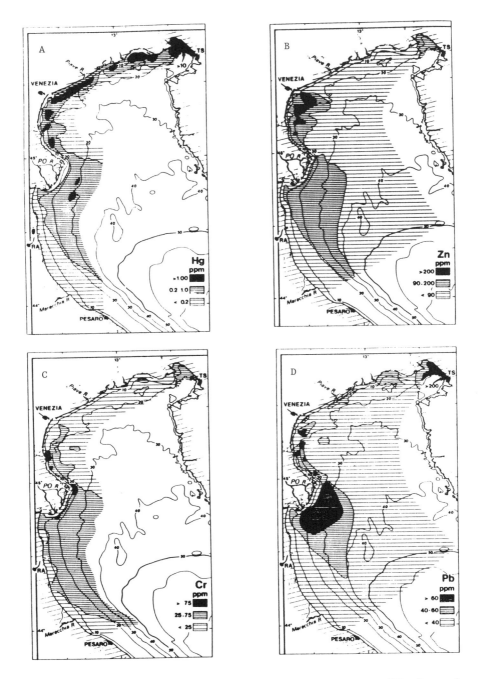

Figure 4.3   Distribution of (A) Hg, (B) Zn, (C) Cr and (D) Pb (mg/kg, dry weight) in surface sediments [Frascari, F., et al, 1988]

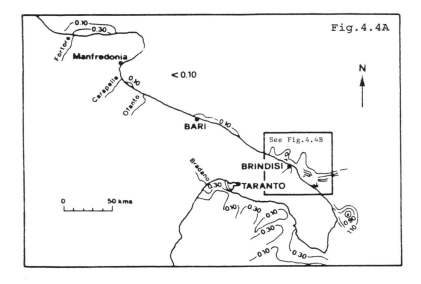

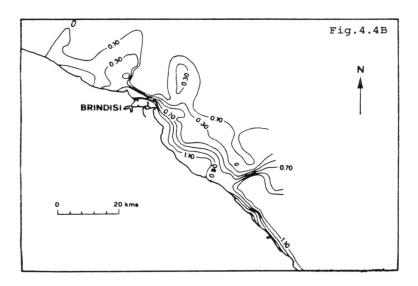

Figure 4.4   (A) Apulian region and (B) area of Brindisi: Distribution of mercury (mg/kg) in surficial sediments (<0.063 mm) [Viel, M., et al., 1985]

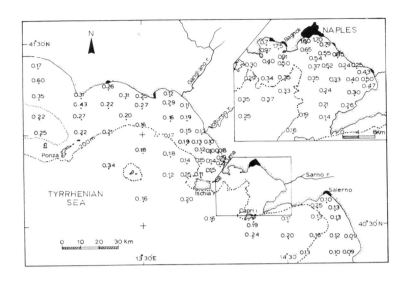

Fig. 4.5   Mercury concentrations (ug/g dry wt)
in the top 3 cm of sediment
[Baldi, F., et al., 1983]

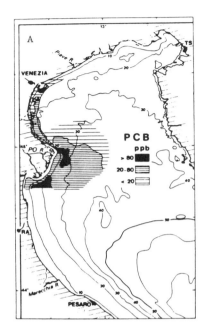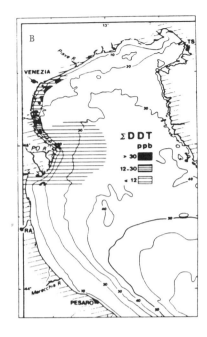

Figure 4.6 Distribution of (A) PCBs and (B) total DDT in surface sediments
[Taken from Frascari et al (1988)]

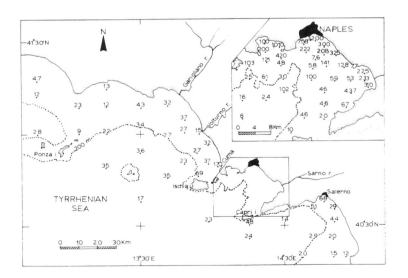

Figure 4.7 PCBs concentrations (ng/g dry wt) in the top 3 cm of sediment [Baldi, F., et al., 1983]

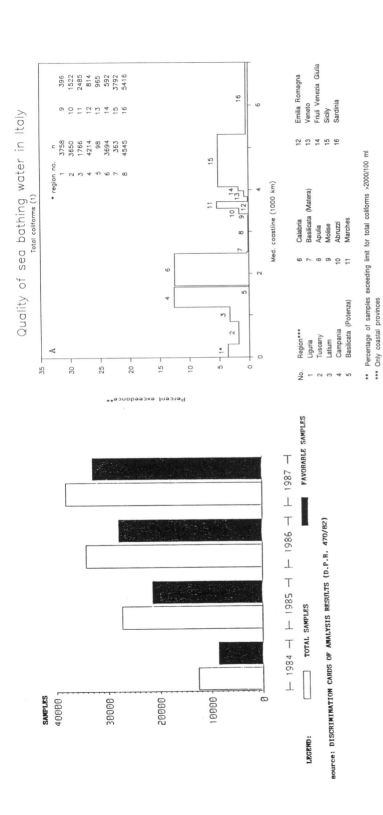

Fig. 4.8 Quality of marine bathing water: total and favorable samples between 1984 and 1987

Fig. 4.9 % Non-compliance of marine water quality of the Mediterranean Regions with existing quality requirements for (A) total and (B) fecal coliforms, (C) fecal streptococci, (D) salmonella, (E) pH, (F) colour, (G) mineral oils, (H) detergents, (I) phenols, (J) transparency and (K) oxygen

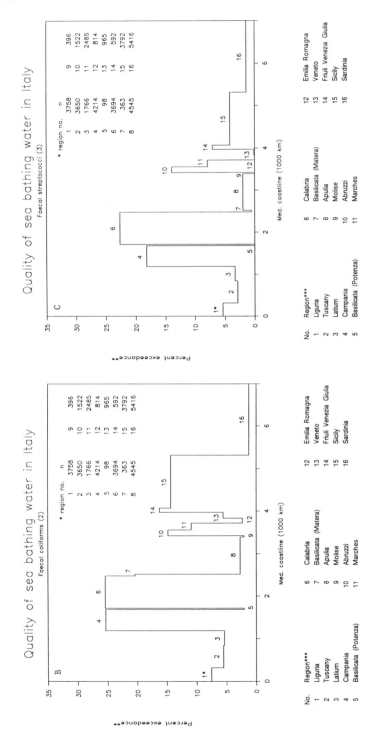

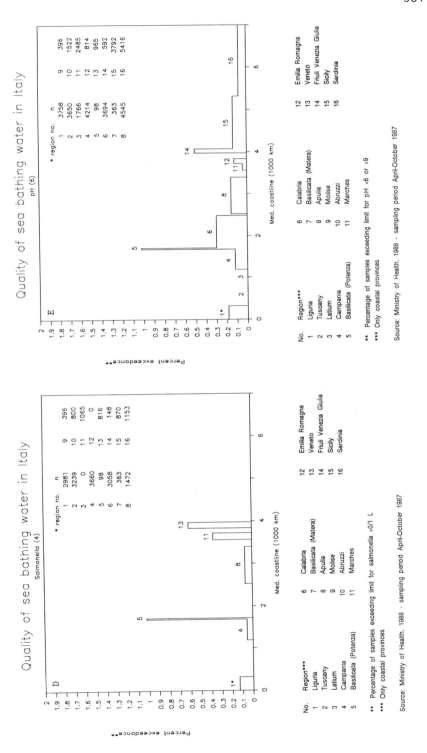

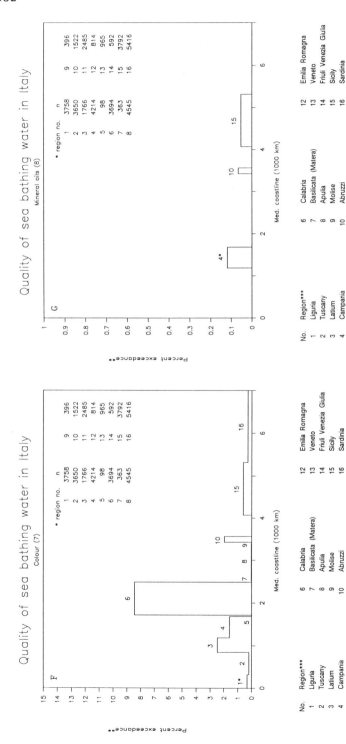

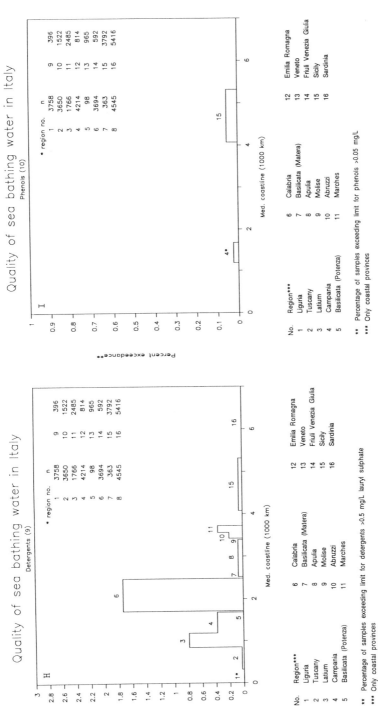

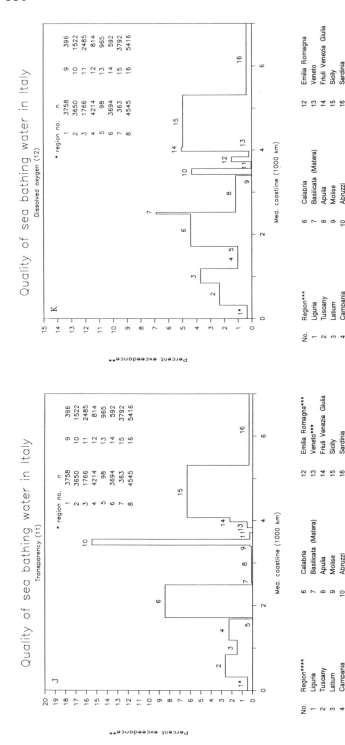

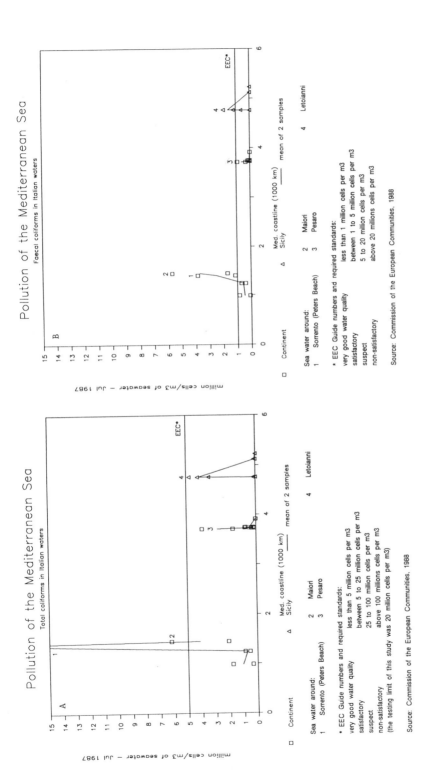

Figure 4.10 Bacteria pollution along the Italian coast: (A) total coliforms, (B) faecal coliforms and (C) faecal streptococci

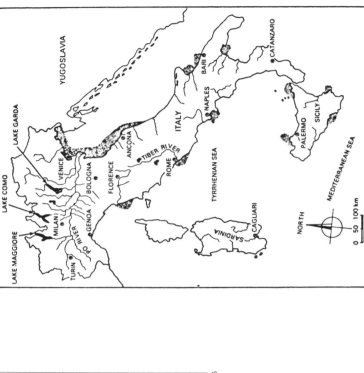

Fig. 4.11 Water resources of Italy showing major areas of coastal eutrophication (stippled areas) [Marchetti et al, 1985]

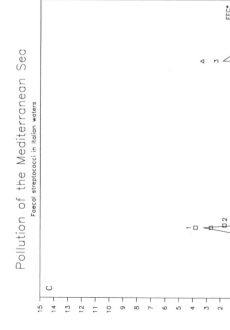

Figure 4.10 continued

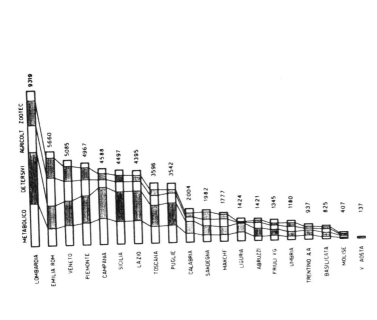

Fig. 4.12 Regional distribution of absolute phosphate discharge (ton/year) depending on source
[Marchetti, R., Passino, R., 1978]

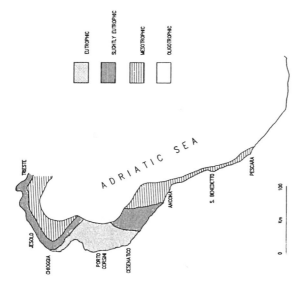

Fig. 4.13 Scheme of trophic conditions in the coastal zone of the Adriatic sea, according to IRSA
[UNEP, MAPTRS No.21, 1988]

388

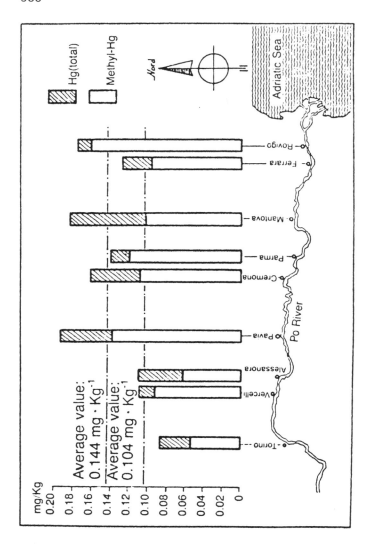

Figure 4.14　Total mercury and methyl mercury in Po River fish [Marchetti et al., 1985]

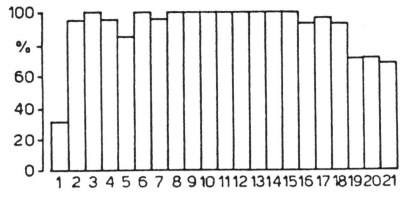

1 Crissolo	8 V. Foce Tanaro
2 Moncalieri	9 M. Foce Ticino
3 Torino	10 V. Foce Ticino
4 Chivasso	11 M. Foce Lambro
5 Verolengo	12 V. Foce Lambro
6 Trino	13 V. Foce Adda
7 Valenza	14 Cremona

15 Casalmaggiore
16 Guastalla
17 Borgoforte
18 V. Foce Secchia
19 Gaiba
20 Pontelagoscuro
21 Polesella

Figure 4.15   River Po: Percentage of non-conformity of the bathing water quality to the limit value of 1000 coli/l for 21 stations
[Ingegneria Ambientale, 15 (1), 37-40, 1986]

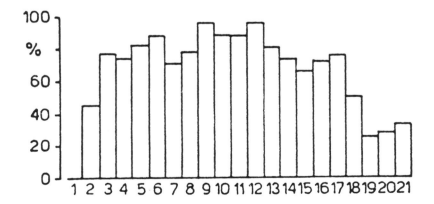

Figure 4.16   River Po: Percentage of non-conformity of the drinking water quality to the CEC limit value for 21 stations
[Ingegneria Ambientale, 15 (1), 37-40, 1986]

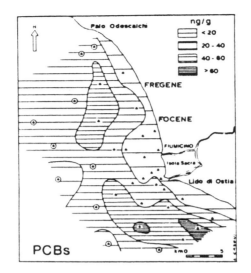

Figure 4.17　　Areal distribution of PCB concentrations in bottom sediments (ug/kg) [Frignani, M., et al., 1984]

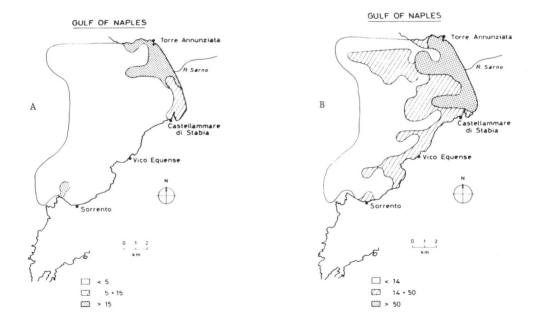

Figure 4.18　　(A) Fecal coliforms and (B) fecal streptococci distribution in superficial sediments; the reticles used to show the distribution patterns relate to n g^{-1}　　[Volterra, L., et al., 1985]

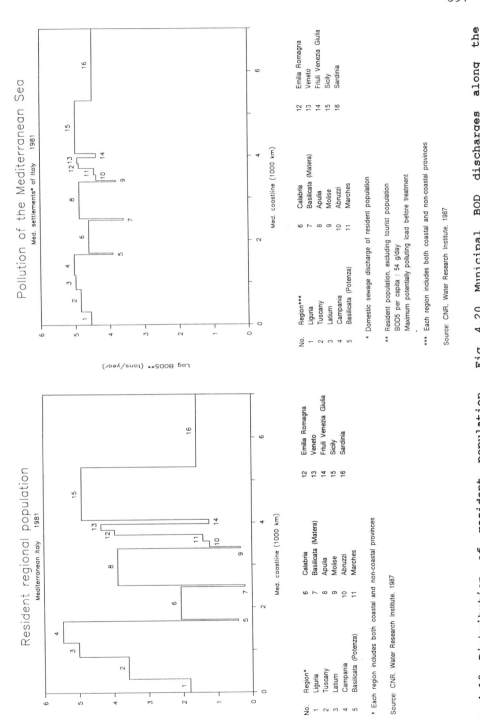

Fig. 4.19 Distribution of resident population along the Italian coast

Fig. 4.20 Municipal BOD discharges along the Italian coast

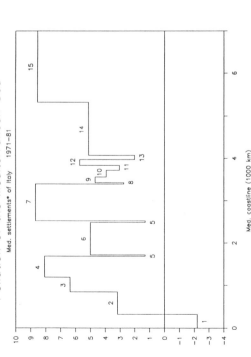

Fig. 4.21 Domestic pollution growth in 1981 compared to 1971 along the Italian coast

Fig. 4.22 A: Industrial organic load produced per worker for the 49 main industrial sectors; B: industrial BOD discharges for the 49 main industrial sectors along the Italian coast

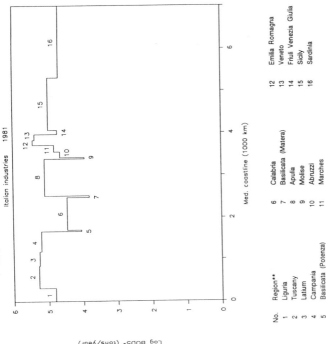

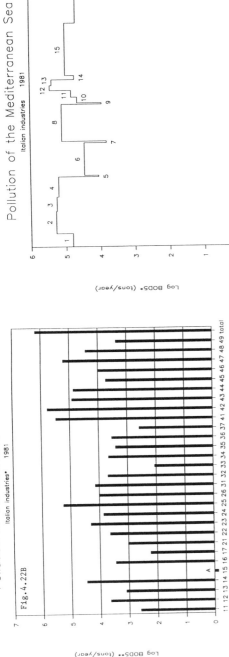

Pollution of the Mediterranean Sea
Italian industries* 1981

ISTAT Code	Industry
11	Fuels: extraction and storage
12	Coal
13	Petroleum, natural gas
14	Oil refining
15	Nuclear power
16	Electric power generation, distribution; gas
17	Water: collection, treatment and distribution
21	Metals: extraction and preparation
22	Metals: production and first processing
23	Non-ferrous metals: extraction
24	Non-ferrous metals: processing
25	Chemicals
26	Synthetic fibres
31	Metal products
32	Machinery: construction and installation
33	Office equipment: construction, installation and repair
34	Construction and installation
35	Automobiles
36	Alternative transportation
37	Precision instruments
41	Food products
42	Sugar, beverages, tobacco
43	Textile
44	Leather
45	Shoes, clothing, linen
46	Wood, furniture manufacturing
47	Paper, printing, editing
48	Rubber processing, plastics
49	Other manufactured goods

* Industrial discharge from the Mediterranean regions including the islands of Sicily and Sardinia Each region includes both coastal and non-coastal provinces

** Population equivalent
BOD5 per capita : 54 g/day
Maximum potentially polluting load before treatment, based on industrial indices related to population equivalents per industrial worker

A. Negligible

Source: CNR, Water Research Institute, 1987

Figure 4.22 continued

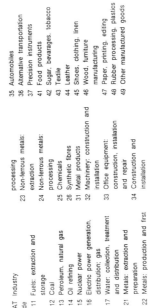

No.	Region**		
1	Liguria	6	Calabria
2	Tuscany	7	Basilicata (Matera)
3	Latium	8	Apulia
4	Campania	9	Molise
5	Basilicata (Potenza)	10	Abruzzi
		11	Marches
12	Emilia Romagna		
13	Veneto		
14	Friuli Venezia Giulia		
15	Sicily		
16	Sardinia		

* Population equivalent
BOD5 per capita : 54 g/day
Maximum potentially polluting load before treatment, based on industrial indices related to population equivalents per industrial worker

** Each region includes both coastal and non-coastal provinces

Source: CNR, Water Research Institute, 1987

Fig. 4.23 Industrial BOD discharges along the Italian coast

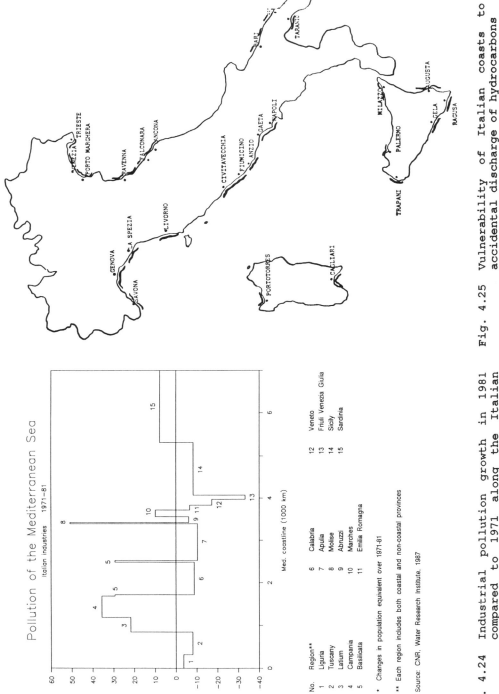

Fig. 4.24 Industrial pollution growth in 1981 compared to 1971 along the Italian coast

Fig. 4.25 Vulnerability of Italian coasts to accidental discharge of hydrocarbons

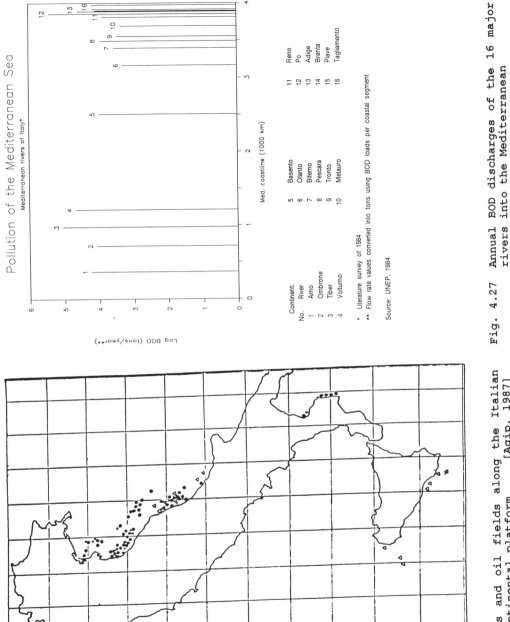

Fig. 4.26 Gas and oil fields along the Italian continental platform [Agip, 1987]

Fig. 4.27 Annual BOD discharges of the 16 major rivers into the Mediterranean

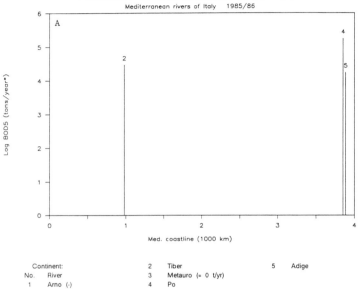
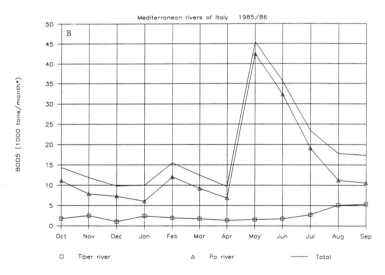

Figure 4.28   (A) Annual and (B) monthly (Oct. 1985-Sept. 1986) BOD discharges of the 5 major rivers into the Mediterranean

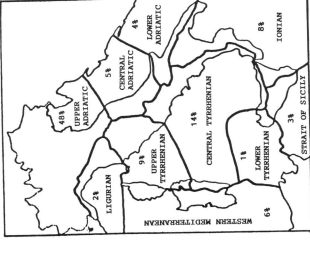
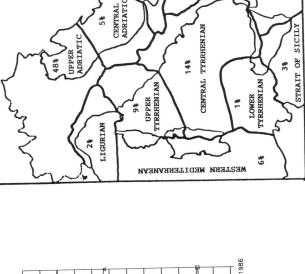

Fig. 4.29 Change of BOD discharges of Tiber, Po and the sum of five Italian rivers between 1982 and 1986 (same rivers as previously)

Fig. 4.30 Relative release of phosphorus into 7 Italian sea basins [Taken from Marchetti, R. (1987). L'eutrofizzazione. Collana scientifica Angeli]

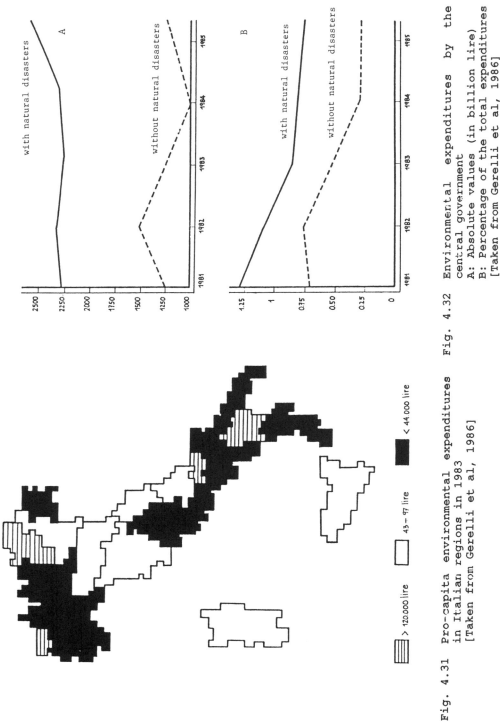

Fig. 4.31 Pro-capita environmental expenditures in Italian regions in 1983 [Taken from Gerelli et al, 1986]

Fig. 4.32 Environmental expenditures by the central government
A: Absolute values (in billion lire)
B: Percentage of the total expenditures
[Taken from Gerelli et al, 1986]

Table 4.1   Total concentrations of heavy metals in the coastal water of Ancona Province (ug/l) [Benetti, et al., 1979]

	Zn	Cu	Pb	Cd	Ni	Hg
1	17.53	5.31	1.20	0.75	2.66	0.03
2	18.12	7.07	1.18	0.91	2.91	0.04
3	15.32	7.12	1.21	0.60	2.35	0.03
4	14.44	6.34	1.31	1.01	2.48	0.03
5	17.20	6.57	1.16	0.65	3.02	0.03
6	17.55	5.80	1.25	0.62	2.71	0.03
7	16.07	5.61	1.54	0.94	2.83	0.03
8	16.16	7.01	1.56	0.90	2.16	0.04
9	18.10	7.24	2.22	1.11	3.05	0.04
10	18.35	6.90	2.50	1.21	2.94	0.04
11	17.72	7.15	2.37	1.05	3.03	0.04
12	19.86	8.25	2.01	0.86	3.60	0.04
13	32.05	10.16	2.51	0.98	3.77	0.04
14	95.15	15.44	3.53	1.41	4.15	0.05
15	111.71	17.16	3.55	1.06	5.25	0.04
16	90.84	9.92	1.91	0.77	4.02	0.05
17	35.30	10.20	1.64	0.96	4.11	0.03
18	15.26	6.33	1.33	0.70	2.85	0.04
19	15.42	5.91	1.34	0.66	2.74	0.03
20	17.03	5.50	1.46	0.75	2.59	0.02
21	16.22	5.11	1.25	0.83	2.16	0.03
22	16.11	6.03	1.20	0.80	2.24	0.03
23	17.18	6.27	1.50	0.77	2.18	0.03
24	17.61	6.30	1.44	0.72	2.33	0.02
25	18.75	7.01	1.29	0.75	2.18	0.03
26	45.60	8.46	1.32	0.71	3.24	0.03
27	62.93	10.18	1.49	0.88	3.56	0.03
28	21.27	6.99	1.31	0.66	2.20	0.03
29	15.12	6.05	1.17	0.68	2.25	0.02
30	15.10	6.16	1.20	0.70	2.22	0.02

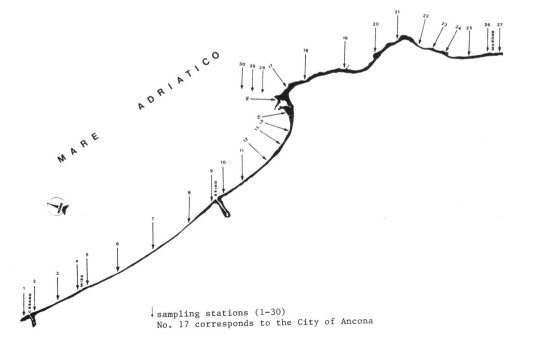

↓ sampling stations (1-30)
No. 17 corresponds to the City of Ancona

Table 4.2　　Heavy metal concentrations in the coastal waters from Punta Ala to Piombino (ng kg^{-1}) [Breder, R., 1987]

	$Hg_{tot.}$	$Hg_{react.}$	$Cd_{diss.}$	$Pb_{diss.}$	$Cu_{diss.}$
1	12.1	0.7	44	165	440
2	13.6	1.1	10	135	360
3	17.8	1.6	14	120	640
4	11.8	1.1	10	110	850
5	12.6	0.5	15	110	740
6	9.6	0.9	11	70	430
7	8.1	0.9	8	60	490
8	6.2	0.7	8	60	530
9	7.3	1.0	10	75	540
10	14.7	3.3	10	405	520
11	8.2	1.4	11	520	400
12	6.9	1.1	13	70	420
13	6.6	0.8	12	75	320
14	5.2	1.5	8	75	530
15	7.2	1.0	9	105	300
16	6.0	1.1	10	465	300
17	7.5	1.5	8	70	280
18	13.4	3.5	10	710	300

Samples obtained on 13 July 1980. Sampling stations 1, 150 m north of Punta Ala port; 18, 150 m east of Piombino port

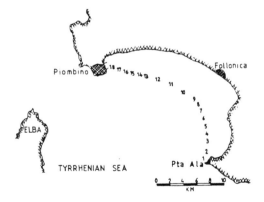

Section of coast from Punta Ala to Piombino (Gulf of Follonica). 1–18: water sampling stations.

Table 4.3   Specific surface area (SSA), clays (>4 um), $CaCO_3$ and trace metal contents in the sediment [Bargagli, et al., 1985]

Sample No.	SSA ($m^2 g^{-1}$)	Clay (%)	$CaCO_3$ (%)	Fe (%)	Mn ($\mu g\, g^{-1}$)	Zn ($\mu g\, g^{-1}$)	Pb ($\mu g\, g^{-1}$)	Cu ($\mu g\, g^{-1}$)	Ni ($\mu g\, g^{-1}$)	Co ($\mu g\, g^{-1}$)
1	11	12	28	6·2	831	99	38	27	111	19
2	20	50	19	7·6	896	205	67	48	120	25
3	33	83	18	8·8	1 446	184	75	72	81	29
4	26	55	20	6·6	1 326	215	85	77	100	22
5	18	25	25	8·2	901	131	88	43	112	20
6	9	20	37	3·5	605	129	60	54	80	20
7	24	64	17	7·6	1 627	115	44	43	81	20
8	17	66	36	9·9	1 347	153	68	65	75	26
9	3	7	52	1·2	307	60	19	7	31	6
10	4	5	46	1·5	763	38	17	12	42	11
11	31	85	43	7·7	830	102	38	26	66	14
12	6	13	23	3·2	711	106	41	14	45	10
13	10	20	50	4·9	438	96	36	24	39	13
14	8	15	12	2·9	388	104	43	18	38	10
15	10	30	27	4·7	495	165	66	36	40	6
16	13	30	46	3·6	382	127	55	26	25	11
17	10	17	56	1·9	370	85	35	18	26	9
18	10	19	33	3·5	500	182	68	33	37	14
19	7	36	30	2·8	562	150	54	29	45	10
20	7	38	28	3·1	434	164	40	31	46	17
21	20	50	13	5·7	615	187	49	38	61	15
22	27	63	13	7·9	661	185	54	37	64	17
23	18	32	20	7·5	647	176	54	41	57	21
24	13	33	30	4·5	739	91	33	20	39	9
25	3	5	78	3·9	882	84	24	14	41	10
26	16	35	47	6·4	533	100	42	25	40	19
27	22	40	46	8·9	750	160	51	48	50	13
28	22	45	19	6·5	973	178	40	44	61	15
29	13	55	15	2·3	654	158	45	40	57	16
30	18	37	16	7·0	661	106	52	34	43	13
31	32	90	17	7·8	887	130	37	45	50	15
32	34	83	13	8·2	2 990	204	13	58	51	13
33	32	84	17	8·0	2 417	133	56	47	66	21
34	30	5	16	8·5	2 681	185	52	54	61	20
35	3	5	17	1·4	563	40	8	4	24	3
36	3	14	3	7·5	1 290	176	100	41	13	27

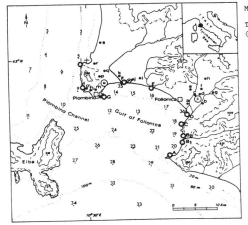

Marine sediments (•), molluscs (○), reed leaves (♦).

The study area and sampling sites. Ⓜ Metallurgical, Ⓟ power and Ⓒ chemical plants.

Table 4.4 Mean, minimum and maximum levels of heavy metals in fish from different areas (mg/kg wet weight) [Region Friuli Venezia Giulia-Regional water reclamation plan - 1985]

Fishing Areas	Copper	Zinc	Cadmium	Mercury	Lead
LIGURE SEA	0.12 / 0.57 / 2.60	2.15 / 4.66 / 13.52	0.025 / 0.04 / 0.20	0.01 / 0.23 / 0.94	0.25 / 0.37 / 0.87
TYRRHENIAN SEA North Centre	0.12 / 0.61 / 2.12	2.68 / 4.53 / 9.74	0.05 / 0.067 / 0.10	0.01 / 0.46 / 1.30	0.25 / 0.745 / 1.25
TYRRHENIAN SEA South Centre	0.12 / 0.30 / 0.75	1.94 / 3.95 / 8.00	0.05 / 0.053 / 0.10	0.10 / 0.28 / 0.68	0.25 / 0.48 / 1.25
TYRRENIAN SEA Sardinian Coast	0.15 / 0.53 / 2.04	2.92 / 5.64 / 13.88	0.017 / 0.045 / 0.117	0.045 / 0.29 / 0.691	0.27 / 0.48 / 1.29
TYRRHENIAN SEA Sicilian Coasts	0.12 / 0.85 / 2.10	2.47 / 7.01 / 14.05	0.01 / 0.047 / 0.10	0.03 / 0.33 / 1.28	0.1 / 0.45 / 1.12
TYRRHENIAN SEA (SARDINIA)	0.16 / 0.73 / 3.30	2.89 / 5.74 / 14.30	0.020 / 0.049 / 0.167	0.05 / 0.19 / 0.65	0.36 / 0.665 / 1.055
AFRICAN SEA	0.16 / 0.49 / 1.23	3.42 / 6.40 / 10.78	0.02 / 0.047 / 0.09	0.11 / 0.49 / 1.44	0.10 / 0.296 / 1.00
JONIAN SEA	0.29 / 1.17 / 4.38	2.43 / 4.75 / 9.53	0.01 / 0.0375 / 0.08	0.04 / 0.42 / 1.62	0.10 / 0.54 / 1.05
ADRIATIC SEA North Centre	0.10 / 0.81 / 1.85	0.50 / 5.20 / 10.00	—	0.06 / 0.46 / 1.06	0.62 / 1.04 / 2.10
ADRIATIC SEA Slavic Coasts	0.06 / 0.46 / 1.81	2.17 / 5.89 / 10.89	0.06 / 0.111 / 0.20	0.05 / 0.54 / 1.41	0.12 / 0.65 / 1.45
ADRIATIC SEA South Centre	0.05 / 0.40 / 1.12	0.50 / 4.25 / 9.45	0.01 / 0.044 / 0.09	0.02 / 0.33 / 1.31	0.10 / 0.611 / 1.14

Table 4.5 Range of concentrations (mg/kg wet weight) of metals in fish and crustaceans from the Trieste Gulf [Region Friuli Venezia Giulia-Regional water reclamation plan - 1985]

	Cd	Cu	Pb	Zn
Gobius niger	0.16÷0.22	0.07÷0.38	0.43÷1.03	5.9÷7.8
Mugil cephalus	0.13÷0.26	0.57÷1.31	0.49÷1.14	8.1÷10.2
Sepia officinalis	0.12÷0.41	4.99÷27.9	0.41÷1.16	12.3÷26.3
Belone belone	0.14÷0.29	0.87÷1.55	0.76÷2.82	23.6÷51.3
Engraulis eucrasicolus	0.34÷0.39	1.15÷1.18	1.40÷1.44	17.6÷20.6
Squilla mantis	0.46÷0.65	15.8÷29.8	0.38÷0.60	14.9÷25.2
Mullus barbatus	0.14÷0.17	0.65÷0.76	0.58÷0.68	6.1÷8.5

Table 4.6 Concentrations (mg/kg fresh weight) of heavy metals in marine organisms from the Gulf of Trieste [Region Friuli Venezia Giulia-Regional water reclamation plan - 1985]

Metal	Mytilus galloprovincialis		
	$\overline{M}$	min	max
Cd	.18	.3	.23
Cr	.23	.18	.26
Cu	2.95	2.4	3.5
Hg	.06	.05	.08
Pb	1.5	1.1	2.0
Zn	24.1	14.7	31.3
As	1.75	1.1	2.4

Metal	Chamelea gallina		
	$\overline{M}$	min	max
Cd	.14	.05	.20
Cr	.42	.34	.51
Cu	2.2	1.4	3.1
Hg	.27	.11	.47
Pb	1.2	.8	2.0
Zn	15.9	10.4	18.7
As	1.7	.9	2.2

Table 4.7 Chlorinated hydrocarbon content (mean ± SD, ug/kg wet weight) of marine organisms collected near Venice, Ancona and La Spezia from June 1976 to December 1979 [UNEP/MAP, TRS no. 3, 1986]

	Venice		Ancona		La Spezia	
Mytilus galloprovincialis						
Samples No.	15		14		9	
Wet wt/Dry wt	5.80	± 1.08	5.46	± 1.21	5.50	± 1.48
ECM% wet wt	1.73	± 0.58	1.89	± 0.66	1.89	± 0.64
Σ PCB	41	± 27	59	± 29	150	± 123
Σ DDT	12.9	± 6.9	18.5	± 7.8	24.7	± 16.9
Σ BHC	1.2	± 1.0	1.4	± 1.3	1.1	± 0.7
Dieldrin	0.7	± 0.8	0.9	± 1.0	1.3	± 1.7
Mullus barbatus						
Samples No.	9		13		14	
Wet wt/Dry wt	3.84	± 0.34	3.95	± 0.66	4.06	± 0.54
ECM% wet wt	5.74	± 2.03	5.11	± 2.50	4.34	± 2.67
Σ PCB	122	± 49	121	± 53	669	± 885
Σ DDT	39.6	± 22.1	42.3	± 22.8	94.0	± 86.0
Σ BHC	4.2	± 2.3	3.7	± 2.9	2.4	± 2.6
Dieldrin	1.0	± 1.3	1.6	± 1.9	5.2	± 9.3
Carcinus mediterraeneus						
Samples No.	14		13		0	
Wet wt/Dry wt	4.68	± 0.82	4.75	± 0.67		
ECM% wet wt	0.33	± 0.12	0.32	± 0.08		
Σ PCB	62	± 35	90	± 44		
Σ DDT	6.6	± 3.5	8.7	± 6.1		
Σ BHC	0.7	± 0.5	0.5	± 0.5		
Dieldrin	0.5	± 0.5	0.5	± 0.6		

Table 4.8 3,4-benzopyrene (BaP) and perylene (Pe) content of *Mytilus* soft parts from the Laguna Veneta and the Adjacent Adriatic Sea. Values represent means ± SD of nine determinations made from July, 1975 to November, 1976 [Fossato et al., 1979]

Stations	Wet weight to dry weight ratio	BaP (μg/kg dry weight)	Pe (μg/kg dry weight)	BAP to Pe ratio
A	5.7±0.2	51.8±14.7	6.3±1.5	6.9±1.3
B	5.9±0.5	54.2±9.3	10.2±2.6	6.3±0.7
C	6.3±0.5	109.5±27.4	15.9±4.7	7.9±0.8
D	5.6±0.3	96.7±18.1	16.2±4.3	7.7±1.0
E	5.2±0.4	16.7±3.9	4.6±1.0	4.4±0.9
F	5.8±0.4	135.1±31.9	16.9±7.1	10.4±1.1
G	5.6±0.3	43.5±20.9	4.9±1.6	8.3±1.6
H	5.3±0.3	29.7±15.6	5.5±1.8	6.0±1.0
I	6.0±0.4	67.1±14.4	9.3±2.1	7.7±1.2
L	4.7±0.3	12.0±4.2	1.5±0.3	7.1±0.9

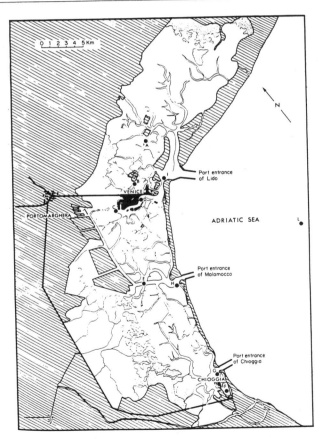

The Laguna Veneta, showing sampling stations

Table 4.9 Quality of marine bathing waters in Italy, 1987
[Ministry of Health]

REGION AND PROVINCE	PERCENTAGE OF DATA COMPLYING WITH ACCEPTABILITY LIMITS	REGION AND PROVINCE	PERCENTAGE OF DATA COMPLYING WITH ACCEPTABILITY LIMITS
1. ABRUZZO		**10. PUGLIA**	
- Chieti	72.9	- Lecce	99.7
- Pescara	76.4	- Brindisi	98.0
- Teramo	60.7	- Taranto	98.4
		- Bari	90.0
2. BASILICATA		- Foggia	95.5
- Matera	78.5		
- Potenza	94.9	**11. TOSCANA**	
		- Grosseto	87.0
3. CALABRIA		- Pisa	60.9
- Reggio Calabria	71.8	- Livorno	93.8
- Catanzaro	88.2	- Lucca	95.9
- Cosenza	67.9	- Massa Carrara	92.9
4. CAMPANIA		**12. VENETO**	
- Salerno	72.7	- Rovigo	99.3
- Napoli	73.0	- Venezia	92.3
- Caserta	82.5		
		13. FRIULI VENEZIA GIULIA	
5. EMILIA ROMAGNA		- Trieste	76.3
- Forlì	99.7	- Gorizia	70.3
- Ravenna	94.3	- Udine	95.8
- Ferrara	91.0		
		14. SARDEGNA	
6. LAZIO		- Oristano	99.3
- Latina	96.1	- Cagliari	97.4
- Roma	84.0	- Nuoro	98.6
- Viterbo	87.6	- Sassari	96.6
7. LIGURIA		**15. SICILIA**	
- La Spezia	86.1	- Siracusa	98.2
- Genova	87.4	- Ragusa	84.1
- Savona	96.5	- Catania	100.0
- Imperia	95.2	- Caltanisetta	40.2
		- Agrigento	90.0
8. MARCHE		- Messina	75.6
- Ascoli Piceno	83.6	- Palermo	60.5
- Macerata	71.2	- Trapani	86.7
- Ancona	92.4		
- Pesaro	99.1		
9. MOLISE			
- Campobasso	97.7		

Table 4.10   Red tides south of the river Po. The coast of Emilia-Romagna   [GESAMP/UNEP, 1987]

Date	Organism	Biomass (10^6 cell/l)	Observations
1969 May	Peridinium depressum, 95%	Very dense.	One week. 8 km. Fish and mollusc kills for lack of oxygen
1975 Sept.7	algal bloom	Very dense.	Massive fish and mollusc kills
1978 March July-August	Skeletonema costatum Gonyaulax polyedra, Gymnodinium coril	100	-- -- -- --
1979 end winter	S. costatum G. polyedra	30	-- --
1981 February June August	S. costatum Glenodinium lenticula Massartia rotunda G. coril	25 30 10 25	Chlorophyll up to 1 mg l^{-1}
1982 March May Summer	S. costatum Glenodinium lenticula G. polyedra + unidentified dinof.	-- -- 10	O_2 less than 1mg l^{-1} over 900km^2
1984 March-April June Aug.Sep.	S. costatum Prorocentrum micans G. polyedra, G. coril, Massartia	-- -- --	Spreading 3 km from shore
1985 February October	S. costatum + other diatoms G. polyedra	Heavy growth Impressive bloom	-- --
1986 January-May	S. costatum	Heavy growth	

Table 4.11   Coastal erosion in Italy
[CNR, Atlante delle spiagge italiane, 1985]

Region	Length of monitored coasts	High coasts		Low coasts		Advancing coasts		Regressing coasts		Stable coasts	
	km	km	%	km	%	km	%	km	%	km	%
Liguria	80.0	44.0	73	16.5	27	0.5	3	4.0	24	12.0	73
Toscana	440.0	220.5	50	219.5	50	35.5	16	93.5	43	90.5	41
Lazio	287.5	63.5	22	224.0	78	10.0	4	125.5	56	88.5	40
Campania	72.0	6.0	8	66.0	92			13.0	20	53.0	80
Calabria	271.5	35.5	13	236.0	87	4.5	2	191.0	81	40.5	17
Sicilia	47.5	5.0	11	42.5	89	4.5	11	6.0	14	32.0	75
Basilicata	56.5	18.0	32	38.5	68			38.5	100		
Molise	34.0			34.0	100			26.0	76	8.0	24
Abruzzo	126.0	14.0	11	112.0	89	2.5	2	57.5	51	52.0	47
Marche	170.0	4.5	3	165.5	97	6.5	4	62.5	38	96.5	58
Emilia Romagna	150.0			150.0	100	18.5	12	60.0	40	71.5	48
Veneto	156.0			156.0	100	14.0	9	66.0	42	76.0	49
Friuli	100.5	14.0	14	86.5	86			2.5	3	94.0	97
Total	1972.0	425.0	22	1547.0	78	96.5	6	746.0	48	704.5	48

Table 4.12   Area (ha) of forest of each Italian Region destroyed by fires, due to different causes, in 1982   [Annuario di statistica forestale, 1983]

Regions	Natural	Intentional	Area (ha) Unintentional						Total	Unknown	Total
			Recreation	Forestry	Agriculture	Waste fires	Matches	Other			
Piemonte	1	212	28	1	160	12	120	136	457	120	790
Valle D'Aosta	—	—	—	—	3	—	—	—	3	5	8
Lombardia	2	293	1	64	134	20	79	132	430	354	1.079
Trentino-Alto Adige	4	90	2	—	26	—	2	4	34	30	158
Bolzano	—	—	1	—	—	—	—	1	2	1	3
Trento	4	90	1	—	26	—	2	3	32	29	155
Veneto	—	192	25	—	52	45	89	50	261	5	458
Friuli-Venezia Giulia	642	136	6	2	1	4	410	31	454	420	1.652
Liguria	101	3.115	1	—	643	56	61	137	898	1.129	5.243
Emilia-Romagna	—	45	—	—	44	5	50	85	184	34	263
Toscana	7	1.797	4	150	141	12	503	310	1.120	236	3.160
Umbria	—	15	—	—	1	3	41	14	59	172	246
Marche	16	3	—	—	10	8	37	44	99	14	132
Lazio	40	588	32	9	46	1	335	236	659	44	1.331
Abruzzi	—	18	25	—	14	17	18	10	84	7	109
Molise	—	21	—	—	165	—	20	39	224	25	270
Campania	1	2.890	—	7	26	13	1.700	927	2.673	124	5.688
Puglia	—	1.191	—	—	42	5	318	133	498	—	1.689
Basilicata	2	217	—	—	348	5	763	25	1.141	68	1.428
Calabria	53	4.075	38	170	334	156	1.167	301	2.166	43	6.337
Sicilia	116	2.706	—	—	646	14	283	1.282	2.225	510	5.557
Sardegna	—	12.080	—	—	769	4	—	162	935	2	13.017
ITALIA	985	29.654	162	403	3.605	380	5.996	4.058	14.604	3.342	48.615

Table 4.13 Area of damaged wood due to forest fires and restoring cost for each Italian Region [Annuario di statistica forestale, 1983]

Regions	N.	High forests				Coppice				Total	Total woods	Damages	
		Resinous	Broadleaf	Resinous and broadleaf	Total	Simple	Composite	Seriously degradated	Total			Wood damaged	Restoring cost
Piemonte	154	112	74	26	212	396	85	97	578	790	205.472	260.645
Valle D'Aosta	5	6	1	—	7	1	—	—	1	8	7.000	11.300
Lombardia	192	190	17	48	255	599	104	121	824	1.079	484.369	826.453
Trentino-Alto Adige	55	17	—	—	17	140	—	1	141	158	29.265	111.800
Bolzano	3	2	—	—	2	—	—	—	1	3	3.400	1.200
Trento	52	15	—	—	15	140	—	—	140	155	25.865	110.600
Veneto	104	67	20	28	115	315	3	25	343	458	46.171	53.499
Friuli-Venezia Giulia	197	612	281	21	914	244	356	138	738	1.652	400.863	1.540.561
Liguria	1.169	1.750	297	108	2.155	2.662	335	91	3.088	5.243	1.304.468	1.366.909
Emilia-Romagna	89	24	3	9	36	223	1	3	227	263	73.700	83.546
Toscana	626	650	190	53	893	2.043	144	80	2.267	3.160	1.637.357	1.876.431
Umbria	94	15	—	17	32	205	9	—	214	246	198.861	139.325
Marche	55	19	—	5	24	104	4	—	108	137	182.704	436.550
Lazio	304	400	24	71	495	745	63	28	836	1.331	621.639	1.319.205
Abruzzi	32	6	31	7	44	54	9	2	65	109	49.977	164.534
Molise	45	18	11	—	29	158	78	5	241	270	61.848	327.925
Campania	1.121	407	147	200	754	4.494	45	395	4.934	5.688	2.047.527	2.973.928
Puglia	244	671	146	247	1.064	321	105	199	625	1.689	865.819	2.533.751
Basilicata	199	370	581	197	1.148	268	9	3	280	1.428	432.941	2.068.494
Calabria	898	921	1.070	754	2.745	3.211	289	92	3.592	6.337	4.653.205	6.213.062
Sicilia	216	702	1.789	2.099	4.590	826	12	129	967	5.557	2.408.318	6.992.676
Sardegna	335	937	8.942	367	10.246	1.621	519	631	2.771	13.017	18.062.150	15.353.450
ITALIA	6.134	7.894	13.624	4.257	25.775	18.630	2.170	2.040	22.840	48.615	33.593.654	44.624.044

Area (ha), cost (thousands lire).

Table 4.14 Water quality indicators for a number of rivers[a] [ISTAT, Statistiche Ambientati, 1984]

INDICATOR	MEASUREMENT UNIT	1975	1976	1977	1978	1979	1980
Po							
Flow rate	m^3/s	1.737	1.760	2.571	1.752	....	....
BOD_5	mg/l	7,3	7,3	7,1	7,3	....	....
Nitrate(NO_3)	mgN/l	1,35	1,35	1,34	1,35	....	....
Total phosphorus	mgP/l	0,231	0,234	0.238	0,234	....	....
COD	mg/l	13,9	13,7	13,8	13,9	....	....
Dissolved solids	mg/l	359	401	812	180	294	317
Cadmium[b]	µg/l	....	....	0.160	0,070	0,055	0.045
Lead[c]	µg/l	0,40	0,61	0,16	0,12	0,15	0.55
Cromium[c]	µg/l	....	....	0,50	1,10	0,90	0,60
Copper[c]	µg/l	0,60	0,92	0,60	1,44	0,45	0,85
pH		7,5	8,6	7,8	8,4	8,3	7,8
Adige							
Flow rate	m^3/s	217	171	287	208	....	....
BOD_5	mg/l	4,0	3,8	4,1	4,0	....	....
Nitrate(NO_3)	mgN/l	0,879	0,961	0,804	0,892	....	....
Total phosphorus	mgP/l	0,191	0,161	0,220	0,186	....	....
COD	mg/l	8,6	8,5	9,5	8,6	....	....
Dissolved solids	mg/l	....	....	....	144	183	196
Cadmium[b]	µg/l	....	....	....	0,08	0,18	0,30
Lead[c]	µg/l	....	....	....	0,08	0,18	0,30
Cromium[c]	µg/l	....	....	....	0,68	0,55	0,70
Copper[c]	µg/l	....	....	....	0,82	0,50	0,85
pH		....	....	....	8,1	7,8	7,7
Tiber							
Flow rate	m^3/s	....	243	....	....	....	....
Dissolved oxygen	mg/l	....	....	3,5	4,6	4,9	5,3
Nitrate(NO_3)	mgN/l	....	1,51	2,02	0,9	1,19	....
Ammonia(NH_3)	mgN/l	....	1,58	1,08	0,79	0,78	....
Phosphorus orthoph.	mgP/l	....	0,26	0,33	0,26	0.199	....
Dissolved solids	mg/l	865	994	997	568	828	....
Cadmium[b]	µg/l	....	0,18	0,20	0,30	0,09	....
Lead[c]	µg/l	0,21	0,46	0,21	0,26	0,12	....
Cromium[c]	µg/l	....	....	1,00	0,70	0,10	....
Copper[c]	µg/l	0,50	0,81	0,70	0,96	0,30	....
pH		7,5	7,5	7,4	7,7	8,0	....

[a] At stations Pontelagoscuro, Boira Pisani and Ostia Antica for the Po, Adige and Tiber, respectively;
[b] total concentration;
[c] dissolved concentration

Table 4.15 Number of plant and animal species endangered of extinction in different regions

YEAR / REGION	(1982) plants	(1981) birds	(1983) mammals	(1977) reptiles	(1983) amphibians
PIEMONTE	1	25	5	3	7
VALLE D'AOSTA	1	25	5	3	7
LOMBARDIA	-	22	3	1	6
TRENTINO ALTO ADIGE	-	37	6	6	7
VENETO	4	37	6	6	7
FRIULI VENEZIA GIULIA	6	37	6	6	7
LIGURIA	1	25	5	3	7
EMILIA ROMAGNA	1	27	3	3	6
TOSCANA	1	21	7	3	5
UMBRIA	-	15	6	4	4
MARCHE	-	9	3	4	4
LAZIO	1	15	6	4	4
ABRUZZO	1	11	6	4	4
MOLISE	-	11	6	4	4
CAMPANIA	5	11	4	3	4
PUGLIA	2	19	3	5	3
BASILICATA	-	10	5	5	5
CALABRIA	-	10	5	5	5
SICILIA	9	15	3	9	1
SARDEGNA	8	30	4	5	3
ITALY	41	61	12	16	15

Table 4.16 Organic loads and residual quota to purify in Po basin (total thousand Equivalent Inhabitants per year)
[Taken from the report of the "Conferenza Permanente Interregionale per la tutela ed il risanamento del fiume Po". 1988]

	CIVIL*	INDUSTRIAL	ZOOTECHNICAL	TOTAL
1. GENERATED LOAD	17.923	58.579	60.066	138.568
2. PURIFIED QUOTA IN PUBLIC SYSTEMS	8.605	9.016	...	17.621
3. PURIFIED QUOTA IN PRIVATE SYSTEMS	...	29.730	57.084 (**)	86.822
4. NON-PURIFIED QUOTA AS IN MAY 88 (% ON GENERATED LOAD)	9.318 (52%)	19.825 (34%)	4.982 (8%)	34.125 (23%)
5. ADDITIONAL QUOTA WHICH WILL BE PURIFIED WITH THE REALIZATION OF INTERVENTIONS FIO 1986-1988.	835	1.190	130	2.155
6. RESIDUAL QUOTA TO PURIFY (% ON GENERATED LOAD)	6.693 (37%)	18.635 (32%)	4.852 (8%)	30.180 (22%)

* does not include the quota to attribute to the tourist population.
** or rather on the ground.

Table 4.17 Synthesis of produced and released loads in the basin of Po River
[Taken from the report of the "Conferenza Permanente Interregionale per la tutela ed il risanamento del fiume Po". 1988]

DESCRIPTION	NITROGEN (tonn/year)	PHOSPHORUS (*) (tonn/year)
A. GENERATED LOADS		
A1. CIVIL	85060	14730
A2. PRODUCTIVE	28350	1470
A3. CIVIL+PRODUCTIVE (A1+A2)	113410	16200
A4. ZOOTECHNICS	363770	60690
A5. FERTILIZERS	280380	84470
TOTAL	757560	161360
B. APPLIED LOADS ON AGRICULTURAL GROUND		
B1. ZOOTECHNICS (95% of A4)	345600	57660
B2. FERTILIZERS (100% of A5)	280380	84470
TOTAL	625980	142130
C. RELEASED LOADS BY GROUND IN SURFACE WATER BODIES		
C1. ZOOTECHNICS (washed away quota by ground)	58760	1720
C2. FERTILIZERS (washed away quota by ground)	56050	2530
C3. RELEASED QUOTA BY UNCULTIVATED GROUND	6720	340
TOTAL	121530	4590
D. RELEASED TOTAL LOADS IN SURFACE WATER BODIES		
D1. CIVIL+PRODUCTIVE	103930	15440
D2. ZOOTECHNICS (5% of A4)	18170	3020
D3. GROUND (TOTAL = C)	121530	4590
TOTAL	243630	23050

(*) The value of phosphorus load has been rounded to ton.

Table 4.18 Amounts of nutrients discharged into the lagoon of Venice according to different evaluation methods [Veneto Region]

	PHOSPHOROUS (t/year)(%)				NITROGEN (t/year)(%)			
	domest.	indust.	agric.	Total	domest.	indust.	agric.	Total
Method IRSA	513.5 (75.3)	57.8 (8.5)	111.0 (16.2)	682.3 (100)	2 648 (40)	1 000 (15)	2 960 (45)	6 608 (100)
Bendoricchio (85)	1 658 (63)		967 (37)	2 634 (100)	4 268 (45)		5 231 (55)	9 499 (100)
Cossu (1985)	466 (30.7)	50 (3.3)	1 000 (66)	1 516 (100)	1 061 (16.2)	1.000 (15.2)	4 500 (68.6)	6 561 (100)
Perin (1980)	780 (56.4)	541 (39.1)	51 (3.7)	1 348 (100)	2 827 (21.4)	7 409 (56.1)	2 642 (20)	13 202 (100)
Zingales (1980)			250				2 500	
Rinaldo (1981)			1 157				9 897	

Table 4.19 Special and toxic-noxious waste (TPA) produced in the provinces forming part of the basin draining into the lagoon of Venice [Veneto Region]

	indust. waste		biolog. sludge	hospital + patholog. waste	power station ash	TOTAL
	special	toxic-noxious				
PADOVA	432 000	95 000	35 000	1 000	14 000	577 000
TREVISO	510 000	94 000	38 000	700	-	622 700
VENICE	393 000	90 000	76 000	1 000	200 000	760 000
TOTAL	1 335 000	279 000	129 000	2 700	214 000	1 959 700

Table 4.20 Costs of algal blooms in the lagoon of Goro in 1987

Factor	Cost (Billion of liras)
1. Deaths of marine organisms in 1987	6.3
2. Loss of fishing activities in 1988	8.4
3. Disposal of marine organisms	0.5
4. Works to increase water circulation in the lagoon	2.5
5. Removal of algae from the lagoon	2.5
Total	20.2

Table 4.21 Regional laws concerning pollution (A) and general environmental (B) matters during the last 3 legislatures
[Taken from "Nota preliminare alla relazione sullo stato dell'ambiente". Ministero dell'ambiente 1987]

	I legislature		II legislature		III legislature*		Total 3 legislature	
	A	B	A	B	A	B	A	B
Piemonte	3	3	7	14	5	8	15	25
Lombardia	6	10	6	8	14	19	26	37
Liguria	5	6	2	4	9	11	16	21
Veneto	9	9	9	9	4	5	22	23
E. Romagna	4	8	4	7	5	5	13	20
Toscana	2	2	10	11	8	8	20	21
Umbria	-	3	5	8	2	5	7	16
Marche	-	4	-	-	4	5	4	9
Lazio	2	2	4	6	9	11	15	19
Abruzzo	-	-	-	2	5	6	5	8
Molise	-	-	3	3	3	3	6	6
Campania	2	2	4	5	1	1	7	8
Puglia	3	5	4	7	5	7	12	19
Basilicata	-	-	1	1	4	5	5	6
Calabria	-	-	-	-	1	1	1	1
Total	36	54	59	85	79	100	174	239

* Until 1985

Table 4.22   Marine reserves with possible limitations of human activities

1) Golfo di Portofino;
2) Cinque Terre;
3) Secche della Meloria;
4) Arcipelago Toscano;
5) Isole Pontine;
6) Isola di Ustica
7) Isole Eolie;
8) Isole Egadi;
9) Isole Ciclopi;
10) Porto Cesareo
11) Torre Guaceto;
12) Isole Tremiti;
13) Golfo di Trieste;
14) Tavolara, Punta Coda Cavallo;
15) Golfo di Orosei, Capo Monte Santu;
16) Capo Caccia, Isola Piana;
17) Isole Pelagie;
18) Punta Campanella;
19) Capo Rizzuto;
20) Penisola del Sinis, Isola di Mal di Ventre.

Note: So far only two such reserves (Island of Ustica and Miramare in the Gulf of Trieste) have been established (D.M. 12.11.1986), whereas the others are expected to follow shortly.

Table 4.23   Main international (environmental) conventions ratified by Italy

- London Convention of 1954, regulating the prevention of pollution of the sea by oil, ratified in 1961;
- Genoa Convention of 1973, regulating exploitation of the sea bed and offshore installations;
- London Convention of 1972, regulating dumping of wastes and other matters at sea, ratified in 1983;
- London Convention of 1973, regulating pollution from navigation, ratified in 1980;
- Barcelona Convention of 1976, regulating the protection of the Mediterranean Sea against pollution, ratified in 1979;
- Agreement between Italy and Yugoslavia, regulating the protection, ratified in 1977;
- Agreement with France and Monaco, regulating the protection of Mediterranean Waters, ratified in 1981;
- Agreement with Greece, regulating the protection of the Ionian Sea;

Table 4.24 Resources (billion Italian Lire and percentage) provided by Fio to Italian regions

REGIONS	YEARS			
	1982-1984		1986-1988	
	ABSOLUTE VALUE	PERCENTAGE	ABSOLUTE VALUE	PERCENTAGE
ABRUZZO	124.2	4.3	172.4	5.8
BASILICATA	72.8	2.6	384.3	13.0
CALABRIA	105.0	3.7	56.2	1.9
CAMPANIA	53.1	1.8	267.2	9.0
EMILIA ROMAGNA	253.2	8.8	190.8	6.5
FRIULI	89.9	3.1	85.7	2.9
LAZIO	66.6	2.3	128.4	4.4
LIGURIA	69.6	2.4	140.5	4.8
LOMBARDIA	255.6	8.9	315.6	10.7
MARCHE	189.9	6.6	103.9	3.5
MOLISE	135.6	4.7	26.9	0.9
PIEMONTE	307.5	10.8	192.2	6.5
PUGLIA	139.6	4.9	131.4	4.4
SARDEGNA	100.7	3.5	81.8	2.8
SICILIA	144.5	5.0	54.5	1.8
TOSCANA	332.7	11.6	118.3	4.0
TRENTINO	-	-	-	-
UMBRIA	41.9	1.6	35.8	1.2
VALLE D'AOSTA	62.7	2.2	40.0	1.4
VENETO	318.4	11.2	428.0	14.5
TOTAL	2863.5	100.0	2957.1	100.0

Table 4.25 Regional distribution of the resources (million Italian Lire) allocated by the Fio in 1986-1988 to different sectors

REGION	WATER PROTECTION AND TREATMENT *	WASTE DISPOSAL	OTHER SECTORS **	TOTAL
ABRUZZO	48 891	41 103	82 381	172 375
BASILICATA	61 482	18 535	304 325	384 342
CALABRIA	32 892	/	23 340	56 232
CAMPANIA	152 059	58 226	56 893	267 178
E. ROMAGNA	96 551	44 268	50 000	190 819
FRIULI	66 000	19 734	/	85 734
LAZIO	85 466	/	42 991	128 457
LIGURIA	120 176	20 320	/	140 496
LOMBARDIA	147 365	94 789	73 498	315 652
MARCHE	51 836	17 320	34 766	103 922
MOLISE	/	/	26 856	26 856
PIEMONTE	148 304	43 824	/	192 128
PUGLIA	54 075	20 109	57 251	131 435
SARDEGNA	46 906	34 872	/	81 778
SICILIA	/	/	54 469	54 469
TOSCANA	83 654	34 673	/	118 327
UMBRIA	23 066	12 774	/	35 840
V. D'AOSTA	43 037	/	/	43 037
VENETO	173 653	80 403	173 965	428 021
TOTAL	1 435 413	540 950	980 735	2 957 098

* includes sewerage, purification and reclamation
** includes forestation, idraulic works, soil and environmental protection.

Table 4.26 Expenditures in 1988 by the central government of Italy through Fio 1986-1988, law 441/1987 and law 67/1988 for water protection and treatment and waste disposal

AREA	EXPENDITURES (BILLION ITALIAN LIRE)			TOTAL
	FIO 1986-1988 AND ART. 1 LAW 441/1987	ART. 17 AND 18 LAW 67/1988	ART. 1 BIS AND 1 TER LAW 441/1987	
CENTRAL AND NOTHERN ITALY	1,473	608	646	2,727
SOUTHERN ITALY	778	297	364	1,439
TOTAL ITALY	2,251	905	1,010	4,165
PROV. OF NAPLES	191	67*	*	258
BORMIDA BASIN	38	30*	*	68
LAMBRO, OLONA AND SEVESO BASIN	119	63*	*	182
PO BASIN	686	500*,**	380*	1,566
TEVERE BASIN	108	5*	*	113
ARNO BASIN	161	5*	*	166
ADIGE BASIN	94	5*	*	99
LAGOON OF VENICE	71	-	*	71

* Projects to be covered with these resources have not been defined yet.
** According to previsions of the reclamation plan of the Lambro, Olona and Seveso basin approved by the Council of Ministries on 18 July 1988, 82 billion lire of these resources will be allocated for the Lambro, Olona and Seveso basin which is part of the Po basin.

Table 4.27 Regional environmental expenditures (billion Italian Lire) in 1982-1984
[Taken from "Nota preliminare alla relazione sullo stato dell'ambiente". Ministero dell'ambiente (1987)]

	Natural disasters			Soil protection			Forestation			Nature conservation			Water protection			Hunting and fishing			TOTAL		
	1982	1983	1984	1982	1983	1984	1982	1983	1984	1982	1983	1984	1982	1983	1984	1982	1983	1984	1982	1983	1984
Piemonte	20.9	17.8	19.5	5.6	9.7	11.7	10.1	9.4	12.8[1]	21.6	46.1	48.2	12.4	4.0	8.7	4.3	4.1	5.8	83.0	91.1	106.7
Val D'Aosta	0.3	0.1	0.4	18.2	19.8	22.6	6.4	9.0	22.8[1]	0.1	0.1	0.03	4.5	4.9	11.8	0.01	0.01	0.01	29.5	33.9	57.6
Lombardia	n.p.	n.p.	n.p.	n.p.	n.p.	n.p.	15.9	10.9	10.2	n.p.	n.p.	n.p.	56.3	46.0	73.7	5.9	5.9	6.0	78.1	62.8	89.9
Liguria	4.5	7.2	2.7	5.4[2]	7.5[2]	6.3[2]	6.1	4.7	9.4	1.6	1.9	2.2	18.2	11.0	61.7	1.6	1.8	1.7	37.4	33.4	84.0
Veneto	5.2	5.8	4.9	27.9[2]	64.4[2]	70.6[2]	19.8	31.5	22.0	0.3	0.4	0.8[3]	0.4	8.5	16.1	3.6	5.3	5.1	57.2	115.9	119.5
Prov. Bolzano	5.6	10.3	30.6	15.5	19.8	25.3	3.5	4.1[4]	4.8	6.2	14.6	28.1[3]	14.3	13.3	33.1	0.2	0.2	0.3	45.3	62.3	20.4
Prov. Trento	3.9	5.4	4.1	22.7	25.1	26.6	1.6[4]	1.9[4]	2.2[4]	0.6	1.1	0.9	31.9	43.2	52.8	1.1	0.6	0.6	61.8	77.3	87.1
Friuli V.Giulia	2.7[5]	3.6[5]	3.6[5]	16.0	22.3	35.9	13.8	7.9	21.8	2.7	3.6	3.6	11.2	16.8	16.5	0.6	0.6	0.9	47.0	54.8	82.3
Emilia Romagna	40.5[5]	32.9[5]	59.2[5]	19.0	35.4	24.8	9.7	9.6	6.3	1.5	2.3	1.6	71.6	45.6	65.2	8.0	12.3	8.7	150.6	139.1	165.8
Toscana	3.9	5.4	2.7	15.7	18.7	16.6	37.2	51.3	51.9	0.6	0.8	0.5	49.1	39.4	54.8	5.0	8.3	10.4	111.5	123.9	136.9
Marche	10.1	10.5	27.6	11.2	44.4	15.1	1.1	1.0	0.9	0.5	0.4	0.9	-	-	-	1.1	0.7	1.6	24.0	57.0	46.1
Umbria	1.0	0.8	12.3	16.4[2]	5.6[2]	28.1[6]	12.9	14.1	20.4	0.9	1.1	1.1	2.4	0.5	7.5	1.8	2.4	3.8	35.4	24.5	73.2
Lazio	17.3	20.4	19.2	2.9[2]	7.2[2]	3.8	7.0	5.3	5.6	0.9	1.5	1.9	23.4	30.9	67.9	2.0	2.0	5.3	53.5	67.3	103.7
Abruzzo	4.2	4.0	3.7	6.0	4.6	4.2	3.4	35.4[1]	4.9	18.1	0.01	7.1	4.0	7.7	-	1.5[7]	1.3[7]	1.8[7]	36.2	53.0	21.7
Molise	1.4	0.7	19.3	8.2	7.7	9.5	4.1	2.2	4.5	1.7	10.9	0.7	8.0	9.5	8.1[8]	0.5	0.7	0.2[7]	23.9	31.7	42.3
Campania	58.2	258.1	56.2	100.3	122.8	84.0	57.2	64.9	76.4	4.0	4.1	5.1	74.3[8]	60.4[8]	45.6[8]	4.0	5.2	5.7	298.0	515.7	273.0
Puglia	24.1	94.7	51.7	3.0	3.0	3.0	22.7	22.4	16.1	19.3	2.0	1.3	5.0	5.5	5.0	3.6	3.6	3.7	77.7	131.2	80.8
Basilicata	1.7	0.8	10.9	9.9	5.1	2.4	0.03	0.03	0.03	3.7	3.8	3.2	8.1	10.4	6.5	0.4	0.6	0.8	23.8	20.7	23.8
Calabria	55.7	46.7	37.7	8.2	10.9	7.1	200.0	239.7	260.3	0.2	0.6	-	24.4	8.2	5.3	3.6	1.9	2.7	292.1	308.0	313.1
Sicilia	n.p.	n.p.	n.p.	n.p.[9]	n.p.[9]	n.p.[9]	n.p.	n.p.	n.p.	n.p.	n.p.	n.p.	n.p.	n.p.	n.p.	n.p.	n.p.	n.p.	n.p.	n.p.	n.p.
Sardegna	35.0	40.0	12.2	9.6	47.3	14.6[6]	41.3	53.8	83.4	14.6	52.5	22.2	25.4	25.3	48.8	3.6	11.7	22.9	129.5	230.6	204.4
ITALIA	304.6	566.4	378.5	321.7	481.3	410.3	472.8	579.0	636.7	98.9	146.9	133.3	444.9	391.1	589.1	52.4	69.2	88.0	1695.3	2233.0	2232.4

[1] Includes FIO funds.
[2] Includes funds for purification.
[3] Includes waste disposal.
[4] Includes nature conservation.
[5] Includes natural disasters.
[6] Includes special funds for housing consolidation.
[7] Indicates actual expenditures.
[8] Includes water purification.
[9] Includes special funds for groundwater recovery.

Table 4.28 Projections concerning manufacturing industry in Italy according to different scenarios
[Taken from the "Blue Plan" (1987)]

SCENARIO YEAR	Aggravated continuing scenario		Moderate continuing scenario		Aggregated alternative scenario	
	Local Units	Workers	Local Units	Workers	Local Units	Workers
1985	638 000	5 043 000	638 000	5 043 000	638 000	5 043 000
2000	526 000	4 264 000	740 000	4 733 000	761 000	4 950 000
2025	433 000	3 637 000	574 000	4 193 000	804 000	4 825 000

Table 4.29 Projections concerning total and coastal receptive structures* in Italy according to different scenarios
[Taken from the "Blue Plan" (1987)]

SCENARIO YEAR	Aggravated continuing scenario		Moderate continuing scenario		Aggregated alternative scenario	
	Total Beds	Coastal Beds	Total Beds	Coastal Beds	Total Beds	Coastal Beds
1985	19 041 000	8 979 000	19 041 000	8 979 000	19 041 000	8 979 000
2000	18 022 000	8 110 000	20 499 000	9 751 000	22 448 000	11 118 000
2025	16 576 000	6 630 000	21 283 000	10 235 000	24 586 000	12 785 000

* Include hotels, accomodations other than hotels and vacation houses

Table 4.30 Projections concerning arrivals and average stays in Italy according to different scenarios [Taken from the "Blue Plan" (1987)]

SCENARIO YEAR	Aggravated continuing scenario		Moderate continuing scenario		Aggregated alternative scenario	
	Arrivals	Average stay (days)	Arrivals	Average stay (days)	Arrivals	Average stay (days)
1985	57 400 000	5.88	57 400 000	5.88	57 400 000	5.88
2000	66 500 000	5.07	77 300 000	5.20	83 100 000	5.46
2025	73 500 000	4.59	87 600 000	5.00	106 600 000	5.19

Table 4.31 Projections concerning transportation in Italy according to different scenarios: carriers [Taken from the "Blue Plan" (1987)]

SCENARIO

	Aggravated continuing		Moderate continuing		Aggregated alternative	
	2000	2025	2000	2025	2000	2025
Motor cars (thousand)	26.000	31.000	24.000	29.000	23.000	25.000
Trucks (thousand)	2.350	2.800	2.150	2.550	1.950	2.250
Railway wagons (passengers)	16.000	18.500	18.000	21.000	22.000	28.000
Railwai goods wagons (units)	105.000	120.000	115.000	135.000	135.000	180.000
Merchant ships (tonnage, thousand)	9.100	11.000	9.500	12.000	11.500	15.000

Table 4.32   Costs (billion Italian Lire) of measures proposed in the reclamation plan of the Lambro, Olona and Seveso basin

	TO BE BORNE BY THE STATE (1/1)	TO BE BORNE BY LOCAL AUTHORITIES AND A LIMITED COMPANY* (1/2)	TOTAL (1/1)+(1/2)
Drinking water supplies	300	600	900
Pollution abatement in surface waters	404	1,350	1,754
Protection of water against contamination	500	-	500
Solid urban waste	240	256	800
Solid industrial waste	-	200	200
Sewage sludge	-	100	100
Management of table and ground water	200	-	200
Abatement of atmospheric pollution	3	-	3
Abatement of acoustic nuisance	3	-	3
High-risk productive enterprises	6	-	6
Protected areas	100	-	100
Planning and control system	140	-	140
Personel training	65	-	65
Public information and education concerning the environment	29	-	29
TOTAL	1,990	2,810	4,800

N.B. (*): An appropriate limited company can be formed by the State and the Region to contribute towards the punctual implementation of the measures provided for in the Plan by 1992 by ensuring adequate financing for the investment to be carried out by the local authorities and the limited company itself.

Table 4.33   Financing for measures (billion Italian Lire) provided for under the reclamation plan of the Lambo, Olona and Seveso basin

	1988	1989	1990	1991	1992	TOTAL
(A) TO BE BORNE BY THE STATE						
Law n. 67 of 11.3.1988 (Art. 18, 1st sub-para. (a))	63					
Law n. 67 of 11.3.1988 (Art. 18, 1st sub-para. (b))	73					
Law n. 67 of 11.3.1988 (Art. 18, 1st sub-para. (c))	8					
TOTAL (A)	144	463	416	466	501	1,990
(B) FINANCING						
"Serie speciale Lambro" (Lambro special series) Bonds	144	340	300	220	120	1,124
EIB loans	144	340	300	220	120	1,124
Cassa Depositi e Prestiti	72	170	150	110	60	562
TOTAL (B)	360	850	750	550	300	2,810
TOTAL (A)+(B)	540	1,313	1,166	1,016	801	4,800

Table 4.34　　　　　Reclamation of the Lambro, Olona and Seveso basin: operating and debt servicing costs (billion Italian Lire per annum)

	OPERATING COSTS (4/1)	DEBT SERVICING (4/2)	TOTAL (4/1)+(4/2)
Drinking water supplies	49	66	115
Surface water pollution abatement	85	149	234
Protection of water resources	5	-	5
Solid urban waste	83	62	145
Solid industrial waste and sewage sludge	18	33	51
Soil management	2	-	2
Abatement of atmospheric pollution	-	-	-
Abatement of acoustic nuisance	-	-	-
High-risk enterprises	-	-	-
Protected areas	1	-	1
Planning and control system	2	-	2
Personal training	-	-	-
Public information and education concerning the environment	-	-	-
TOTAL	245	310	555

N.B. (*) Regulatory or planning measures.

Table 4.35　　　　　Reclamation of the Province of Naples: cost of measures provided for in the plan [Ministry of the Environment, 1988]

- measures against pollution of water	Lit	374 billion
- protection of drinking water supplies	Lit	30 billion
- disposal of solid waste	Lit	97 billion
- measures against atmospheric pollution	Lit	39 billion
- measures against acoustic nuisance	Lit	55 billion
- measures against soil pollution	Lit	240 billion
- measures in respect of high-risk industrial installations	Lit	6 billion
- measures against fire risks	Lit	36 billion
- protected areas	Lit	72 billion
- planning and supervision	Lit	68 billion
- public information and education regarding environmental matters	Lit	20 billion
TOTAL	Lit	1 027 billion

Table 4.36   Costs (billion Italian Lire) of the interventions foreseen in the 10-year reclamation plan of the lagoon of Venice

INTERVENTION	1988	1989	1990	1991	1992	1993	1994	1995	1996	1997	1998	TOTAL
1) Sewerage purification and connection systems	185	235	235	275	275	115	115	115	115	100	100	1,865
2) Hydraulic regulation of rivers debouching into the lagoon	-	-	60	60	60	60	-	-	-	-	-	240
3) Irriguous transformations	-	-	40	50	50	50	-	-	-	-	-	190
4) Hydraulic regulation of Brenta, Bacchiglione and Sile rivers	-	-	-	-	-	2	5	5	3	-	-	15
5) Additional works of hydraulic regulation	-	-	-	-	-	9	9	9	9	9	11	56
6) Protected areas	-	-	10	10	10	10	-	-	-	-	-	40
7) Environmental monitoring	-	3	3	4	-	-	-	-	-	-	-	10
8) Waste disposal	200	-	-	-	-	-	-	-	-	-	-	200
9) Harvesting and disposal of algae	10	15	15	15	10	10	8	8	8	8	8	115
10) Vivifying works	12	25	25	25	25	25	25	25	25	25	25	262
11) Reduction of agricultural pollution	-	10	10	10	10	10	-	-	-	-	-	50
TOTAL	407	288	398	449	440	291	162	162	160	142	144	3,043

Table 4.37   Funds needed (billion Lire) for sewage purification and aqueduct systems according to regional water treatment plans

REGION	SEWERAGE AND PURIFICATION	AQUEDUCT	TOTAL
1) VALLE D'AOSTA	70.7	-	70.7
2) TRENTO	59.3	-	59.3
3) BOLZANO	145.6	-	145.6
4) FRIULI VENEZIA GIULIA	216.2	-	216.2
5) LIGURIA	1153.0	401.3	1554.3
6) UMBRIA	528.1	147.3	675.4
7) MARCHE	No approved plan available		
8) LAZIO	2680.0	1270.5	3950.5
9) ABRUZZO	221.8	-	221.8
10) MOLISE	49.6	-	49.6
11) CAMPANIA (without the Province of Naples)	639.5	263.2	902.7
12) PUGLIA	991.3	164.6	1155.9
13) BASILICATA	No approved plan available		
14) CALABRIA	No approved plan available		
15) SICILIA	2773.8	-	2773.8
16) SARDEGNA	853.4	410.5	1263.9
TOTAL	10382.3	2657.4	13039.7

Table 4.38 Financial resources (billion Italian Lire) needed for adapting to needs the solid waste disposal systems in Italy

INTERVENTION	FINANCIAL RESOURCES NEEDED PER YEAR				
MUNICIPAL WASTES	1989	1990	1991	1992	TOTAL
1. Improvement of existing plants	450				450
2. Contruction of new plants	900	900	700	-	2 500
3. Improvement of waste collection systems in large cities	150	150	-	-	300
4. Reclamation of abandoned dumps and contaminated areas	300	500	500	300	1 600
INDUSTRIAL WASTES					
5. Contributions to industry for reducing amount and toxicity of wastes produced	600	600	600	-	1 800
6. Rotative fund for the construction of polivalent platforms for disposal of wastes	300	300	-	-	600
T O T A L	2 700	2 450	1 800	300	7 250

Table 4.39 Estimation of funds needed (billion Italian Lire) in some environmental sectors in the period 1989-1992

SECTOR	REQUIREMENT PER YEAR				
	1989	1990	1991	1992	TOTAL
1. Lambro, Olona and Seveso Basin (comprehensive solution)	1 313	1 166	1 016	810	4 305
2. Po Basin (partial solution)	1 000	1 000	1 000	1 000	4 000
3. Province of Naples (comprehensive)	308	272	218	162	960
4. Lagoon of Venice (partial)	288	398	449	440	1 575
5. Water tratment (rest of Italy) (comprehensive)	3 000	3 000	3 000	3 000	12 000
6. Waste disposal (rest of Italy) (comprehensive)	2 700	2 450	1 800	300	7 250
7. Soil conservation (partial)	1 000	1 000	1 000	1 000	4 000
T O T A L	9 609	9 286	8 483	6 712	34 090

5. GREEK MEDITERRANEAN ENVIRONMENT

The country is located in the northeastern part of the Mediterranean Sea. The coastline of continental Greece is 4,078 kilometers, including Peloponnesos, while that of the islands is 10,942.9 kilometers. This makes a total of 15,020.9 kilometers (National Statistical Service, 1970). Greece has an area of 132,000 km^2 and a population of 9.7 million (population density of 73 inhabitants per km^2). Although the average population density is relatively low, the majority of those involved in economic activities is concentrated in urban areas (58% of the population).
The largest urban centers are located near gulfs (e.g. Athens, Heraklion, Patras, Thessaloniki, Volos). Due to the high urban and industrial waste load in these areas, the pollution of the environment has become a major problem.

Because of lack of adequate space in Greece, there is a strong competition between industry, agriculture and tourism for land in flat areas, especially near the coasts. The above antagonism leads to an irreversible environmental deterioration, which also affects the natural resources and represents a serious problem for the future development of the country. Only a limited degree of planning and resource management has occurred during the last two decades.

In this report, the extent of marine pollution in the Greek seas is investigated. The measures already taken or being planned to control the problem are also described. In particular, the state of the Gulfs of Saronikos, Thermaikos, Pagassitikos and a few others is examined in detail and comparisons are made with other polluted and unpolluted areas. Both programs for sewage treatment plant construction and reception facilities for ship wastes are examined. The chapter further discusses the institutional and legal issues.

5.1.   State of the environment

5.1.1.   State of the sea

Most of the Greek territory borders on the sea, i.e. the Aegean Sea between Greece and Turkey and the Ionian Sea between Greece and Italy.

The surface currents have in general an anticlockwise direction, northwards in the eastern Aegean and Ionian and southwards in the western Aegean, following the general Mediterranean trend. It must be noted that many enclosed bays and gulfs (where the main economic activities occur) have local currents which are wind driven and deviate from the general pattern.

Greek marine waters are generally oligotrophic. Due to low influx of nutrients from land drainage and the particular circulation pattern of the eastern Mediterranean Sea, there is a small phytoplankton crop and a low rate of primary production. However, in some areas (e.g. polluted gulfs) nutrient enrichment induces dense phytoplankton crops. Table 5.1 (Balopoulos and Friligos, 1986) gives the ratio of nutrients for polluted gulfs to Aegean Sea background values. Elefsis Bay, the most industrialized area in Greece, has the highest concentration of most nutrients, especially ammonia (NH_4^+-N) (up to 15.8 times background). The Bay of Thessaloniki, contaminated by sewage and industrial effluents, shows the highest phosphate (PO_4^{3-}-P) values (up to 5.3 times background). The inner Saronikos (influenced by sewage effluents) and western Saronikos Gulfs (having the greatest depths) have equally enriched phosphate values. The western Thermaikos Bay and the Alexandroupolis Gulf have high silicate (SiO_4^{4-}-Si) values due to the discharge of rivers. The North Euboikos Gulf experiences high nitrate (NO_3^--N) and silicate concentrations from upwelling currents from great depths and underwater springs.

Generally, the quality of the receiving waters, with respect to nutrients, is a function of the different sources of nutrients, as well as the morphology of the area and the water circulation.

Heavy-metal concentrations along the Greek coast reveal some high levels in specific areas, as recorded by the European Community (EC) in 1987 (CEC, 1988) (Figure 5.1). Significant levels of mercury were detected near the Xanthi industrial area, Thassos and Ialyssos (Rhodes). Increased cadmium concentrations were observed mostly near the island of Tinos. In the Dodecanese area high levels of zinc were detected. The Athens coastal area was found systematically polluted with mercury, cadmium and lead.

Oil pollution is a major problem for the Greek coasts (tar on beaches), especially in the northern Ionian Sea with a mean tar concentration of 150 µg/m^2 (Golik et al., 1988).

Deballasting of oily waters from ships and release of oily compounds into the sea were permitted until 1978 on both sides of South Crete. However, after 1980, measurements of pelagic tar in the eastern Mediterranean indicate a reduction in quantity.

During the period from July 26 to August 1, 1987, 58 Greek beaches were surveyed for bacteriological bathing-water quality by the EC (CEC, 1988), as shown in Figure 5.2. Of the total 58 beaches examined, 39 were in compliance with the EC guide number (recommended criteria limit), 18 exceeded the EC guide number but were in compliance with the required standard, while 1 (Corfu) exceeded the required standard.

5.1.1.1. Saronikos Gulf/Elefsis Bay

Since 1951 the population of the Greater Athens metropolitan area has risen from 1,400,000 to 3,500,000 inhabitants.

This growth continues, though more slowly, causing acute problems. The Saronikos Gulf is polluted, due to domestic and industrial effluents.

The quality of the sea bottom has undergone a great change with respect to heavy metals and organic carbon in the following areas:
- Drapetsona Bay (wastes from a large fertilizer factory)
- Keratsini Bay (outfall of the main sewage discharge of the Athens municipality)
- Port of Piraeus entrance
- Elefsis Bay (industrial zone).

In the above areas, concentrations of various heavy metals have risen considerably above background levels (Voutsinou-Taliadouri, 1981). The bottom-sediment concentrations are shown in Table 5.2 and indicate that the heavy-metal concentrations in Drapetsona Bay have increased 1,000 times for copper (Cu), 3,500 times for iron (Fe) and 700 times for zinc (Zn) compared to the reference values (Voutsinou-Taliadouri, 1981). In Keratsini Bay organic carbon levels have increased 30 times above normal levels. The total affected area in the Saronikos Gulf is estimated at 40 km^2. Total, particulate and dissolved chromium in the water column was measured in samples collected at 23 stations during three cruises (March, July and October 1984) in the upper Saronikos Gulf (Grimanis et al., 1985).
Maximum values of chromium were found near the Athens sewage outfall (ASO). Total chromium concentrations ranged from 0.45 to 18 µg/l. About 5 times more total chromium was found in seawater samples near ASO than in samples at stations situated several miles away. Dissolved and particulate chromium concentrations ranged from 0.33 to 12 µg/l and 0.10 to 12 µg/l, respectively. The predominant chromium form was particulate chromium (40-70% of total) at all stations that were under the immediate influence of the ASO. Dissolved chromium is the predominant form (80-99% of total) at stations not directly influenced by the ASO.

Decreasing total and dissolved chromium concentrations with depth were observed at some stations close to ASO.

Very high insecticide dichlorodiphenyl trichlorethane (DDT) and polychlorinated biphenyl (PCB) concentrations of 1,900 and 775 µg/kg dry weight, respectively (Fowler, 1985), have been measured in sediments near the Athens outfall in the Saronikos Gulf. The above values decrease to below 20 µg/kg in a distance of about 5 kilometers, while average Mediterranean values have a range of 0.8-9.0 µg/kg dry weight. It should be noted that while the use of DDT has been prohibited since 1970 it is still present in the environment.

As part of a monitoring program to estimate the damage inflicted by PCBs and DDT, specimens of the striped mullet (Mullus barbatus) have been obtained from the Saronikos Gulf since 1975 (Voutsinou-Taliadouri and Satsmadjis, 1982a). This fish species was chosen because of its commercial interest. It also feeds on organisms living on the bottom of the sea where pollutants, if present, accumulate. As striped mullets are detected throughout the Mediterranean Sea, comparisons are possible throughout the basin.

Five sampling areas were selected in the Saronikos Gulf (Table 5.3). Area A comprises the site of the sewage outfall at Keratsini Bay and Piraeus Harbor, as well as the eastern section of Elefsis Bay. Area B adjoins area A to the south, but it is not directly exposed to intense pollution. Area C, further to the south at the entrance to the gulf, lies in the main path of the outgoing stream. On the other hand, areas D and E are partially sheltered from the major sources of pollution due to the predominant currents. PCBs and the common chlorinated pesticides were investigated.

Table 5.3 indicates, for each parameter and area, the average of the results of all analyses ("mean"), their maximum ("high") and minimum ("low"), the standard deviation ("σ") and the ratio of the mean in that area to the background mean in area E ("A/E", "B/E", etc.). The samples from the most polluted area, in the vicinity of Piraeus Harbor, contained on the average 15 times as much PCBs and DDTs as those from the cleanest area. The ratios A/E, B/E etc. show that the concentrations of the major chlorinated hydrocarbons (PCBs, DDE, DDT and DDD) decrease sequentially from area A to B, to C, to D and E; most pronounced from A to B. This shows that striped mullets take in the toxic compounds either through the gills or from food at the sea bottom to reach a concentration parallel to the sediment content.

The decrease with increasing distance from the source depends on the constituent. PCBs used in industries will originate primarily from the central sewage discharge. The PCB levels drop sharply from A to B (2.6 times); from B to C (2.5 times); and from C to D (1.9 times). The pesticides DDE, DDT and DDD also originate from the countryside. The values of DDD decline even more pronounced than those of PCBs: 5.3 times from A to B; and 3.0 from B to C. On the other hand, DDT and DDE decrease stepwise. Their concentrations diminish 3.8 and 2.4 times, respectively, from A to B, do not change significantly from B to C, then decrease 2.7 and 2.2 times, respectively, from C to D. This suggests perhaps that DDE has a greater resistance to degradation in the sea or within the striped mullet than DDD. Amongst the minor constituents (benzene hexachloride (BHC), heptachlor epoxide, dieldrin and endrin) only the last two show a declining trend.

Another feature of the major chlorinated hydrocarbons is the large spread of the concentrations in a given area. Thus, in areas A, B, C, D and E the means of the high to low ratios are 4.7, 5.7, 18.3, 12.6 and 10.5 respectively. This leads to the inference that the striped mullet does not always remain in the same area, but travels across the Saronikos Gulf. Fish heavily contaminated, migrating within a few weeks to less foul waters, would still contain high levels of the toxic chemicals since the detoxification process is a slow one.

It must be noted that except for several specimens caught

near the sewage outfall the concentrations of the organochlorine residues stay well below the health limits.

The main sources of discharged nutrients in the Saronikos Gulf are the following (Friligos, 1981a):
1. Factories around the northeast and west side of Elefsis Bay.
2. The shipbuilding yard in Perama Bay.
3. The electricity-generating plant at Keratsini Bay (oil fired) which recirculates seawater for cooling purposes at a rate of 5 m^3 per sec, causing some nutrient release and temperature increase.
4. The sewage outfall at Keratsini Bay discharging at a rate of 350,000 m^3 per day.
5. The fertilizer factory near Drapetsona Bay with a daily production of 700 tons of fertilizers and 200 tons of phosphoric acid.
6. Piraeus Harbor activities.

The situation is particularly serious at Keratsini Bay, which is only 25 to 35 meters deep and receives all the sewage from Athens. Table 5.4 shows the main characteristics of the sewage effluent in this area. For the Saronikos Gulf, the observed nutrient concentrations in the 24-hour composite sewage effluent are within the limits specified by the Greek Ministry of the Environment, except for ammonia (Ministry of the Environment, 1987).

It was determined that the fertilizer factory at Drapetsona Bay is the second main nutrient source in the Saronikos Gulf after the sewage outfall at Keratsini. It contributes large amounts of nitrate and silicate and smaller quantities of phosphate and ammonia, resulting in high ambient concentrations in Drapetsona Bay (Friligos, 1981a).

The increased nutrient concentrations in the waters near the east coast of the Saronikos Gulf have caused the occurrence of the "red tide" phenomenon especially from dinoflagellates. This was observed for the first time ever in this area during the summer of 1978 (Satsmadjis and Friligos, 1983). The nitrates appear to be the main limiting factor to this proliferation of phytoplankton organisms.

Elefsis Bay lies at the north end of the Saronikos Gulf with depths of about 30 meters. The bay receives industrial effluents, mainly from oil refineries and steel works, along its northern shore and domestic wastes through the east channel from the Keratsini Athens outfall. These inputs generate severe ecological stresses, the most acute being the summer anoxic conditions (Friligos, 1983).

Water samples were collected at station K (indicated by * in Table 5.3), the deepest location in Elefsis Bay, from 1, 10, 20 and 30 meter(s) depth at seasonal intervals between 1973 and 1976. Temperature, salinity, dissolved oxygen and inorganic nutrients were measured (Friligos, 1983).
The salinity range was 38-39°/$_{oo}$. There was one temperature cycle during the year. Cooling of the upper layers reached a minimum in February-April, with temperatures of about 12-14° C. The water mass warms up and eventually reaches a temperature of about 25° C in August. Summer temperature differences of 10° C between sea surface and bottom result in the development of a strong stratification, which persists for about 4 months and causes anoxic conditions below 20 meters. Biodetritus in the water column, terrigenous detritus from run-off, and accumulated detritus in the sediments are the most likely deoxidants. The water-column biodetritus also consumes oxygen since it is supplemented continually by algal growth in the surface water resulting in sinking dead algae. The bottom sediments are effective deoxidants, as the water-sediment interface, with the associated decomposition microorganisms, oxidizes solubles that diffuse out of the sediments.
The silicate concentrations range from 28 to 280 µg/l in the top 20 meters. Below 20 meters, the values were 140-560

µg/l in 1974 and 140-980 µg/l in 1975. Silicate maxima occurred at the bottom of the water column at times of stratification and deoxygenation. This silicate is produced by the biochemical degradation of biodetritus, either in the water column or at the sediment interface.

Phosphate maxima occurred at the bottom of the water column at times of stratification and deoxygenation. The values of phosphate, down to a depth of 20 meters, ranged from 3.1 to 31 µg/l. Below 20 meters, the values were 9.3-15.5 µg/l in 1974 and 31-217 µg/l in 1975. The highest values of up to 217 µg/l occurred during anoxic conditions in 1975. Under oxic conditions phosphorus accumulates in the bottom sediments (chemical precipitation and sedimentation of detritus) while under anoxic conditions phosphate is released from the sediments.

The highest nitrate values of 14-70 µg/l were observed in the winter. The nitrate decreased to below 1.4 µg/l due to denitrification and biological uptake during the period from July to October 1975. The nitrite varied between 1.4-7.0 µg/l in the top 20 meters and between 7.0-14 µg/l below 20 meters. The corresponding ammonia concentrations were 14-210 µg/l and 210-280 µg/l. It should be noticed that nitrite exceeded nitrate concentrations in June 1975, with the onset of denitrification. Higher ammonia values, similar to silicate and phosphate, occurred in 1975 during prolonged anoxic periods. The release of ammonia and its subsequent oxidation to nitrate via nitrite proceeds throughout the winter and spring. These nutrients become evenly distributed within the entire water column after destratification.

The subsurface waters of Elefsis Bay contain very high levels of ammonia, phosphate and silicate, especially during the end of the anoxic period of up to 280 µg N/l, 217 µg P/l and 980 µg Si/l respectively. Ammonia is the predominant form of inorganic nitrogen throughout the water column.

Measurements at station K in Elefsis Bay showed mean concentrations of ammonia (NH_4^+-N), nitrate (NO_3^--N), nitrite (NO_2^--N), phosphate (PO_4^{3-}-P) and silicate (SiO_4^{4-}-Si) in the water 15.4, 6.7, 3.1, 6.2 and 4.6 times the mean levels in the outer Saronikos Gulf, respectively (Table 5.5). The predominance of ammonia over all the forms of inorganic nitrogen was the result of the anoxic conditions prevailing below the stable thermocline and the direct additions of large loads of ammonia from both domestic and industrial effluents.

The bathing-water quality around the heavily populated Attika peninsula has been monitored for relatively long periods. Between 1976-1983 and beyond, EC criteria for bathing waters were not met near the sewage outfall of Keratsini and as a consequence the beaches were closed to the public.

5.1.1.2. Gulfs of the northern Aegean Sea

The northern part of the Aegean Sea has semi-enclosed, relatively shallow and fairly small water bodies.

A survey of DDT, PCBs and other organochlorines in marine organisms started in December 1975 and continued through December 1979, and covered the three main gulfs, Thermaikos, Strymonikos and Kavala. The above gulfs receive domestic, industrial and agricultural wastes and the discharge of six important rivers.
Samples of the mussel Mytilus galloprovincialis showed significant PCB increases during the period of investigation, as shown in Table 5.6 (Kilikidis et al., 1981). The concentration of PCBs in mussels doubled in 1979, reaching 580 µg/kg wet weight, as compared to 1975-1976. Organochlorine pesticides had smaller changes as shown in Figure 5.3. The northern Aegean Sea is characterized by high values of PCBs in mussels, while the values of chlorinated insecticides in the same organisms are low compared to other Greek areas.

Thermaikos Gulf

The Thermaikos Gulf (Figure 5.4) in the northern part of Greece is fairly long and shallow with a depth not exceeding 50 meters. Three rivers discharge into the gulf: the Axios (containing large quantities of industrial effluents), the Loudias, and the Aliakmon. The innermost section of the gulf receives the effluents of about 240 factories and the domestic wastes of the Greater Thessaloniki area (1,200,000 inhabitants).

Samples of surface sediments of the Thermaikos Gulf were collected at a network of 56 stations closely spaced in the innermost section and at the mouth of the Axios River and analyzed for cadmium (Cd), lead (Pb), chromium (Cr), copper (Cu), nickel (Ni), cobalt (Co), zinc (Zn), manganese (Mn) and iron (Fe) (Voutsinou-Taliadouri and Leondaris, 1986). The most polluted section is located near the sewage outfall to the east of the Bay of Thessaloniki and the industrial zone to the west (Table 5.7). Next comes the area around the mouth of the Axios River. The highest pollution ratios (mean divided by reference value) are exhibited by cadmium, zinc, lead and copper (17.8, 16.0, 11.8 and 7.9, respectively). Zinc, lead and copper have a similar pattern. Cadmium has its highest levels close to the Axios River. The Aliakmon River causes only moderate nickel, cobalt and chromium pollution.

A two-year investigation (Table 5.8) found considerable concentrations of lead (Pb), cadmium (Cd) and mercury (Hg) in marine organisms (Vasilikiotis et al., 1983). Mercury concentrations in molluscs were 86 µg/kg wet weight near the industrial zone and 29 µg/kg in the unpolluted zone representing a concentration factor of 3.0. The concentrations have not yet reached permissible levels. The main source of mercury is one factory using chlor-alkali electrolysis in its manufacturing process.

The most voluminous organic matter discharges occur in the

eastern part of the Bay of Thessaloniki, receiving the city sewage effluent. This part of the bay is characterized by high organic carbon in the sediment and low dissolved oxygen near the sea bottom, particularly during the summer (Sobemap, 1981). The western section of the bay is affected primarily by discharges from the industries. The outer part of the Bay of Thessaloniki receives the outflow of three rivers (the Axios, the Loudias and the Aliakmon) and is least affected.

Three municipal discharges (Table 5.9) exceeded the acceptable level of 80 mg/l for organic matter as biochemical oxygen demand (BOD) while all municipal discharges and one industrial discharge exceeded the acceptable level of 250 mg/l for chemical oxygen demand (COD). For the entire Thermaikos Gulf a discharge of 2,800 l/sec, 45,000 kg of BOD per day and 59,500 kg of COD per day was measured.

The limits for BOD and COD adopted by the Greek Ministry of the Environment for Thermaikos Gulf are higher than the corresponding limits set for Saronikos Gulf by a factor of 2 and 1.5, respectively. The different limits for the two gulfs reflect their different absorbing capacities.

The distribution of nutrients (phosphate, nitrate, nitrite and ammonia) in water taken at 11 sampling stations in the Thermaikos Gulf (Figure 5.4) was studied over a period of 1 year in 1985-1986 (Samanidou et al., 1987).

Stations 1 and 5 were located in an area where slaughterhouse and municipal wastes are discharged. Stations 2 and 11 are near an industrial area and a harbor respectively. Station 4 is near the Axios estuary, while 3, 9 and 10 are in recreational and mussel cultivation areas. Station 8 was located in the middle of the gulf. Stations 6 and 7 are in areas which do not receive any wastes and which are used for fishing and recreation.

Strong eutrophication conditions prevail at stations 1 and 5. Their high nutrient values are due to the proximity to slaughterhouse and municipal sewage outfalls.

In the winter a high total inorganic nitrogen content was

found throughout the study area, followed by a minimum in the summer (Figure 5.5). The concentration of ammonia (NH_4^+-N) at station 1 was extremely high (6.25 µg-at/l or 87.5 µg/l) due to anoxic conditions. The nitrite (NO_2^--N) concentration was nearly five times higher than the average.

A significant increase in nitrate (NO_3^--N) (5.31 µg-at/l or 74.3 µg/l) was observed in the estuarine waters of the Axios River. During heavy rainfall the nutrients (especially nitrate) are carried by run-off from the surrounding agricultural area after fertilizer application. Plotting the monthly values of nitrate concentrations for station 4 versus time, a curve with a maximum nitrate concentration occurs at the end of the winter and during the spring, while minima occur in the summer. The peak concentrations coincide with heavy rainfall.

The concentration of orthophosphate (PO_4^{3-}-P) increased slightly during the winter and decreased during the summer. Significant seasonal fluctuations were observed for ammonia, with a decrease in the summer.

The lowest ratio of atomic nitrogen to phosphorus (N/P) of 2.71 was observed at station 5 near the sewage outfall. In unpolluted oceanic waters a ratio of 16 : 1 is typical. The results show an increase in the ratio of nitrogen to phosphorus with increasing distance from the sewage outfalls, due to the relative rapid phosphorus decrease.

The nutrient data show that the Bay of Thessaloniki can be divided into two principal regions. The inner bay is characterized by high nutrient concentrations reflecting strong anthropogenic influences. The outer bay is characterized by lower concentrations that are similar to literature values for slightly polluted areas.

Pesticides and PCB (Ministry of the Environment, 1985) enter the bay through the three rivers. The U.S. Environmental Protection Agency (EPA) has adopted a maximum allowable concentration of 1 ng/l of PCB in coastal seawater. PCB concentrations of 2.1-4.76 ng/l (Table 5.10) in the Thermaikos Gulf exceed this limit. The pesticide concentrations mentioned in Table 5.10 are within the limits adopted by the U.S. EPA (Lamb, 1985).

Kavala Gulf

The Kavala Gulf is situated close to Kavala City in the north of Greece which is rapidly developing due to exploitation of offshore oil wells.

Water samples were collected twice a month at four sampling stations in the Kavala Gulf during a period of two years from 1981 and 1982. Analyses were conducted for organochlorine pesticides (the organochlorine biocide lindane, p,p'-DDT and dieldrin), PCB, polycyclic aromatic hydrocarbons (PAH) and volatile halogenated hydrocarbons (tetrachloroethene and trichloroethene) (Fytianos et al., 1985).
Pesticides were found in trace amounts (Table 5.11). PCB exceeded the U.S. EPA permissible level of 1 ng/l. The highest concentrations of pesticides (except p,p'-DDT) and PCB were measured near the offshore oil platform. The p,p'-DDT maximum occurred near the industrial area. The second highest concentration was measured in the harbor area.

PAH ranged from 27 in the middle of the gulf to 80 ng/l near the offshore oil platform (followed by the values in the harbor area). These values are comparable to those reported in the literature for unpolluted and slightly polluted areas, respectively.

The tetrachloroethene and trichloroethene concentrations varied between 0.35 and 3 µg/l, with the highest near the offshore oil platform, followed by the values in the harbor area. The concentrations in the waters near the offshore platform, the harbor and the industrial area exceed the EC permissible level of 1 µg/l.

5.1.1.3. Pagassitikos Gulf

The pollution in the Pagassitikos Gulf comes primarily from the sewage of the municipality of Volos (107,000 inhabitants) and the adjoining industrial zone (Figure 5.6). During the rainy season, some waste water from the industrial city of Larissa (102,000 inhabitants) is discharged into streams which enter the gulf at Volos.

High PCB concentrations have been measured (Satsmadjis et al., 1988) in the muscle of the striped mullet *Mullus barbatus* (Table 5.12). Significant increases have been found during the rainy season. The Pagassitikos Gulf (1986-1987) is not as heavily polluted as the Saronikos Gulf (1975-1976).

5.1.1.4. Euboikos Gulf

High chromium (Cr) concentrations (up to 41,000 mg/kg) were measured in sediments of the North Euboikos Gulf (Voutsinou-Taliadouri and Varnavas, 1985).
The study area is of special interest because it receives the slag from the "LARKO" iron-nickel alloy smelting plant. The maximum concentrations occurred 8 kilometers northeast of the smelting plant, directly near the area of discharge where dumping is allowed. The degree of enrichment in chromium is 1,000 relative to normal pelagic sediments. However, the concentrations of chromium decrease to background values within a few kilometers from the site, suggesting that the discharged slag is retained in the area. A comparison with other Greek regions (Table 5.13) shows that chromium in the North Euboikos Gulf is about 100 times higher.

5.1.1.5. Patraikos Gulf

In August 1980 and February 1981, sediment samples were collected at 21 stations in the Greek Gulf of Patraikos

(Voutsinou-Taliadouri and Satsmadjis, 1983).

The heavy-metal concentration tends to rise in the central part of the gulf where the sediment is finer (high adsorption capacity) and the depth greater. Neither the port of Patras nor the river deltas perceptibly disturb this sedimentation pattern. The range of mean concentrations of heavy metals between August 1980 and February 1981 in mg/kg (except for iron in $°/_{oo}$) are shown in Table 5.14.

The levels of lead (Pb), cobalt (Co), nickel (Ni), chromium (Cr) and iron (Fe) are similar to those in the eastern Aegean Sea while copper (Cu), zinc (Zn) and manganese (Mn) are about two times higher as shown in Table 5.15 (Voutsinou-Taliadouri et al., 1987).

5.1.1.6. Aegean Sea

Sediment samples from the eastern Aegean were analyzed for trace metals in March 1980 (Voutsinou-Taliadouri and Satsmadjis, 1982b). Lead (Pb) levels in the eastern Aegean Sea were found to be 2.7 times lower and zinc (Zn) levels were 3.2 times lower than those in two polluted Aegean Gulfs (Table 5.16).

The distribution of trace elements in the surface waters of the eastern Aegean Sea has also been investigated (Scoullos and Dassenakis, 1986) and compared to other seas (Table 5.17). The Aegean Sea has rather low copper (Cu) and rather high lead (Pb) and zinc (Zn) concentrations. Metal concentrations in the eastern Aegean appear to be higher than the Mediterranean average by a factor of 2.4 for Zn, 1.9 for Cu and 6 for Pb. This may be due to the influence of the Sea of Marmara, a high-salinity landlocked sea.

The nutrient content in the waters of the eastern Aegean has also been studied (Friligos, 1986). Nutrient levels were much higher (25 times for nitrates and 4.4 times for silicates) in the northeastern as compared to the south-eastern Aegean. The maximum values shown in Table 5.18 are mostly from the northeastern Aegean. The nutrient richness in the north may be due to waters originating from the Sea of Marmara and the Black Sea.

Dissolved/dispersed hydrocarbons were measured in 1980 in the eastern Aegean (Gabrielides and Hadjigeorgiou, 1986). High concentrations in the range of 3.99-14.81 µg/l (in Kuwait crude oil equivalents) were observed, indicating significant pollution by petroleum hydrocarbons (mainly light fuel oil). For comparison the observed hydrocarbon concentrations in the Saronikos Gulf were in the range of 1.6-5.6 µg/l in 1980-1981 (MAP, 1987). The situation is worst in the northern Aegean due to heavy shipping traffic to and from the Black Sea as well as natural oil seepage from the oil fields near the island of Thassos.

5.1.1.7. Ionian lagoons

Sediment samples were collected at different points in the Messolonghi region (Palaiopotamos, Aitolikon, Messolonghi and Kleissova lagoons) on 7-9 June 1983 and 26-28 March 1984. Except for the Aitolikon Lagoon the depth is in the range of 0.2-2.0 meters, characterizing the region as brackish lagoons.
Organic carbon rose (6.9-4.4%) at the relatively deeper (0.7-1.2 meters) points and fell (1.8-2.4%) at the shallowest (0.2-0.4 meters) locations (Voutsinou-Taliadouri et al., 1987). The mean value of 3.39% is ten times the value measured in the eastern Aegean Sea (0.35%).
Nutrient-rich run-off water, evaporation, relative isolation from the open sea and shallow depth cause the accumulation of organic matter in the sediments.

5.1.2. State of the freshwater

Water deficits result from the unequal rainfall in time and space, the unequal distribution of activities and, consequently, of consumption patterns in some regions of the country. The western part of the country has the largest run-off. Water inter-basin transfer is used to solve the drinking-water supply to Athens. A similar but much more costly work is planned for the water supply of the plain of Thessalia using the waters of the upper Acheloos River. This multipurpose hydraulic work includes three dams for electricity and will cost more than 200 billion drachmae. Due to financial difficulties the above project has not yet started.

A general picture of the availability of water resources in Greece is presented in Table 5.19. Data are based on a publication on long-term perspectives in the EC region (UN/ECE/WATER 26/1981). The total water flow in the country was 62.9×10^9 m^3 per year (in 1980). The total water abstraction in 1980 was 6.95×10^9 m^3, the total water consumption was 5.38×10^9 m^3, while the amount of collected waste water was 0.67×10^9 m^3 (Table 5.20). The difference between abstracted and consumed amount of water has been due to recycling and reuse, especially in agriculture (Table 5.21). A 36% increase in the demand for water is expected by 2000. Thus more investments are needed in order to secure the availability of the required amount of water.

Rivers in Greece are relatively short (not more than 300 kilometers) and four of them are shared with the neighboring countries (the Evros, Nestos and Strymon rivers with Bulgaria and the Axios River with the former Yugoslavian states). Table 5.22 presents the main Greek rivers, their drainage area and average outflow of 34.6×10^9 m^3 per year. Most of the main rivers discharge to the Aegean Sea while the Acheloos, Alfios, Arachthos, Kalamas and Louros rivers discharge to the Ionian Sea.

An extended monitoring network of the Ministry of Agriculture controls the quality of surface waters used for irrigation. Another parallel network by the Ministries of Environment and Agriculture controls the quality of the six main rivers (the Nestos, the Strymon, the Axios, the Aliakmon, the Pinios and the Acheloos). The quality of these rivers has been in accordance with the EC directives with the exception of the bacteriological quality, due to domestic sewage discharges.

The average concentration of fecal coliforms was 4,288/100 ml in the Axios River in 1985 and that of fecal streptococci as high as 2,179/100 ml. In the Strymon River the values were 6,518/100 ml and 3,580/100 ml, respectively (CEC, 1987).

The BOD discharge from 7 major Greek rivers into the Mediterranean is shown in Figure 5.7. The Acheloos discharged the largest BOD load.

5.1.3. State of the coastal strip

The coastal zone in Greece extends over a length of 15,021 kilometers and is almost equally divided between the mainland and the islands. Over 57% of the population lives in this area. The five largest urban centers are located in the coastal zone (Athens-Piraeus, Thessaloniki, Patras, Heraklion, Volos). Over 80% of all industrial activities and 90% of all tourist and recreational activities are carried out in this area.

The main pollution sources (Camhis and Coccossis, 1982) in the coastal strip are uncontrolled, unsanitary municipal refuse and hazardous waste disposal sites.

The use of sanitary landfills is very limited in Greece. Even in the Greater Athens area, only two are sanitary landfills out of 30 designated disposal sites and both of them do not function properly. Approximately 65% by weight of the municipal refuse is dumped in 1,420 places which

have a license from the local authorities. The remainder is dumped in 3,430 places without a license. The production of leachates with BODs above 20,000 mg/l, the production of flammable and explosive gases (methane, mercaptans) from anaerobic decomposition and foul smells present real problems (Ministry of the Environment, 1987). Due to the porocity of the soil, groundwater contamination occurs in many sites. Surface waters are contaminated adjacent to landfills.
The situation is worst in Volos and Thessaloniki where municipal wastes are co-disposed with industrial wastes.

A serious pollution problem is caused by the disposal of pigsty and olive-mill wastes in septic tanks at the islands of Crete, Corfu, Thassos and also around Amvrakikos Gulf and South Peloponnesos. Domestic wastes are also frequently dumped in septic tanks. Tourism development and the construction of vacation homes exacerbate the problem, particularly at the islands.

The coastal strip also faces an erosion problem, mainly at the islands (except for Andros) and in the area of East Macedonia and Thrace. The problem is attributed to lack of vegetation, and overgrazing. These areas also suffer from frequent fires in the summer season. The acreage anually destroyed by fire ranges from 22 km^2 in 1959 to 1,054 km^2 in 1985. The area destroyed each year has increased between 1955 and 1986. It is generally believed that most fires are deliberately set to destroy coastal habitat and to make land available for tourist activities.

Another serious problem found all along the Greek coast is saltwater intrusion in aquifers caused by coastal groundwater overpumping (Doxiadis Centre for Ecistics).

The following eight coastal sites are considered Wetlands of international importance by the RAMSAR Convention (UNEP/WG.163/4, 1987):

- Evros Delta, Nestos Delta, Lake Mitrikou, Lake Vistonida (Thrace);
- Axios Delta (Macedonia);
- Amvrakikos Gulf, Messolonghi Lagoon (eastern central Greece);
- Kotykhi Lagoon (northeastern Peloponnesos).

Domestic waste waters, run-off from cultivated land and wastes from animal breeding units have damaged the above biotopes. For example, Vistonida's fish harvest has declined to one-seventh of the 1973 harvest, while mass fish deaths have been reported in Messolonghi.

The Specially Protected Areas Protocol of the Barcelona Convention designated existing "marine and coastal protected areas of the Mediterranean region" having an ecological and biological value. The following four are located in Greece:
- Gorge Samaria National Park (Crete)
- Vai Aestetic Forest (Crete)
- Pefkias-Xylokastron Aesthetic Forest (northern Peloponnesos)
- Sounio National Park (Attika Prefecture).

Fires and overuse by tourists are the most serious problems for the above four areas.

Sixteen coastal sites in Greece have been selected as Coastal Historic Sites of Common Mediterranean Interest by the Contracting Parties to the Barcelona Convention (UNEP/IG.74/5, 1987).
They are the following:
- Athens, Delphi, Meteora (central Greece);
- Mount Athos (Macedonia);
- Epidaurus and Nauplion, Mycenae, Olympia, Tiryns (Peloponnesos);
- Cnossos, Phaestos (Crete);
- Delos, Paros and its quarries, Rhodes, Thera, Thassos and its quarries (Aegean Islands);
- Corfu (Ionian Islands).

Most of the damage to the monuments in Athens is caused by acid deposition (Ambio, 1977). Data on the quality of the environment, specifically the atmosphere, are necessary for possible investments by the Mediterranean Action Plan or the United Nations Educational, Scientific and Cultural Organization (UNESCO).

5.1.4. Air pollution

Air pollution problems, especially the smog formation ("Nefos"), are severe in Athens, where 70% of the economic activity and almost 40% of the population is concentrated. Air pollution increased rapidly in Thessaloniki (Salonika) which is the second urban and industrial centre in the country. Smog formation begins to appear in a number of other towns (e.g. Heraklion, Kavala and Patras).

Traffic is the main air pollution source in Athens, contributing 75% on a yearly basis in comparison with all other sources. Industry contributes about 20% and domestic sources 5%. The contribution of the different pollution sources to nitrogen oxides (NO_x) and volatile organic compounds (VOC) emissions in Athens is presented in Table 5.23.

During recent years severe air pollution episodes occurred in Athens. During air pollution episodes, industrial activities and traffic are greatly reduced in accordance with the existing emergency plan. Photochemical smog episodes that occur are comparable to those in Los Angeles (Figure 5.8).

Figure 5.9 presents the results of measurements from the PERPA network after statistical analysis, compared with the international air quality standards. Smoke, total solid particulates (TSP), nitrogen dioxide (NO_2) and carbon monoxide (CO) are the main pollutants exceeding the ambient air quality standards. There is a severe smoke problem in

Athens city center with annual average values of up to 145 $\mu g/m^3$, compared with the limit of 80 $\mu g/m^3$, set by the EC. The U.S. Environmental Protection Agency limit of 260 $\mu g/m^3$ for TSP (not to be exceeded as an hourly average more than once per year) was surpassed in 5, 8, 14 and 20% of the measurements in Elefsis, Athens city center (at two of the sampling stations), Rentis and Drapetsona, respectively.
The NO_2 hourly limit of 200 $\mu g/m^3$ (for 98% of the measurements in one year), set by the EC, was exceeded in 1% of the measurements at two sampling stations in Athens city center. The World Health Organization's CO limit of 10 mg/m^3 for 8-hour averages was surpassed in 30% of the measurements at one sampling station in Athens city center.

Sulfur dioxide air pollution in Athens has been reduced due to the restrictions placed on fuel oil used in the central heating in the Greater Athens area. The reduction in the sulfur content to 0.3% in diesel for traffic and central heating had a similar effect. The reduction in the lead content in gasoline, from 0.84 to 0.15 g/l, has resulted in a decrease in lead concentrations in the air. The ambient lead concentration of 0.8 μg per m^3 of air, as shown in Figure 5.10, is within the guideline set by the World Health Organization at an annual mean of 0.5-1.0 mg/m^3.

Because of high taxation, the majority of the cars are old and small. One-third of the imported cars, registered in Athens in 1986, were second hand (L. Patas, private communication). The application of EC regulations with lower taxes for new modern cars will decrease purchases of old cars, resulting in lower emissions.

The estimated costs of the air pollution control investments in Athens are as follows:
1. $1,100 million (U.S.) for the construction of the metro over a six-year period.
2. $220 million (U.S.) for traffic control improvements over a five-year period.

The power plants constitute the largest air pollution source of sulfur dioxide in Greece. They are located near Ptolemais, Megalopolis, in Attica and on Euvoia Island.

An account of the air-pollutant emissions from the different sources is shown in Table 5.24. The power plants contribute 57% and industry 27% to the 582,000 tons per year of sulfur dioxide (SO_2) emissions. Mobile sources are responsible for 59% of the nitrogen oxides (NO_x) emitted to the atmosphere.

The use of lignite among the primary sources of energy in the generation of electricity is increasing (Figure 5.11), while the relative use of oil is decreasing, thereby reducing the dependency on oil imports. The mining of lignite is expected to increase from 35 million to 65 million tons per year within 10 years. Fly ash emissions, already affecting the mining sites, will further increase. The installation of electrostatic precipitators is a necessity.

5.2. Pressures on the environment

5.2.1. Human settlements

The geographical characteristics of the country (more than 70% is mountainous) and historical socioeconomic reasons have promoted the development of human settlements in the coastal areas. Thus the five largest Greek cities, Athens, Thessaloniki, Patras, Heraklion and Volos, have developed in the coastal zone. Between 1951 and 1981 the urban population increased by 96%, while in the same period the Greek population as a whole increased by only 28%. The corresponding rates for the period between 1928 and 1951 are 49% and 23%, respectively (Table 5.25). This rapid increase created a number of environmental problems because the necessary infrastructure in the coastal cities was not developed accordingly. Most of the municipal wastes have been disposed of in the sea without any treatment. The total organic load originating in Greece and discharged to the Aegean and Ionian Seas amounts to approximately 23 million population equivalents, of which about 20% is domestic and 80% industrial and agricultural (Table 5.26).

The mild climate, the landscape, the extended sandy beaches and the historic sites of Greece attract a large number of foreign tourists. The tourist industry has become a major economic activity. Between 1962 and 1982 the number of foreign tourist arrivals in Greece increased by 890% to more than 5 million per year. Internal tourism has also increased significantly (8.8 million overnight stays in a total of 44 million for 1986). Approximately 90% of all tourism activities occur in the coastal areas, resulting in additional environmental degradation. Some indicators of environmental pressures in Greece are shown in Table 5.27.

Tourist zones already saturated are located in Attika, Thessaloniki, Corfu, Rhodes, Kos and northeastern Crete. Areas under development are situated in North Peloponnesos, western Greece, Chalkidiki and Korinthia (Tourist Organization of Greece).

5.2.2.   Industrial activities

Industrial activities have increased at a high rate (Table 5.27). Almost 80% of the Greek industry is located in the coastal zone. Furthermore, most of the industrial plants have been concentrated in the Greater Athens and Thessaloniki areas (the exception being the power stations). In particular, all four existing refineries, three of the four fertilizer plants and all of the metallurgical industries are located in these areas (Figure 5.12). Industrial waste effluents are often discharged untreated into the sea, hence creating significant water pollution.

Industrial estates were established in semi-urban areas as a means of decongesting the capital and major cities.

Industrial estates are areas especially organized for the establishment and operation of modern manufacturing plants. The Greek legislation provides for the establishment of industrial estates in all 52 prefectures of the country. In 6 of the industrial estates, wastewater treatment units are in operation. The above six are in the following prefectures:
- Achaia
- Heraklion
- Thessaloniki
- Kilkis
- Xanthi
- Komotini.

The total wastewater load discharged annually by the above estates is 300,000 population equivalents and its designed treatment efficiency is 90-95%.

Since 1965 large efforts have been made and considerable funds have been invested by ETBA (Hellenic Industrial Development Bank S.A.) into these projects.

5.2.3. Agricultural activities

Agriculture has also affected the environment significantly with the transition from the traditional type of farming to the intensive modern farming of today. Coastal agricultural area (about 1×10^6 hectares) covers one-third of the whole Greek coastal zone (within 10 kilometers from the coastline). Agriculture contributes to the pollution of the coastal seas, lagoons, lakes and other fragile aquatic ecosystems through run-off and leaching of fertilizers and pesticides. Major coastal agricultural areas are located in Crete, western Peloponnesos, Pieria, Magnisia, Phthiotida, Preveza and Kavala (National Plan for Greece). In 1985, 2.2 million tons of fertilizers and 2,800 tons of pesticides were added to agricultural soil (Table 5.28).

5.2.4.   Fisheries and aquaculture

The fisheries production in Greece was 119,000 tons in 1980 and increased to 138,200 tons in 1986 (Table 5.29). The trawling and coastal fishing effort for cephalopods (squids, octopods and cuttlefish, all very important commercial species) in recent years has been significantly higher than the optimum (Stergiou, 1988). This clearly indicates that cephalopod resources are overfished and authorities must take immediate protection measures. The same applies to gadoid resources (blue whiting, poor cod and especially hake) (Stergiou and Panos, 1988).
The multispecies nature of commercial fish in the Mediterranean poses certain difficulties in developing uniform measures for the protection of the resource. Measures that are favorable for some species may not be so for other. Experimental closing and opening of different areas and/or seasons and license restriction may be used beneficially for the protection and management of the overexploited trawl and coastal fishery resources in Greek waters. A compulsory mesh size of 40 mm (stretched) in trawlers as opposed to 28 mm currently used is also essential for the protection of demersal resources.
Some of the spawning and nursery grounds are affected by domestic pollution or agriculture. The same is also true for the fishing areas identified in a study by the Oceanographic and Fisheries Institute of Athens (Figure 5.13).

During the past few years aquaculture has become an important economic activity. The total area covered by aquaculture is estimated at 131,700 hectares of which 40,000 hectares are lagoons, 54,800 hectares natural lakes, 31,300 hectares artificial lakes and 5,600 hectares rivers. There remains between 50,000 and 70,000 hectares for future development which can reduce the country's dependence on fish imports, currently at 27%.

The most productive area, the Amvrakikos Lagoon, is located

in western Greece. The northern part of the lagoon is covered by the International RAMSAR Convention. It has a total area of 405 km^2 and it is used for fishing and recreation. It is surrounded by agricultural land (585 km^2), forest and pastures (1,006 km^2) and human settlements (two towns with 13,700 inhabitants and 170 villages with a total of 5,000 inhabitants). Near the settlements are 19 olive oil processing units (10,000 tons per year), 50 milling factories (13,000 tons per year) and 8 slaughterhouses.

Two rivers (the Arachthos and the Louros) discharge an average of 61 m^3/sec. A comprehensive development program has been formulated for the above area.

5.3.   Institutional/legal framework

Greece is one of the few countries that have included an article on environmental protection in their Constitution. Article 24 states that the protection of the natural and cultural environment constitutes an obligation of the State and that the State is responsible for taking special preventative or enforcement measures towards its conservation. Articles 21, 22 and 106 deal with the protection of workers, public health, urban development and physical planning.

The application of these constitutional orders was made possible by the introduction and adoption by the Parliament of Law 360/1976 on environmental and physical planning. Certain articles of this Law have been subsequently revised by Law 1032/1980.

According to the Law 360/1976 the policy for physical planning and environment is shaped and monitored by a National Council (NCPPE) chaired by the Prime Minister and composed by the Ministers of National Economy, Finance, Agriculture, Sciences and Culture, Industry and Energy,

Social Affairs, Public Works, Regional Planning and Environment, and Mercantile Marine.

In order to strengthen the implementation and control procedures, a new Ministry of the Environment, Regional Planning and Public Works was created by Law 1032 in 1980 (Figure 5.14). From a small number of administrators in 1972 the new Ministry has now more than 250 employees and regional representation in all prefectures of the country. This Ministry is the main executive body carrying out the environmental policy of the Government.

In addition, there are administrative units in almost every Ministry that have been assigned responsibility for environmental protection (Figure 5.15), such as the Ministry of Agriculture (forest, wildlife, pesticides), the Ministry of Mercantile Marine (pollution by oil and other harmful substances from ships and land-based sources) and the Ministry of the Interior (solid wastes, municipal waste water).

The latest completed five-year plan of 1983-1987 recommended the adoption of a general law, concerning the protection of the environment. This was fulfilled by ratification of the 1650/1986 Environmental Act.

Law 1650 of 1986 determines the overall framework. The Law distinguishes 3 industrial categories, based on their level of pollution. The enforced standards are defined at a regional level (prefectures). They depend on the receiving environment and the nature of the industry. Normally, prefectoral decrees should conform to EC directives.

Law 1069 of 1980 has compelled each municipality of more than 10,000 inhabitants to create a municipal enterprise for water supply, wastewater collection and treatment. In parallel it has defined the basis for project financing:
- the cost of investment has to be covered through State subsidies (up to 40%), a tax on housing (3% of the tax

on property returns to the municipality) and water tariffs (80% or less);
- the tariff should cover operational costs;
- the regional Prefect should check for discrepancies between cost and tariff.

Law 1298/90 supports industry's antipollution investments by giving incentives, from 40% (Athens) to 50% (near the borders) of the cost. The differentiation has been set on a regional basis, with the objective of decentralization.

An important legislative act is the Law 743/1977 on the "protection of marine environment from pollution caused by vessels and coastal industrial activities". This Law provides, among others, for the organization of a marine pollution abatement service in the Ministry of Mercantile Marine and for fines imposed on violators of this Law. Certain provisions of this Law were later harmonized with those of the Convention for the Protection of the Mediterranean Sea Against Pollution (Barcelona Convention) and its two Protocols on the "prevention of pollution in the Mediterranean Sea from dumping by vessels and aircrafts" and on "cooperation for preventing pollution in the Mediterranean Sea from hydrocarbons and other harmful substances in cases of emergency", which were ratified by Greece in 1978.

Two other Protocols for the "protection of the marine environment from land-based sources" and for "specially protected marine areas" have been ratified in 1986. Greece has also ratified the MARPOL, STCW and Civil Liability Conventions.

Oil pollution is the biggest pollution problem in the Mediterranean Sea and presents a danger of national dimension to the Greek coast. That is the reason why Greece has developed a contingency plan, since the early 1970s has created regional oil combating units, has eleven skimmers and multipurpose ships, and two airplanes. Oil pollution equipment has been installed in 40 ports all over the country. Moreover, the fines which were imposed on violations by ships or land sources were $2.2 million (U.S.) for the period 1978-1982.

Reception facilities will be created at the main ports of the country, according to the latest five-year plan. This project will be partly financed by the EC.

Another important development is the introduction of prefectorial regulations on effluent standards in coastal areas for "land-based sources of pollution of the marine environment" to establish maximum limits of disposal of various substances (Table 5.30).

To promote regional development in Greece, the establishment of any industry in the area of Attica around Athens was prohibited in the 1980s for a period of five years. Establishing industries outside this area was encouraged by tax and custom exemptions. Since 1982 a series of laws have come into force dividing the country into four development areas with different incentives.

The prefectorial services are charged with the control of the proper application of laws and permits.

By the fundamental Law 1650/86 the installation of environmental quality control units (KEPE) in any prefecture has been regulated. The Prefect can nominate enforcement functionaries from the prefecture personnel, including representatives of the Ministry of the Environment. Representatives of the local authorities can also participate. The units have all the necessary rights for proper supervision by entering premises, sampling, suggesting measures or sanctioning.

Administration fines, after Prefectorial Decision, can be imposed up to 10 million Greek drachmae or 60,000 ECU. In severe cases (e.g. accidental spill with victims) the

Minister has the right to increase the fines up to 6 million ECU. Temporal or permanent withdrawal of a permit is possible together with fines of up to 6,000 ECU per day of illegal work.

The Penal Code is applied to violations of environmental legislation. Complaints can be made by those who are affected by actions that cause environmental damage.
For violation of the environmental legislation, imprisonment for up to 2 years and penalties can be levied.

New environmental impact assessment legislation harmonizes Greek law with the relevant EC directive.

The international conventions ratified by Greece are shown in Table 5.31.

5.4. Past and future trends

5.4.1. Sewage treatment plants

In 1981 Greece had only three municipal sewage treatment plants with a total capacity of less than 50,000 population equivalents (Sobemap 1981, 1982). These plants served three small provincial municipalities, hence the small capacity. The two largest Greek cities (Athens and Thessaloniki) had limited sewer networks but no treatment plants. Since that time, plans have been made and treatment plants have been constructed in many Greek towns (Table 5.32).

In 1978 the Greek government received a loan from the World Bank to help the financing of the sewage treatment projects for Volos and Thessaloniki. It also paid for technical assistance and training for the Ministry of Public Works and related technical studies. Additional financing was provided by the European Investment Bank (EIB), and after 1985 the Volos project was completed. The Thessaloniki project is complete but not in operation due to effluent disposal problems.

The sewage treatment project for Athens has been financed through a loan of $62 million by the EIB (Goola, 1986). The project includes the construction of a treatment plant at Psitalia (a small island in the Saronikos Gulf near the coast) and at Metamorphosis which will serve both Athens and Piraeus. The construction of the treatment plant at Psitalia was under way in 1990. The Metamorphosis treatment plant has been constructed, but there have been problems with its operation. The EIB and other sources have provided loans for sewage treatment projects in the following cities: Agios Nikolaos (Crete), Chalkis, Ioannina, Kastoria, Larissa and Ptolemais. The Ioannina part is ready but not yet in operation, as local authorities do not accept the discharge of effluent into the Kalamas River.

Another source of finance (both loans and subsidies) for sewage and water supply projects has been the European Fund for Regional Development. Until the end of 1984, the Fund has provided up to 11.5 billion drachmae covering 40% of the cost in many projects. Recently, treatment plants are being constructed in the industrial zones of Thessaloniki (10,000 m^3 per day), Patras (5,000 m^3 per day) and Heraklion (4,000 m^3 per day).

The investment of the public sector predominates (Table 5.33). About 37% of the total cost estimated in 1983 for water supply and wastewater treatment has been invested as of 1986. There remains about $1.5 billion (U.S.) (estimation based on 1987 prices) for projects under construction or planned. Treatment plants have been installed in 42% of all firms in the agro-food industry (Table 5.34).

5.4.2. Port reception facilities

Greece has ratified the International MARPOL Convention

with Law 1269 passed in 1982 for the establishment of port facilities to receive oil, chemicals and wastes (MEDWAVES No 9, 1987).

The Greek Ministry of Mercantile Marine has selected ten ports on the basis of their traffic, in which reception facilities must be established as a matter of priority (Ekonomikos Tachydromos, 1988). These ports are:
- Piraeus/Elefsis (common facilities)
- Thessaloniki
- Kavala
- Patras (the municipality of Patras has not accepted the establishment of a port facility and the solution of a permanent floating facility has been adopted)
- Chalkis
- Corfu
- Syros
- Rhodes
- Heraklion
- Volos.

The total cost of establishing the above facilities is estimated to be over 2 billion drachmae (excluding the Patras facility which will cost 47 million drachmae and 50% of that will be covered by the Mediterranean Strategy and Action Plan of the European Community). The management of the above facilities, when they are constructed, will be the responsibility of the local harbor authorities.

The above projects have not yet been included in the State budget for public works because of lack of funds. It is expected that construction contracts will be given to companies in the private sector, since no state company has the necessary equipment and technical expertise. Recently, there has been a change in the Governmental Decision. The number of reception facilities that will be constructed has decreased to eight and all of them will be floating facilities.

Until the above facilities are constructed, alternative temporary solutions have been adopted. In the ports of Piraeus, Thessaloniki, Syros and Chalkis floating reception facilities have been established. In the rest of the selected harbors, wastes are loaded on special trucks and transported to appropriate installations for processing.

The most important floating facility is that at Piraeus, which also serves Elefsis Harbor. It comprises a floating separator and collecting ships. It is run by a private company on a ten-year contract with the Piraeus port authority. The total cost of setting up this facility was 350 million drachmae. In 1987 it received 400,000 tons of oil residues and produced 15,000 tons of oil products (primarily for export).

To abate accidental oil release, contingency plans have been developed. Coastal antipollution centers are equipped with the necessary means and equipment. A number of oil spillages occurred during the past years (two exceeded 5,000 tons of oil). Law 743/77 imposes fines and strict measures against ships or land-based sources for oil spillages.

5.5.   Conclusions

The greatest urban centres in Greece are located near gulfs (e.g. Athens, Thessaloniki, Patras, Volos). Due to the high urban and industrial waste load in these areas, the pollution of the marine environment has become a serious problem. More treatment plants and improved efficiency in operation of the existing ones are necessary for control of both municipal and industrial discharges. The estimated cost is $1.5 billion (U.S.).

A number of ports has been selected in which reception facilities are established as a matter of priority. The total cost of the above facilities is estimated to be around $13 million (U.S.).

Air pollution is a major problem in Athens. Traffic is responsible for 75%, industry contributes about 20% and domestic sources 5%. Severe air pollution episodes often occur. The estimated cost of the antipollution investments needed in Athens is $1,320 million (U.S.): $1,100 million (U.S.) for the construction of the metro and $220 million (U.S.) for traffic control improvements.

The unequal rainfall in time and space, the unequal distribution of activities and consequently of consumption patterns have caused water deficits in some regions of the country (especially in the Greater Athens area). Further investments are needed in order to secure the availability of the required amount of water.

There are significant problems due to the uncontrolled disposal of municipal refuse and industrial solid waste. Treatment of the above waste must be implemented as soon as possible.

The coastal strip faces an erosion problem. Preventive and remedial measures against deforestation (firefighting equipment, planting) are urgently needed.

At present, financial and other difficulties are causing delays in several major environmental projects.

References

AMBIO, Monuments endangered. Ambio, 1977, 6, 355.

ARCHEIVALA, S. J., Waste water treatment and disposal: engineering and ecology in pollution control. 1981. M. Dekker Inc.

BALOPOULOS, E. Th. and FRILIGOS, N. Ch., Transfer mechanisms and nutrient enrichment in the northwestern Aegean Sea: Thermaikos Gulf. 1986. Rapports et Procès-Verbaux des Réunions. Commission Internationale pour l'Exploration Scientifique de la mer Méditerranée. Vol. 30, no. 2, p. 134.

CAMHIS, M. and COCCOSSIS, H., The national coastal management program of Greece. Ekistics 293, March/April 1982.

CEC, Qualité des eaux douces superficielles. Synthèse 1982-1986. Volume 3: résume statistique. December 1987 B-1400.

CEC, 26.7.1987-1.8.1987 plages propres. Résultats de la campagne de sensibilisation. Brussels 1988.

ECONOMOPOULOS, A., Air pollution in Athens. Critical evaluation of targets and achievements. Technical Chamber of Greece Workshop. Athens, 6-7 June 1988.

EKONOMIKOS TACHYDROMOS (Greek weekly business journal), Mercantile marine: positive steps under the pressure of international conventions. Athens, 17 March 1988.

FOWLER, S. W., Assessing pollution in the Mediterranean Sea. In: Pollutants and their ecotoxicological significance (ed. Nürnberg, H. W.), 1985 pp. 269-287. John Wiley & Sons Ltd. New York.

FRILIGOS, N., Influence of various sources of pollution on the distribution of nutrients in the sea-water of the upper Saronikos Gulf. 1981a. Rev. Int. Océanogra. Méd. 62, 47-61.

FRILIGOS, N., Enrichment by inorganic nutrients and oxygen utilisation rates in Elefsis Bay (1973-1976). 1981b. Mar. Pollut. Bull. 12, 431-436.

FRILIGOS, N., Preliminary observations on nutrient cycling and a stoichiometric model for Elefsis Bay, Greece. Marine Environ. Res., 1983, 8, 197-213.

FRILIGOS, N., Distribution of nutrition salts. In: Biogeochemical studies of selected pollutants in the open waters of the Mediterranean (MED POL VIII). Addendum, Greek oceanographic cruise 1980. MAP Technical Reports Series No. 8, Addendum, pp. 25-28. UNEP, Athens 1986.

FYTIANOS, K., VASILIKIOTIS, G. and WEIL, L., Identification and determination of some trace organic compounds in coastal seawater of northern Greece. Bull. Environ. Contam. Toxicol., 1985, 34, 390-395.

GABRIELIDES, G. P. and HADJIGEORGIOU, E., In: Biogeochemical studies of selected pollutants in the open waters of the Mediterranean (MED POL VIII). Addendum, Greek oceanographic cruise 1980. MAP Technical Reports Series No. 8, Addendum, pp. 29-34. UNEP, Athens 1986.

GOLIK, A., WEBER, K., SALIHOGLU, I., YILMAZ, A. and LOIZIDES, L., Pelagic tar in the Mediterranean Sea. Mar. Pollut. Bull., 1988, 19, 567-572.

GREEK ORGANIZATION OF LOCAL AUTHORITIES, Preliminary study of sectors of water supply and sewage systems. Athens, 5-6 April 1986 (in Greek).

GRIMANIS, A. P., KALOGEROPOULOS, N., ZAFIROPOULOS, D. and VASSILAKI-GRIMANI, M., Chromium in the water column and sediment cores from Saronikos Gulf. 5th International Conference on Heavy Metals in the Environment. Vol. 2, pp. 427-429. Athens 1985.

KILIKIDIS, S. D., PSOMAS, J. E., KAMARIANOS, A. P. and PANETSOS, A. G., Monitoring of DDT, PCBs and other organochlorine compounds in marine organisms from the North Aegean Sea. Bull. Environm. Contam. Toxicol., 1981, 26, 496-501.

LAMB, C. J., Water quality and its control. 1985 pp. 116-118. J. Wiley and Sons Inc.

MAP, Assessment of the present state of pollution by petroleum hydrocarbons in the Mediterranean Sea. UNEP, Athens 1987.

MEDWAVES, Issue No 9, II/1987, Coordinating Unit of the Mediterranean Action Plan, Athens.

MINISTRY OF THE ENVIRONMENT, Regional Planning and Public Works Data. 1985. Greece (in Greek).

MINISTRY OF THE ENVIRONMENT, Regional Planning and Public Works, Report on the environment in Greece, solid wastes. Athens 1987 (in Greek).

NSSG, Statistical year book. 1988.

SAMANIDOU, V., FYTIANOS, K. and VASILIKIOTIS, G., Distribution of nutrients in the Thermaikos Gulf, Greece. The Science of the Total Environment. 1987, 65, 181-189.

SATSMADJIS, J. and FRILIGOS, N., Red tide in Greek waters. Vie et Milieu, 1983, 33, 111-117.

SATSMADJIS, J., GEORGAKOPOULOS-GREGORIADES, E. and VOUTSINOU-TALIA-DOURI, F., Red mullet contamination by PCBs and chlorinated pesticides in the Pagassitikos Gulf, Greece. Mar. Pollut. Bull., 1988, 19, 136-138.

SCOULLOS, M. and DASSENAKIS, M., Dissolved and particulate zinc, copper and lead in surface waters. In: Biogeochemical studies of selected pollutants in the open waters of the Mediterranean (MED POL VIII). Addendum, Greek oceanographic cruise 1980. MAP Technical Reports Series No. 8, Addendum, pp. 45-63. UNEP, Athens 1986.

SOBEMAP, Quantitative and qualitative inventory of sewage treatment plants in EEC. 5: Greece. October 1981.

SOBEMAP, Cost and efficiency of sewage treatment plants in the EEC. 5: Greece. September 1982.

STERGIOU, K. I., Allocation, assessment and management of the cephalopod fishery resources in Greek waters, 1964-1985. 1988. Rapp. Comm. int. Mer Médit., 31, 2 p. 253.

STERGIOU, K. I. and PANOS, Th., Allocation of gadoid fishery in Greek waters, 1964-1985. 1988. Rapp. Comm. int. Mer Médit., 31, 2 p. 281.

UNEP/IG.74/5, Report of the fifth ordinary meeting of the contracting parties to the Convention for the protection of the Mediterranean Sea against pollution and its related protocols. Athens, 7-11 September 1987.

UNEP/PAP, G. Tsekouras project No. PAP-4, mt/5102-83-05. 1984.

UNEP/WG.163/4. Annex I, Directory of marine and coastal protected areas in the Mediterranean. Draft. UNEP, Athens 1987.

VASILIKIOTIS, G., VOULOUVOUTIS, N., FYTIANOS, K. and SAMARA, N., Die Schwermetallbelastung im Nord-Ägäischen Meer (Griechenland) ermittelt anhand von Fisch- und Muscheluntersuchungen. Gwf-wasser/abwasser, 1983, 124, 37-39.

VOUTSINOU-TALIADOURI, F., Metal pollution in the Saronikos Gulf. Mar. Pollut. Bull., 1981, 12, 163-168.

VOUTSINOU-TALIADOURI, F. and LEONDARIS S. N., An assessment of metal pollution in Thermaikos Gulf, Greece. 1986. Rapports et Procès-Verbaux des Réunions. Commission Internationale pour l'Exploration Scientifique de la mer Méditerranée. Vol. 30, no. 2, p. 43.

VOUTSINOU-TALIADOURI, F. and SATSMADJIS, J., Influence of metropolitan waste on the concentration of chlorinated hydrocarbons and metals in striped mullet. 1982a. Mar. Pollut. Bull. 13, 266-269.

VOUTSINOU-TALIADOURI, F. and SATSMADJIS, J., Concentration of some metals in East Aegean sediments. 1982b. Rev. Int. Océanogr. Méd. 66-67, 71-76.

VOUTSINOU-TALIADOURI, F. and SATSMADJIS, J., Distribution of heavy metals in sediments of the Patraikos Gulf (Greece). Mar. Pollut. Bull., 1983, 14, 33-35.

VOUTSINOU-TALIADOURI, F., SATSMADJIS, J. and IATRIDIS, B., Granulometric and metal composition in sediments from a group of Ionian lagoons. Mar. Pollut. Bull., 1987, 18, 49-52.

VOUTSINOU-TALIADOURI, F. and VARNAVAS, S. P., Distribution of Cr, Zn, Cu, Pb and C organic in the surface sediments of northern Euboicos Bay, Greece. 5th International Conference on Heavy Metals in the Environment. Vol. 2, pp. 356-358. Athens 1985.

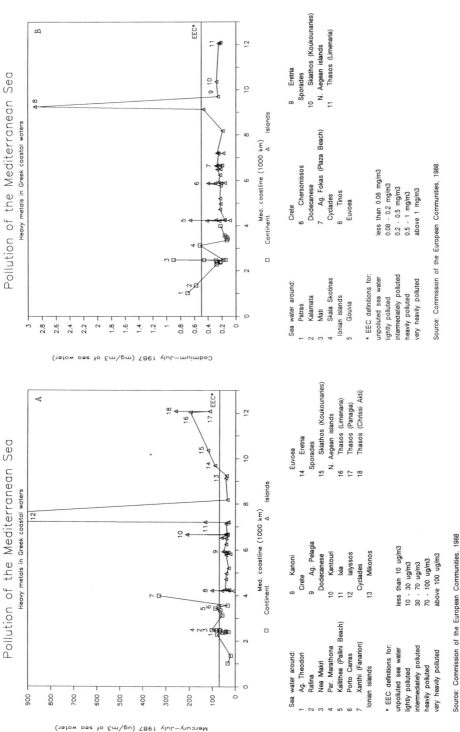

Figure 5.1   (A) Mercury, (B) Cadmium, (C) Lead and (D) Zinc concentrations along the Greek coast

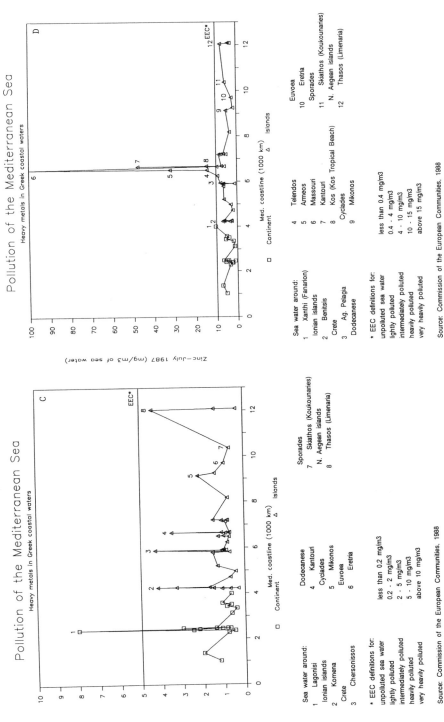

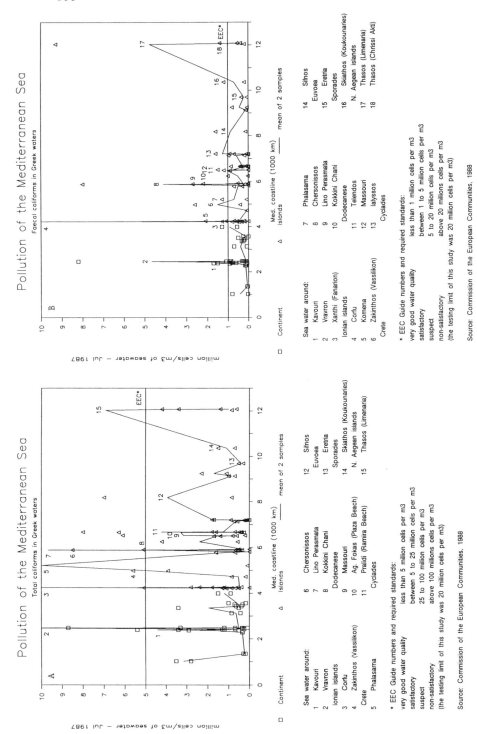

Figure 5.2 Bacterial quality of the bathing waters along the Greek coast
A: total coliforms; B: faecal coliforms, C: faecal streptococci

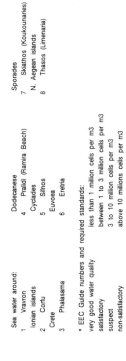

Fig. 5.3　Annual variation of chlorinated hydrocarbons in _Mytilus galloprovincialis_ from the Gulfs of the Northern Aegean Sea [Kilikidis, 1981]

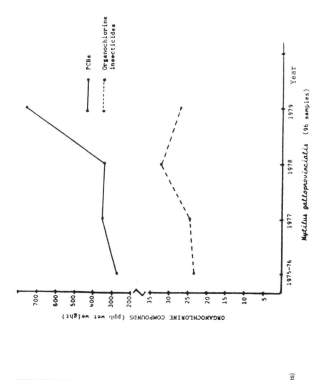

Figure 5.2　continued

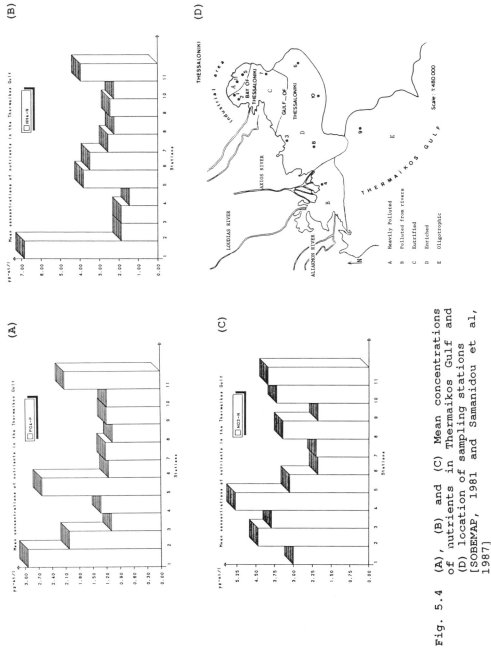

Fig. 5.4 (A), (B) and (C) Mean concentrations of nutrients in Thermaikos Gulf and (D) location of sampling stations [SOBEMAP, 1981 and Samanidou et al, 1987]

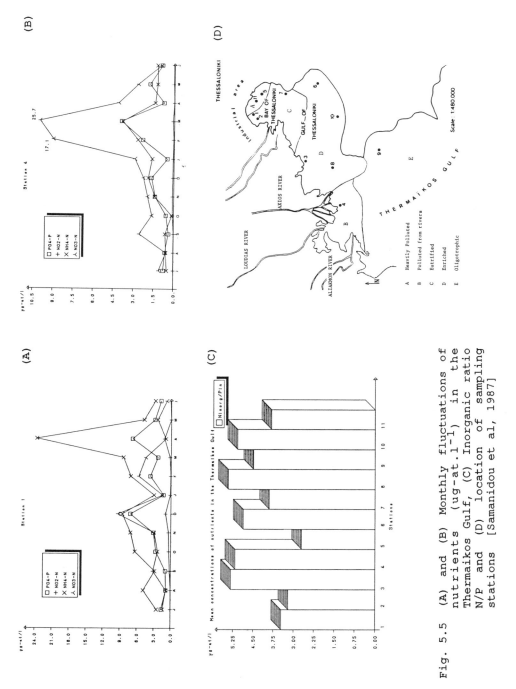

Fig. 5.5 (A) and (B) Monthly fluctuations of nutrients (ug-at.l⁻¹) in the Thermaikos Gulf, (C) Inorganic ratio N/P and (D) location of sampling stations [Samanidou et al, 1987]

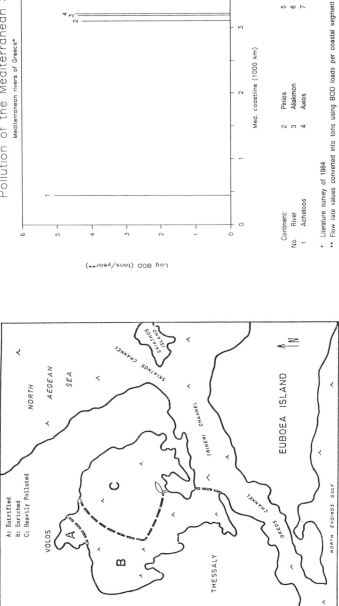

Fig. 5.6 Pagassitikos Gulf [SOBEMAP, 1981]

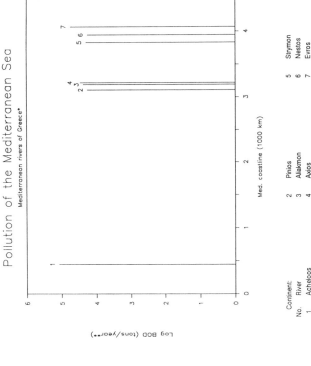

Fig. 5.7 BOD discharges along the Greek coast for the 7 major rivers

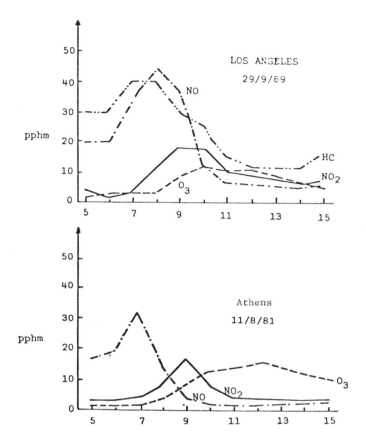

Figure 5.8 Diurnal variation of photochemical pollutants
[Professor Lalas, private communication]

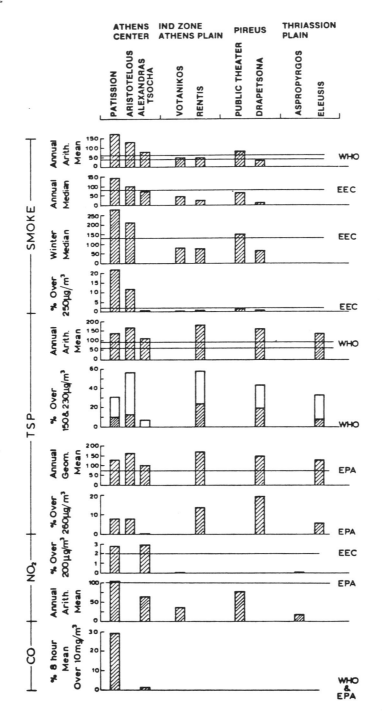

Figure 5.9 Ambient standards violation in Athens
[A. Economopoulos: Technical Chamber of Greece workshop, 6-7 June 1988]

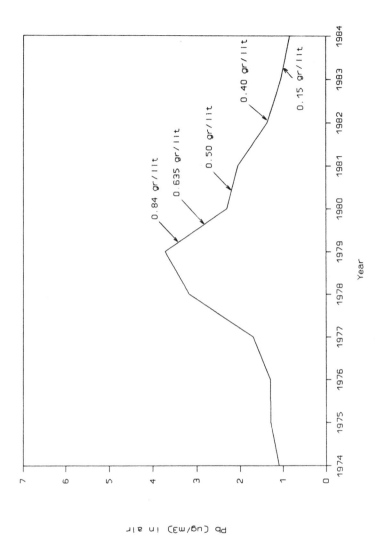

NOTE. in 1988 gasoline with 0.15g/l lead is used in all over the country.

Figure 5.10 Lead content in petrol in Athens and lead concentration in air

474

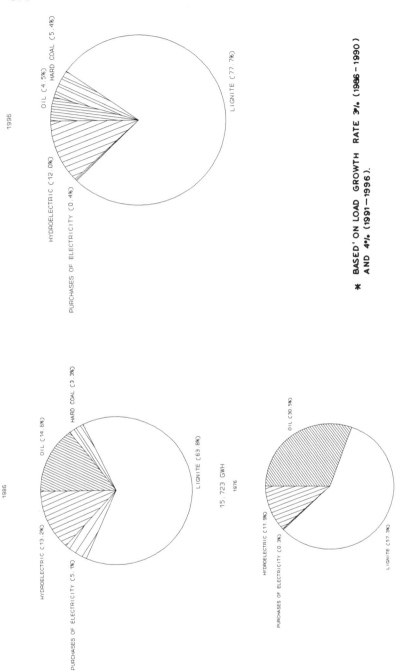

Figure 5.11 Participation of primary sources of energy in the generation of electricity of the mainland system in selected years and purchases of electricity   [Public Electricity Board]

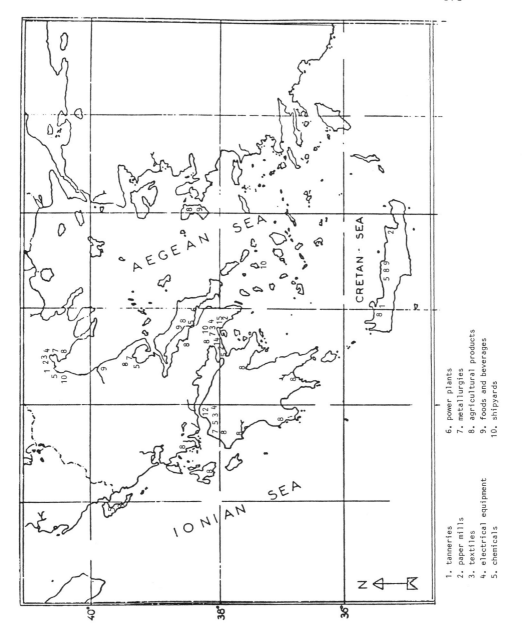

Figure 5.12   Distribution of industrial activities along the coast   [Secretariat for physical planning and the Environment]

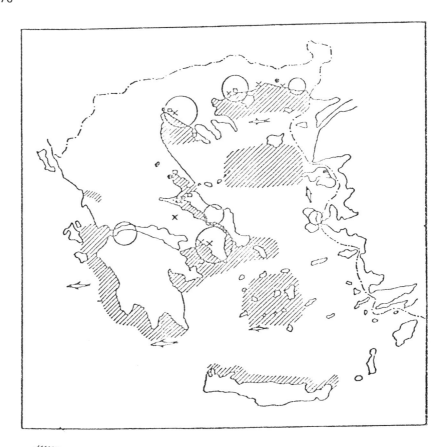

Figure 5.13 Greek fishing areas and related industrial activity [Ministry of Agriculture]

MINISTRY OF ENVIRONMENT, PHYSICAL PLANNING AND PUBLIC WORKS
(P.D. 5/1.2.1988/03 A 19)

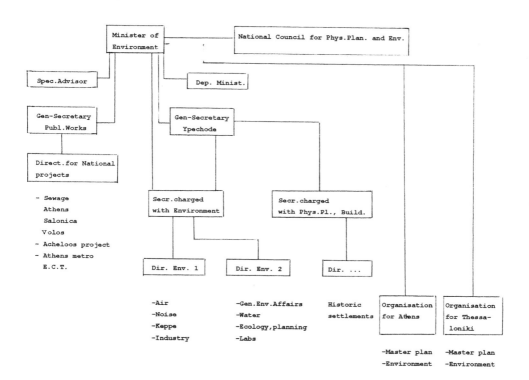

Figure 5.14 Organization of the Ministry of Environment

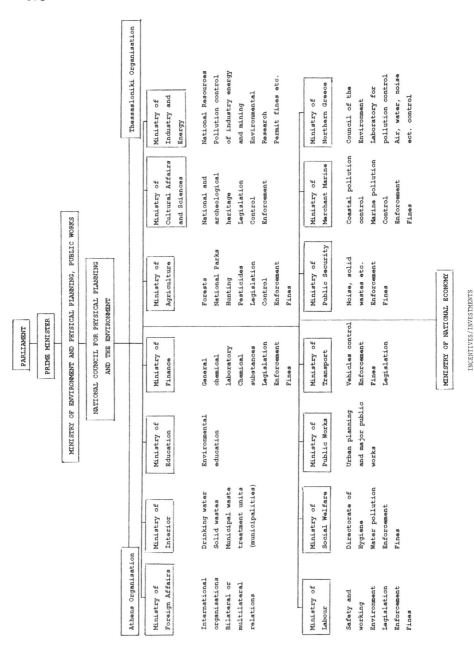

Figure 5.15 Governmental organization of Environmental Affairs

Table 5.1 Ratio of Nutrients for Polluted Gulfs to background values. (Background values are the mean Outer Saronikos Gulf concentrations 1972-1976, Friligos 1981b) [Friligos, 1986b]

Area	PO_4^{-3}-P	SiO_4^{-4}-Si	NH_4^+-N	NO_2--N	NO_3--N	ΣN
Elefsis Bay	5.11	4.15	15.80	3.05	7.00	9.67
Western Saronikos Gulf	2.25	2.95	2.50	1.11	6.39	4.00
Inner Saronikos Gulf	2.50	1.39	4.10	1.55	2.60	3.97
North Euboikos Gulf	2.87	13.20	1.66	0.49	10.20	5.27
Kavala Gulf	1.80	1.49	1.00	0.51	1.41	1.10
Alexandroupolis Gulf	1.32	3.28	1.00	0.65	6.21	3.27
South Euboikos Gulf	1.46	1.41	0.65	0.48	1.17	0.86
Pagassitikos Gulf	1.02	2.80	2.60	1.25	2.36	2.28
Thessaloniki Bay	5.33	3.35	4.58	3.83	3.88	4.14
Western Thermaikos Bay	2.09	3.81	2.91	2.40	3.80	3.22
Eastern Thermaikos Bay	1.18	2.21	1.97	1.53	2.20	2.00

Table 5.2   Observed sediment max. concentrations in Saronikos Gulf in ppm [Voutsinou, 1981]

Area	Element[1]:	Cu	Fe	Zn	%C
Elefsis Bay		8	400	21	1.86
Keratsini Bay		58	710	630	6.22
Drapetsona Bay		1000	7700	720	3.50
Piraeus Harbour		364	3650	625	4.50
Normal Levels		1	2	1	0.2

Notes: 1. Extraction by dilute acid solution
(0.05 N HCL + 0.025 N H_2SO_4)

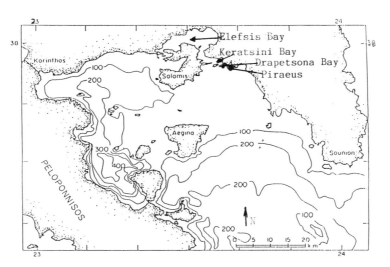

Saronikos Gulf (The most polluted areas
are shown; contour lines in meters)
(Voutsinou, 1981)

Table 5.3 Chlorinated hydrocarbons in Mullus barbatus (striped mullet) from Saronikos Gulf in ug/kg (ppb) wet weight
[Voutsinou, and Satsmadjis, 1982]

Area (No. sampl.)	Parameter	PCBs (ppb)	DDE (ppb)	DDT (ppb)	DDD (ppb)	ΣDDTs (ppb)	BHCs (ppb)	Hept. epox. (ppb)	Dieldrin (ppb)	Endrin (ppb)
A (13)	Mean	460	72	78	66	216	5.6	0.3	17	2.3
	High	1200	151	118	145	400	8.2	0.5	50	3.5
	Low	170	39	48	26	125	3.8	0.2	0.4	0.9
	σ	380	35	24	52	105	1.4	0.2	17	1.5
	A/E	17.56	8.89	15.6	27.5	13.94	3.73	1.5	34	11.5
B (14)	Mean	176	29.7	20.3	12.4	62.4	5.6	0.3	2.4	1.5
	High	340	48	41	20	102	10	0.6	4.7	2
	Low	63	8.6	5.3	5	20.1	1	0.1	0.1	0.5
	σ	87	12	10	8	24	3.5	0.2	2.2	0.8
	B/E	6.72	3.67	4.06	5.17	4.03	3.73	1.5	4.8	7.5
C (18)	Mean	70	31.7	26.6	4.1	62.4	2.5	0.5	1	1.1
	High	124	67	51	11	122	5.5	1.2	1.8	2.3
	Low	16	12.5	9.7	0.2	28.9	0.2	0.1	0.1	0.1
	σ	37	20	12	4	34	2	0.5	0.6	0.9
	C/E	2.67	3.91	5.32	1.71	4.03	1.67	2.5	2	5.5
D (17)	Mean	36.5	14.2	9.7	4.3	28.2	4.6	0.3	1.1	0.9
	High	90	35	18	7.7	58	9.9	0.7	3.1	2.2
	Low	3.8	4	3.5	0.6	8.4	0.4	0.0	0.0	0.3
	σ	36	14	8	3.3	23	4.8	0.1	1.4	0.9
	D/E	1.39	1.75	1.94	1.79	1.82	3.07	1.5	2.2	4.5
E (12)	Mean	26.2	8.1	5	2.4	15.5	1.5	0.2	0.5	0.2
	High	53	20.5	12	5.2	37.7	2.5	0.5	0.9	0.6
	Low	13.9	2.3	1	0.3	3.6	0.3	0.1	0.1	0.0
	σ	13.1	5.9	3.5	1.9	10.8	0.8	0.2	0.3	0.2

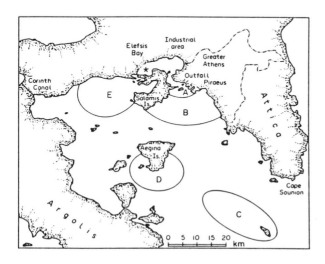

Study area in the Saronikos Gulf. Areas A-E correspond to the above table; * correspond to sampling station K in table 5.5 (Voutsinou & Satsmatjis, 1982)

Table 5.4 Concentration of nutrients in Keratsini and Drapetsona Bays [Friligos, 1981a]

Nutrient	Keratsini Bay Sewage Effluent Concentr.[1] µg/l	Keratsini Bay Seawater Max Concentr.[2] µg/l	Drapetsona Bay Seawater Max Concentr.[2] µg/l	MOE Limits µg/l
NH_4^+-N	21,560	154	154	15,000
NO_2-N	252	21	1.4	600
NO_3-N	1,512	14	84	20,000
PO_4^{-3}-P	2,976	62	62	total P 10,000
SiO_4^{-4}-Si	8,904	56	112	n.a.

Notes : [1] Sampling Date May 1975
[2] Sampling Date November 1973
Sampling Depth 3 m

Table 5.5 Mean Nutrient Concentrations at station K (Elefsis Bay). Integrated Water Column Averages in ug/l [Friligos, 1983]

Date of the Cruise	SiO_4-Si	PO_4-P	NH_4-N	NO_2-N	NO_3-N
December, 1973	87.4	6.51	56.42	9.52	48.58
February, 1974	75.6	4.34	50.12	8.26	59.22
April, 1974	72.2	3.72	21.70	4.76	62.58
June, 1974	184.8	11.47	39.48	2.52	52.78
November, 1974	159.0	21.70	225.96	4.06	17.22
December, 1974	286.4	22.01	183.12	10.64	98.70
February, 1975	72.0	19.53	54.60	10.5	34.30
June, 1975	230.7	28.52	40.60	10.08	5.18
September, 1975	341.6	73.47	105.56	1.96	0.84
December, 1975	133.0	28.52	59.50	7.28	22.68
March, 1976	75.9	5.89	15.96	5.46	29.26
Overall mean	156.2	22.94	77.42	6.86	39.2
Background*	34.2	3.72	5.04	2.24	5.88
Ratio	4.6	6.2	15.4	3.1	6.7

* Background = mean Outer Saronikos Gulf water nutrient concentrations, 1972 - 76 (Friligos, 1981 b).

Table 5.6 Annual Variations in the Level of Organochlorine Compounds in *Mytilus galloprovincialis*, ug/kg wet weight [Kilikidis, 1981]

Compound	1975-1976 (40 samples)	1977 (10 samples)	1978 (23 samples)	1979 (23 samples)
DDE	9	9	11	9
DDD	3	9	11	6
DDT	8	4	8	10
Total DDT	20	22	30	25
HCB	n.d.	2	2	2
Aldrin	3	n.d.	n.d.	0.8
PCBs	270	350	340	580

Table 5.7 Levels of metals* (mg/kg) in various areas of Thermaikos Gulf [Voutsinou, 1986]

Area		Cd	Pb	Cr	Zn	Cu
Sewage Outfall	Mean	4.4	220.0	180.0	770.0	135.0
	Range	2.8-6.0	100-330	140-210	235-1610	100-170
	Pollut.Ratio	14.6	11.8	1.9	16.0	7.9
Industrial Zone	Mean	2.8	165.0	290.0	290.0	80.0
	Range	2.2-3.1	120-245	215-390	220-375	68-85
	Pollut.Ratio	9.3	8.9	3.0	6.1	4.7
Axios River	Mean	5.3	100.0	280.0	200.0	53.0
	Range	2.6-8.7	81-120	230-320	155-250	44-67
	Pollut.Ratio	17.8	5.4	2.9	4.1	3.1
Aliakmon River	Mean	0.3	22.0	280.0	86.0	24.0
	Range	0.3	21-23	215-330	62-135	18-28
	Pollut.Ratio		1.2	2.9	1.8	1.4
Reference Area	Mean	0.3	18.0	95.0	48.0	17.0
	Range	0.3	11-27	66-120	32-74	8-28

Continued

Area		Ni	Co	Mn	Fe
Sewage Outfall	Mean	96	20	480	32,000
	Range	76-115	16-24	295-670	24,000-40,000
	Pollut.Ratio	1.2	1.2	1.0	1.8
Industrial Zone	Mean	93	19	580	35,000
	Range	80-100	16-22	565-660	11,000-46,000
	Pollut.Ratio	1.1	1.2	1.2	1.9
Axios River	Mean	140	25	955	47,000
	Range	110-175	19.29	665-1340	29,000-53,000
	Pollut.Ratio	1.7	1.6	2.1	2.6
Aliakmon River	Mean	240	33	800	34,000
	Range	210-290	19-37	685-1040	22,000-46,000
	Pollut.Ratio	3.0	2.1	1.7	1.9
Reference Area	Mean	81	16	465	18,000
	Range	55-105	14-18	215-740	12,000-22,000

* Extraction by 50% conc. HCl

Table 5.8 Heavy metal content in fish and mollusk samples in ug/kg F.W. in Thermaikos Gulf [Vasilikiotis et al, 1983]

Species		Pb	Cd	Hg
Fish:	Mullus Barbatus average. entire bay	390	124	143
Mollusk:	Mytillus Galloprovincialis from coast near industrial zone	460	103	86
Mollusk:	Mytillus Galloprovincialis from beach where swimming is allowed	305	30	29
Bioconcentration factor in mollusk		1.5	3.4	2.97
Max permissible content in seafood sold in Greece		-	-	700

Table 5.9 Organic Matter in discharges in Thermaikos Gulf [Ministry of Environment, 1985]

Site	Indicator:	pH	COD mg/l	BOD_5 mg/l
Factories				
State Refinery		7.52	255	60
N.G. Chem. Ind.		7.35	68	30
Thes. Petr. Ind.		7.95	120	40
Sindos Cement		8	110	50
Sewage Discharge Points				
Eastern Sindos		7.56	484	80
Western Sindos		7.73	545	90
Malgara		8.37	412	70
Klidion		8.15	720	135
Lefkos Pirgos		7.9	510	270
River Mouths				
Axios		8.83	25	n.a.(1)
Loudias		7.98	100	n.a.
Aliakmon		8.49	50	n.a.
Greek Min. of Environment Max. Allowed Values		8.5	250	80

Notes: 1. n.a. = not available

Table 5.10 Pesticide and PCB concentrations in Thermaikos Gulf (ng/l)   [Ministry of Environment, 1985]

	River Mouth		
Pesticide	Axios	Loudias	Aliakmon
Lindane	.042	.037	.056
Dieldrin	.29	.16	.32
Aldrin	.17	.22	.3
Endrin	.027	.016	.039
Chlordane	.02	.029	.017
Parathion	.99	1.27	.85
PCB	4.76	3.62	2.1

Table 5.11 Organic pollutants in water samples of Kavala Gulf in ng/l (mean values) [Fytianos et al., 1985]

Sampling area	Lindane	P,P'-DDT	Dieldrin	PCB	PAH	Tetrachlorethene ug/l	Trichlorethene ug/l
1 Harbour	0.07	0.9	0.6	2.1	55	2.50	2.10
2 Near the industrial area	0.05	1.2	0.4	1.8	34	1.80	1.70
3 Middle of the gulf	0.02	0.4	0.2	1.2	27	0.39	0.35
4 2 Kms from the offshore oil well platform	0.11	0.8	0.8	2.7	80	3.00	2.80

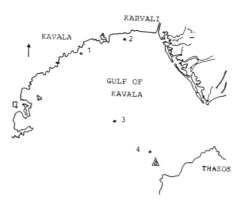

The sampling stations in Kavala Gulf

Table 5.12 Max. concentration (mg/kg F.W.) of organochlorines in red mullet muscle [Satmadjis et al, 1985]

Gulf	PCBs	DDTs	BHCs
Pagassitikos	.321	.026	.004
Saronikos	.46	.216	.006

Table 5.13 Range of Cr in surface sediments of various Greek areas (mg/kg) [Voutsinou and Varnavas, 1985]

Area	Range	Method used for the extraction
N. Euboikos Bay	40-41,000	2 N HCL
Messolongi Lagoon	11-246	2 N HCL
Pagassitikos Gulf	97-395	2 N HCL
Kavala Bay	20-278	2 N HCL
Lesbos Island	40-247	2 N HCL
East Aegean Sea	52-157	2 N HCL
Thermaikos Gulf	31-107	0.5 N HCL
Saronikos Gulf	144-480	0.5 N HCL
Saronikos Gulf	80-1,100	total
Patraikos Gulf	70-210	total

Table 5.14 Ranges of mean concentrations of heavy metals between August 1980 and February 1981, in mg/kg, except for Fe in o/oo, from 21 sampling stations in Patraikos Gulf [Voutsinou and Satsmatjis, 1983]

Metal	Range	Mean	Metal	Range	Mean
Co	11-23	19	Ni	60 - 128	110
Cr	55-119	100	Pb	10 - 20	16
Cu	16-43	35	Zn	43 - 88	72
Mn	750-2610	1420	Fe	16.5-32.0	28.3

(For the extraction of the metals 2 N Hcl was used).

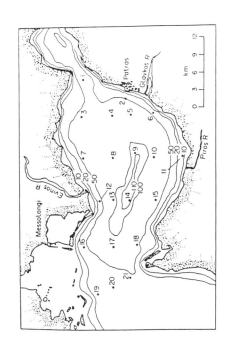

The 21 sampling stations in the Patraikos Gulf (depth contours in meters) (Voutsinou & Satsmatjis, 1983)

Table 5.15 Comparison of levels of heavy metals in sediments from Patraikos Gulf with sediments from East Aegean Sea (determinations carried out with the same methods) [Voutsinou et al., 1987]

Region	Fe (0/00)	Mn (ppm)	Zn (ppm)	Cr (ppm)	Ni (ppm)	Co (ppm)	Cu (ppm)	Pb (ppm)
Patraikos Gulf	28.3	1420	72	100	110	19	35	16
East Aegean Sea	22.5	861	38	86	131	15	17	17

Table 5.16 Average concentrations (mg/kg) of trace metals in sediments from the Aegean Sea (extraction by conc. HNO_3) [Voutsinou et al, 1982]

Constituent metal	East Aegean Sea	Pagassitikos Gulf	Thermaikos Gulf
Ni	143	176	121
Mn	925	1079	965
Cr	92	116	131
Cu	18	34	28
Co	16	28	20
Zn	39	126	104
Pb	17	29	46
Cd	.14	na	.15

Table 5.17 Average concentration (ug/l) of dissolved metals in sea water [Archeivala, 1981; Scoullos and Dassenakis, 1986]

Area	Study Year	Dpth (m)	Zn Range	Zn Mean	Cu Range	Cu Mean	Pb Range	Pb Mean
East Aegean	1980	0	4-8.9	5.6	.5-1	.75	.9-1.8	1.2
Aegean	1975	0	4.7-5.8	5.3	.25-.64	.45	na	
Mediterran.	1973-75	0	.1-7.9	2.3	.06-4.9	.4		.2
NW Meditet.	1974	0	.1-9.9	2.3	.03-21.6	.8	na	
Black Sea	1969	2-2150	.2-42.4	1.9	.04-66.3	1.8	na	
Marmara Sea	1967	0		32.0		18.0		≥10
NE Pasific	1971	5	.8-3.4	2	.4-1.2	.8		.35
NW Atlantic	1962-65	0-200		4.3		1.2		.02

Table 5.18 Nutrient Content (ug/l) in the East Aegean in 1980 [Friligos, 1986]

Component	Range (min - max)
$PO_4 - P$	2.17 - 3.41
$NH_4 - N$	9.1 - 18.9
$NO_2 - N$	0.7 - 22.26
$NO_3 - N$	1.26 - 31.92
$SiO_4 - Si$	21.0 - 91.6

Table 5.19 Water resources [ECE/Water/26 Long term perspective for water use and supply]

		Unit	Year			
			1970	1975	1980	1990
	a	b	c	d	e	f
1	Average annual inflow of surface water into the country	10^9 m^3/year	13.500	13.500	13.500	13.500
2	Average annual run-off originating within the country	"	49.400	49.400	49.400	
3	Average annual outflow of surface water from the country		62.900	62.900	62.900	
4	Guaranteed supply of surface water by 75%	10^9 m^3/year	5.950	5.950	6.980	
5	Guaranteed supply of surface water by 95% (1)	"	4.760	4.760	5.584	
6	All Surface reservoirs with total storage capacity exceeding 10^6 m^3	Number	32	35	37	45
	a. artificial reservoirs	"	9	12	14	22
	b. natural " (lakes)	"	23	23	23	23
7	Capacity of reservoirs in the country. Total:	10^9 m^3	8.700	10.950	12.500	15.500
	a. artificial reservoirs	"	6.700	8.950	10.500	13.500
	b. natural " (lakes)	"	2.000	2.000	2.000	2.000
8	Usable storage Effective:	10^9 m^3	4.500	6.000	7.000	9.150
	a. artificial reservoirs	"	4.100	5.600	6.600	8.750
	b. natural " (lakes)	"	0.400	0.400	0.400	0.400
9	Area of water surface of reservoirs	km^2	820	900	940	1.020
	a. artificial reservoirs	"	150	230	270	350
	b. natural " (lakes)	"	670	670	670	670
10	Total usable groundwater resources (2)	10^9 m^3/year	2.500	2.500	2.500	
11	Of which groundwater resources not connected with surface resources (2)	"	0.800	0.800	0.800	

1) It is estimated the 80% of the item 4
2) Karstic groundwater is not included

Table 5.20 Water balance of Greece
[Reference same as for table 5.19]

		Unit	Year			
	a	b	1970 c	1975 d	1980 e	1990 f
1	Total water abstraction	10^9 m^3	4.254	5.847	6.945	9.587
2	Total water consumption	"	2.959	4.522	5.375	7.462
3	Total return flow	"	1.295	1.325	1.530	2.125
4	- discharged into surface water	"	1.036	1.060	1.256	1.700
5	- discharged into saline water	"	0.259	0.265	0.306	0.425
6	surplus (+) or deficit (-) of water resources for guaranteed level of: 75%	"	+5.232	+3.663	+3.799	
7	" 95%	"	+4.042	+2.473	+2.403	
8	Balance taking account of planned measures for guaranteed level of: 75%	"	+2.900	+1.331	+0.437	
9	" 95%	"				
10	Waste water total (1)	"	0.350	0.530	0.665	1.030
11	- of which polluted waste water (1)	"	0.350	0.530	0.665	1.030
12	Waste water treated	"	0.087	0.130	0.150	
13	Number of treatment station	number	1*	1*	3**	

(1) Only fresh water waste is included which derives from sewages and industrial use.
* Preliminary treatment by screening and grit removal
** Provision for two biological treatment stations

Table 5.21 Water abstractions and consumptions
from different water consumers
[Reference same as for table 5.19]

	Water consumer	Utilization	Unit	Surface water resources					Groundwater				
				Year					Year				
	a	b	c	1970 d	1975 e	1980 f	1990 g	2000 h	1970 i	1975 j	1980 k	1990 l	2000 m
1	Population	abstraction	10^9 m^3	0.220	0.400	0.510	0.830	1.200	0.100	0.190	0.240	0.320	0.390
2		consumption	"	0.220	0.400	0.510	0.830	1.200	0.100	0.190	0.240	0.320	0.390
3	Industry	abstraction	"	0.030	0.035	0.040	0.080	0.160	0.030	0.045	0.050	0.060	0.070
4		consumption	"	0.030	0.035	0.040	0.080	0.160	0.030	0.045	0.050	0.060	0.070
5	Agriculture	abstraction	"	2.810	3.470	4.165	5.900	8.260	0.910	1.435	1.580	1.890	2.270
6		consumption	"	1.700	2.430	2.915	4.150	5.780	0.725	1.150	1.260	1.515	1.820
7	Thermo-electric power	abstraction	"	0.005	0.020	0.040	0.050	0.050	0.0025	0.010	0.015	0.030	0.030
8		consumption	"	0.005	0.020	0.040	0.050	0.050	0.0025	0.010	0.010	0.030	0.030
9	Losses due to evaporation		"	0.102	0.163	0.213	0.303	0.407	0.045	0.079	0.092	0.12	0.144

Table 5.22 Average yearly outflow, drainage area and length of the main Greek rivers

Water district	Rivers	Drainage area (Km^2)	Av. yearly outflow (10^6 m^3)	Length (km)
W. Peloponnese		3374	2100	
	Alfios	3374	2100	110
	Ladonas*	(749)	(635)	
	Loussios*	(117)	(215)	
	Aroanios*	(186)	(82)	
N. Peloponnese		1359	785	
	Pinios	794	441	
	Glafkos	66	39	205
	Silinous	131	70	
	Krathis	79	69	
	Vouraikos	183	116	
	Asopos K.	106	50	
E. Peloponnese			54	
	Evrotas		54	82
W. Sterea		5183	7091	
	Acheloos	4118	5988	220
	Evinos	635	873	
Epiros		4713	5808	
	Aoos	665	854	
	Voidomatis	369	524	80
	Kalamas	1481	1619	
	Louros	343	609	
	Arachtos	1855	2202	
E. Sterea		137	16	
	Charandros	137	16	
Thessaly		7248	2727	
	Pinios	7081	2529	
	Tavropos	167	198	
W. Makedonia		6342	2488	
	Aliakmon	6075	2292	297
	Edesseos	123	105	
	Tripotamos	144	88	
Centr. Makedonia		22450	5031	
	Axios	22450	5031	76
E. Makedonia		10937	3440	
	Strymon	10937	3440	116
Trace		32339	5069	
	Evros	27465	3250	204
	Nestos	4874	1819	130
Total		94082	34609	

* Tributary of Alfios

Table 5.23   Summary of NO_x- and VOC-emissions in Athens
[P. Coddevilles et al 1984/ EEC contract]

Source type	Operating day/y	NOx kt/year	NOx t/day	VOC kt/year	VOC t/day
Industry					
Fuel oil consumption	250	2 - 5	8 - 20	0.1-0.3	0.4-1.2
Refineries	340	3 - 5	9 - 15	3-5	9-15
Printing	250	-	-	4-5	16-20
Fermentation processes	250	-	-	2.2-2.5	8-10
Paint consumption	300 1)	-	-	9-15	30-50
Misc. Processes	250	-	-	0.25-0.3 [2]	1-1.2
Misc.ind.solvents	250	-	-	0.8-1	3-4
total		5-10	17-35	19-29	57-101
Road transport					
gasoline, exhaust	365	11-15	30-40	33-45	90-123
diesel, exhaust	365	8-12	22-33	4-6	11-16
LPG, exhaust	365	0.3-0.5	0.8-1.4	0.3-0.5	0.8-1.4
gasoline, evaporation	365	-	-		20-30
total		19.3-27.5	53-74		122-170
Gasoline marketing					
bulk storage, handling	365	-	-	0.5-0.7	1.4-1.9
filling stations	250	-	-	1.1-1.8	4.4-7.2
total				1.6-2.5	6-9
Other transport	365	4-8	11-22	0.4-1.2	1-3.5
Non-industr. solvents					
chemical cleaning	250	-	-	1.3-1.9 [2]	5-7
automobile repair	365	-	-	0.5-0.7	1.4-2
consumer products	365	-	-	7-12	20-30
total				8.8-14.6	26-39
Natural emissions		-	-		0.5-1.5
Grand total		28-45	80-130		220-320

1) Influenced by non-industrial consumption 2) excl. chlorinated VOC

Table 5.24   Air pollution in Greece 1980 (in 1,000 tonnes = KT) OECD provisional data

Pollutant	Power plants	industry	mobile sources	residential, other uses	agriculture	total
SO_2	335 [1]	160	46	31	10	582 (359 KT/1983)[1]
NO_x	21	35	135	5	34	230 (229)[1]
VOC	> 1	24	55	232 (solvents 17)[2]	10.2	291 (173)[1]

Note (1): 300 KT/1988 Estimated improvement due to government measures to reduce sulphur content in fuel oils.

Sources: 1. EEC/ERL potential for air pollutant emission reduction in certain chemical industries, 1986

2. EEC/Ministry of Environment, control of VOC emissions from architectural surface coating operations, 1985 and EUR 10646/1986 report of EEC.

Table 5.25 Urban, Semi-urban and Rural Population: censuses of 1928, 1951 and 1981   [NSSG, 1988]

	1928	1951	1981
Urban	1,931,937 (31.1%)	2,879,994 (37.7%)	5,659,528 (58.1%)
Semiurban	899,466 (14,5%)	1,130,188 (14,8%)	1,125,547 (11,6%)
Rural	3,373,281 (54.4%)	3,622,619 (47.5%)	2,955,342 (30.3%)
Total	6,204,684 (100%)	7,632,801 (100%)	9,740,417 (100%)

Notes: (1) Urban population =Settlements of 10,000 inhabitants and over
(2) Semiurban population =Settlements of 2,000 -9,999 inhabitants
(3) Rural population =Settlements up to 1,999 inhabitants

Table 5.26   Estimated organic load discharged in Greece into the seas   [SOBEMAP, 1981]

To Ionian Sea	9,500,000 p.e.	41 %
To Aegean Sea	13,500,000 p.e.	59 %
Total	23,000,000 p.e.	100 %
Of which :		
- domestic load	± 5,000,000 p.e.	± 20 %
- industrial and agricultural load	± 18,000,000 p.e.	± 80 %

Ratio (1) $\dfrac{\text{total organic load}}{\text{population (1)}} = \dfrac{23,000,000}{9,200,000} = 2,50$

Ratio (2) $\dfrac{\text{industrial load (2)}}{\text{total load}} = \dfrac{18,000,000}{23,000,000} = 0,80$

Ratio (3) $\dfrac{\text{population linked to sewer (3)}}{\text{population}} = \dfrac{2,350,000}{9,200,000} = 0,25$

number of p.e. = $\dfrac{BOD\,5}{54}$ (with the BOD 5 in gr/day)

(1) Population 1975.
(2) Including agricultural load.
(3) Athens = 2,000,000 p.e., Thessaloniki = 60 % x 570;000 = 350,000 p.e

Table 5.27   Indicators of Environmental Pressures in Greece [UNEP, 1984]

Indicator	Growth Index 1962 = 100			Abs. Val.
	1972	1977	1982	1982
GDP (billion $ 1975)	215	260	290	25.56
Industrial Production	272	362	410	-
Energy Requirements (Mtep)	319	413	468(1981)	16.16
Cement (Ktn/month)	278	463	539	1,025
Construction Permits ($\times 10^6$, m^3/month)	391	393	272	4.47
Electricity (MkWh)	440	636	-	20,681
Vehicle Stock (in 1,000s)	488	910	1260	630
Foreign Tourist Arrivals (in 1,000s)	390	645	891(1981)	5,094

Table 5.28   Fertilizers and Pesticides (in 1,000 tn) [Ministry of Agriculture, The Greek Agriculture, 1987]

	1981	1982	1983	1984	1985
Fertilizers	1,667	1,868	2,033	2,072	2,188
Pesticides	5.7	3.5	4.1	3.2	2.8
Trace elements	0.2	0.3	0.5	0.6	0.8
Copper	2.5	2.5	2.4	2.2	2.2
Sulfur	19	21	22	21	21

Table 5.29   Fish Production in 1,000 t/yr
              [Ministry of Agriculture, 1987]

	1980	1986	% change
a. Sea production	91.0	114.2	+ 25 %
b. International waters	4.1	4.3	+ 4.9%
c. Aquaculture	2.1	2.4	+ 14.3%
Total Greek production	97.3	120.9	+ 24.2%
Overseas production	22.0	17.3	− 19.5
Grand total	119	138.2	+ 15.8%

Table 5.30   Effluent Standards per coastal area
              [Ministry of Environment, 1987]

Environmental Parameter	Saronik. Gulf	N&S Euboic Gulf	Attica thru sewers	Attica directly	Geras Gulf (Lesbos)	Pagasitikos Gulf	Patraikos & Korinth. Gulfs	Thermaikos Gulf[b]	Dodecanese Islands
	CPD17823	CPD19640			Y1052		CPD119731	E2/11300	
pH	6-9	6-9	6-9	6-9	6-9	6.5-8.5	6-9	6.5-8.5	6-9
Temperature (°C)	35	35	35	35	30	40	35	40	35
Color	1:20[c]	1:20[c]	-	1:6[c]	1:20[c]	a	1:20[c]	-	
Floating Matter (diam.>1cm)	0	0	0	0	0	a	0	0[d]	
Settleable Matter ml/l (Imhoff/2h)	0.5	0.5	10	0	-	a	10	a	
Suspended Solids total (mg/l)	40	40	500	50	40	50	40	75	40
Dissolved Solids total (mg/l)	1500	1500	3000	1000		a	1500	a	1500
BOD_5 (mg/l)	40	40	500	40	80	30	40	80	60
COD (mg/l)	150	150	1000	120	120	90	150	250	180
Detergents (mg/l) 80% biodeg.	5	5	50	5	5	a	4	a	5
Fats-Oil (animal+veg.) mg/l	20	20	40	5	20	10	20	20	20
Mineral Oils - Hydrocarbons mg/l	15	15	15	1	15	a	5	15	15
Total Toxic Metals (mg/l)	3	3	3	3	3	a	3	a	
Aluminium (mg/l)	5	5	10	1	1	a	7	20	5
Arsenic (mg/l)	0.5	0.5	0.5	0.1	0.5	a	0.5	2	0.5
Borium (mg/l)	20	20	20	2	a	20	20	20	20
Boron (mg/l)	2	2	10	2	2	a	2	2	2
Cadmium (mg/l)	0.1	0.1	0.5	0.05	0.02	a	0.02	0.5	0.1
Chromium III (mg/l)	2	2	2	1	4	a	2	4	2
Chromium VI (mg/l)	0.2	0.2	0.5	0.2	0.2	3	0.2	0.9	0.2

Table 5.30   continued

Environmental Parameter	Saronik. Gulf	N&S Euboic Gulf	Attica thru sewers	Attica directly	Geras Gulf (Lesbos)	Pagasitikos Gulf	Patraikos & Korinth. Gulfs	Thermaikos Gulf[b]	Dodecanese Islands
	CPD17823	CPD19640				Y1052	CPD119731	E2/11300	
Iron (mg/l)	2	2	15	2	2	30	2	30	2
Manganese (mg/l)	2	2	10	1	1	a	2	4	2
Mercury (mg/l)	0.005	0.005	0.01	0.01	0.005	0.05	0.003	0.05	0.005
Nickel (mg/l)	2	2	10	0.5	2	a	2	4	2
Lead (mg/l)	0.1	0.1	5	0.5	0.1	5	0.1	1	0.1
Copper (mg/l)	1.5	1.5	1	0.2	0.1	30	15	2	1.5
Selenium (mg/l)	0.1	0.1	0.2	0.02	0.03	a	0.03	0.1	0.1
Tin (mg/l)	10	10	10	1	10	a	10	10	10
Zinc (mg/l)	1	7	20	0.5	0.5	a	1	10	1
Cyanides (mg/l)	0.5	0.5	3	0.1	0.5	a	0.5	0.5	0.5
Chlorine (mg/l)	0.7	0.7	-	-	0.2	a	0.2	2	0.7
Sulphides (mg/l)	1	1	-	-	-	a	1	2	1
Sulphites (mg/l)	2	2	1	0.2	1	a	1	2	2
Fluorides (mg/l)	6	6	20	2	-	a	16	30	6
Phosphorus (mg/l)	10	10	10	0.2	-	a	10	30	10
Total Ammonia (mg/l) of NH_4	15	15	25	10	-	a	12	40	15
Nitrogen (mg/l) as N in NO_2	0.6	0.6	4	1	-	a	0.6	3	0.6
Nitrogen (mg/l) as N in NO_3	20	20	20	4	-	a	20	50	20
Phenols total (mg/l)	0.5	0.5	5	0.5	-	0.5	0.5	0.5	0.5
Aldehydes (mg/l)	1	1	-	-	-	a	1	1	1
Aromatic Solvents (mg/l)	0.2	0.2	-	-	-	a	0.2	0.4	0.2
Nitrogenous Solvents (mg/l)	0.1	1	-	-	-	a	0.1	0.2	0.1
Chlorinated Solvents (mg/l)	1	1	-	-	-	a	1	2	1
Total Toxic Substances	e	e	e	e	e	e	e	e	e

Notes:
a: not exceeding the eighty-fold value set for drinking water
b: parameters for Acxios River are set to half the value of the corresponding parameter for Thermaikos
   parameters for Lake Volvi and Gallikos River are set to one third of the corresponding value for Thermaikos
c: dilution under which the color must not be visible in a layer of 10 cm thickness
d: diameter > 0.5 cm
e: the following substances
      As, Cd, CrVI, Hg, Ni, Pb, Cu, Se, Zn, CN^-, C_6H_5OH
   must obey the condition:
      $(Y_1/E_1) + (Y_2/E_2) + \ldots + (Y_n/E_n) \leq 3$
   where $Y_1, Y_2, \ldots Y_n$ are the concentrations of the above substances and
      $E_1, E_2 \ldots E_n$ are the corresponding permitted concentrations.

Table 5.31 International Conventions ratified by Greece

Convention Concerning the Use of White Lead in Painting, Geneva 1926/1926

Convention for the Establishment of a General Fisheries Council for the Mediterranean (as amended), Rome, 1949/1952

Convention for the Establishment of the European and Mediterranean Plant Protection Organization (as amended), Paris, 1951/1951

International Plant Protection Convention, Rome, 1951/1954

International Convention for the Prevention of Pollution of the Sea by Oil, London, 1954 (as amended on 11 April 1962 and 21 October 1969) 1967

Amendments to the International Convention for the Prevention of Pollution of the Sea by Oil, 1954, concerning Tank Arrangements and Limitation of Tank Size, London, 1971/1975

Amendments to the International Convention for the Prevention of Pollution of the Sea by Oil, 1954, Concerning the Protection of the Great Barrier Reef, London, 1971/1975

Convention on the Continental Shelf, Geneva, 1958/1972

Convention Concerning the Protection of Workers Against Ionizing Radiations, Geneva, 1960/1983

Convention on Third Party Liability in the Field of Nuclear Energy, Paris, 1960/1970

Treaty Banning Nuclear Weapon Tests in the Atmosphere, in Outer Space and Under Water, Moscow, 1963/19634

European Convention for the Protection of Animals during International Transport, Paris 1968/1978

International Convention on Civil Liability for Oil Pollution Damage, Brussels, 1969/1976

Convention on Wetlands of International Importance especially as Waterfowl Habitat, Ramsar, 1971/1975

Convention Concerning Protection Against Poisoning from Benzene, Geneva, 1971/1978

Convention on the Prohibition of the Development, Production and Stockpiling of Bacteriological (Biological) and Toxic Weapons, and on their Destruction, London, Moscow, Washington, 1972/1975

Convention Concerning the Protection of the World Cultural and Natural Heritage, Paris, 1972/1981

Table 5.31   continued

Convention on the Prevention of Marine Pollution by Dumping of Wastes and Other Matter, London, Mexico City, Moscow, 1972/1981

Protocol of 1978 Relating to the International Convention for the Prevention of Pollution from Ships, London, 1973/1983

Agreement on an International Energy Programme, Paris, 1974/1977

Convention for the Protection of the Mediterranean Sea Against Pollution, Barcelona, 1976/1979

Protocol for the Prevention of Pollution of the Mediterranean Sea by Dumping from Ships and Aircraft, Barcelona, 1976/1979

Protocol concerning Co-operation in Combatting Pollution of the Mediterranean Sea by Oil and Other Harmful Substances in Cases of Emergency, Barcelona, 1976/1979

Protocol Concerning Mediterranean Specially Protected Areas, Geneva, 1982/1982

United Nations Convention on the Law of the Sea, Montego Bay, 1982/1982

International Convention on the Protection of Archaeological Heritage London 1969/1981

Protocol Concerning Protection of Mediterranean from Land Based Sources of Pollution 1980/1986 and Marine Protected Areas

Convention for the Prevention of Pollution from Ships MARPOL 1973/1982

Convention for the Establishment of an International Fund for Oil Spill Pollution Compensation 1071/1986

Convention on Long Range Transfrontier Air Pollution and the Related Protocol for Financing the Co-operative Monitoring Programme 1979/1988

Greek-Italian Convention for the Protection of Marine Environment of Ionian Sea 1979/1982

Convention for the Conservation of Wildlife and Natural Environment in Europe 19/1983

NOTE: First date entry into force of the convention
      Second date ratification by the Greek Parliament

Table 5.32 Coastal Municipal treatment plants in Greece [SOBEMAP 1981, 1982; MoE 1988]

Locality	Population (x1000)	Type (1)	Status 1981	Status end '87
Agios Nikolaos	8,3	BIO + OUTF	planned	n.a. (2)
Alexandroupolis	35,8	BIO	planned	planned
Argos/Nafplio	31,7	BIO	planned	planned
Athens (3)	3027,0	PRIM+OUTF	planned	under construct.
Chalkis	44,9	BIO+BIO.ST	planned	existing
Chania	47,5	BIO	planned	planned
Egio	21,0	BIO	planned	n.a.
Heraklion	102,4	BIO	planned	under construct.
Chios	24,1	BIO	n.a.	under construct.
Ioannina	45,8	BIO	planned	constructed
Janitsa	24,0	BIO	planned	n.a.
Kalamata	42,1	BIO.ST	existing	existing
Kavala	56,7	BIO	planned	planned
Katerini	40,0	n.a.	n.a.	under construct.
Kerkyra	34,0	n.a.	n.a.	planned
Korinthos	22,7	BIO.ST	existing	existing
Kos	12,0	n.a.	n.a.	planned
Messologi	11,4	BIO	existing	existing
Mytilene	24,0	n.a.	n.a.	planned
Patras	154,6	BIO	planned	constructed
Preveza	13,0	n.a.	n.a.	under construct.
Rethymnon	18,2	BIO	planned	planned
Rhodos	41,4	BIO	planned	constructed
Syros	17,0	n.a.	n.a.	planned
Thessaloniki	706,2	PRIM	planned	constructed
Volos	97,8	PREL+OUTF	planned	constructed

Notes:

1. Type abbreviations
   BIO   : Biological Treatment (e.g. oxidation pond) for municipal and industrial sewage brought by sewer
   BIO.ST : Biological treatment (e.g. oxidation pond) for municipal and industrial sewage brought by septic tank
   PREL   : Preliminary treatment (screening, grit removal, flotation)
   PRIM   : Primary treatment (physicochemical treatment, required because of industrial liquid wastes and toxic substances which are likely to inhibit biological processes

2. n.a. = not available

3. Construction is also under way of a peripheral treatment plant in Metamorphosis

Table 5.33 Program of public investment for water supply and sewage treatment 1980-1988 (payments in mil. Drs)[1] [B. Vardakou, Ministry of Interior, 1988]

Year	Prefectural Programs	Ministry of Interior	Ministry of Public Works	Total	% yearly Change
1980	1,759	-	1,692	3,451	
1981	2,852	1,234	487	4,573	32.5
1982	3,451	1,380	1,372	6,203	35.6
1983	6,060	1,297	2,583	9,940	60.2
1984	6,733	896	5,942	13,571	36.5
1985	9,914	1,130	7,074	18,118	33.5
1986	8,698	1,200	7,799	17,697	2.3
1987	8,788	1,040	9,792	19,620	10.8
1988	9,000[1]	2,400	8,400	19,800	0.9

[1] Estimation

* Water supply costs cover about 1/3 of total costs

Table 5.34 Waste water treatment units of the Agro-Food Industry (Including Stock-Farming) [H. Valassopoulos Agricultural Bank of Greece, 1987]

	NO OF INDUSTRIES	TYPE OF WASTE WATER TREATMENT
a.	3,367 (57.93%)	NO TREATMENT
b.	141 (2.43%)	BIOLOGICAL TREATMENT
c.	1,141 (19.63%)	MECHANICAL TREATMENT
d.	1,126 (19.37%)	OTHER TREATMENT
e.	37 (0.64%)	PLANNING

TOTAL 5,812 (100%)

NOTES:

1. CASES C AND D NOT RELIABLE
2. SUGAR INDUSTRY (100% TREATMENT)
3. MEAT INDUSTRY (19% TREATMENT)

ACKNOWLEDGEMENTS

This study is a compendium of the expertise of many people working on environmental issues. The research was financed by the European Investment Bank of the European Community and the World Bank.

Many colleages have helped by providing suggestions, criticisms and elaborations. We owe a special debt of gratitude to the Mediterranean Action Plan and its constituent unit, the Coordinating Unit in Athens, the many colleagues in the country governments and the consulted experts for their contributions to the research reported here. Several coauthored technical reports with us: Antonio Estevan on the Spanish chapter, Dr. Luis A. Romaña on the French report, Professor Vittorio Silano on the Italian report and Dr. Miltiadis Vassilopoulos on the Greek report. Thanks are especially due to Michel Deleau, George Toregas, Guy Clausse, Gianni Carbonaro and Carl Cavanagh of the European Investment Bank for their comments on the draft of the manuscript. Pierre Balland of the Agence de Bassin Rhône-Méditerranée-Corse in Pierre-Bénite also supplied valuable laboratory data. Others who have provided feedback on the draft are our colleagues Professor Amy Schoener at the University of Washington in Seattle and Professor Egbert K. Duursma.

We acknowledge Robert Glaser at the Environmental Inspectorate for the Province of Zeeland who has commented on the chapter of Greece. Michiel de Vlieger of ECD Computers in Delft has helped in supplying the software to print graphs and in providing technical advice. We also thank Dr. Fausto Maria Spaziani (Water Research Institute, Rome) and Dr. Nicola Sarti (Ministry of Health, Rome) both for providing statistical information. A number of graphs were made based on their insights.

Several of the illustrative materials are reprinted with permission from Ricerca, Ambio, Annals of the New York

Academy of Sciences, American Society for Microbiology, Springer-Verlag, R. Oldenbourg Verlag GmbH, Istituto per l'Ambiente, Etas Periodici S.p.A., Gordon and Breach Science Publishers, Scientia Marina, Elsevier Science Publishers, Pergamon Press PLC, John Wiley & Sons Ltd, Centre d'Études et de Recherche de Biologie et d'Océanographie Médicale, American Fisheries Society, PROGNOS AG, DocTer, World Health Organization, Food and Agriculture Organization of the United Nations, United Nations Environment Programme, Plenum Publishing Corp, Intergovernmental Oceanographic Commission, and Marcel Dekker Inc, and by courtesy of Dr. Anthony J. Parker (University of Bristol).

ABBREVIATIONS

kg	kilogram
km^2	square kilometer
km^3	cubic kilometer
l/h	liters per hour
m^3	cubic meter
m^3/sec	cubic meters per second
mg	milligram
mg/m^2	milligrams per square meter
mg/m^3	milligrams per cubic meter
mg/m^2.year	milligrams per square meter per year
mg/m^3.day	milligrams per cubic meter per day
$\mu g/m^2$	micrograms per square meter
$\mu g/m^3$	micrograms per cubic meter
ng/m^3	nanograms per cubic meter
g/l	grams per liter
mg/l	milligrams per liter
μg/l	micrograms per liter
μg-at/l	microgram atoms per liter
μmoles/l	micromoles per liter
ng/l	nanograms per liter
mg/kg	milligrams per kilogram (in sediment, milligrams per kilogram by dry weight; in living matter, milligrams per kilogram by wet or dry weight)
μg/g	micrograms per gram (in sediment, micrograms per gram by dry weight; in living matter, micrograms per gram by wet or dry weight)
μg/kg	micrograms per kilogram (in sediment, micrograms per kilogram by dry weight; in living matter, micrograms per kilogram by wet or dry weight)
<	less than
≤	equal to or less than
>	greater than

CURRICULA VITAE

ANNABELLA COFIÑO-MOLINA (1961) received her bachelor's degree in chemical engineering at the Universidad del Valle de Guatemala. In addition, she received her postgraduate diploma in environmental science and technology at the International Institute for Hydraulic and Environmental Engineering in Delft. She was a scientist at SCMO-TNO. She currently is Assistant to the General Manager of the chemical factory of La Popular in Guatemala City.

FOPPE DE WALLE (1945) is director of SCMO-TNO in Delft, Holland. He received his master's degree in Environmental Engineering, State University in Wageningen, Holland and a Ph.D. in Civil Engineering, University of Washington in Seattle, USA. He is an adjunct professor in the Department of Environmental Health, at the University of Washington in Seattle. He previously was on the faculty at Stanford University in Menlo Park and University of Illinois in Champaign-Urbana, USA.

ANTONIO ESTEVAN ESTEVAN (1948) received his industrial engineering degree from the Industrial Engineering School of the University of Navarre and a Diploma in Urban Technical Management of the Local Government Institute in Madrid. He is consultant and founder of Gabinete de Economia Aplicada S.A.

WIM J. HEINEN (1956) is a scientist with SCMO-TNO. He received his master's degree in molecular sciences at the Wageningen Agricultural University where he specialized in biochemistry and microbiology. In addition he conducted research on the enzymological aspects of aging at the Medical School of the Erasmus University in Rotterdam. His current specialty is the chemistry and microbiology of the marine environment.

SJAK J. LOMME (1961) received his master's degree in technology and society at the Eindhoven University of Technology, where he specialized in technology assessment. He was a scientist with SCMO-TNO. He currently works in the N.V. SEP (Dutch Electricity Generation Board), Fuel Supply Department in Arnhem on long-term strategies for fuel supply of power stations.

MARIA NIKOLOUPOULOU-TAMVAKLI (1958) is a scientist with SCMO-TNO. She received her bachelor's degree in chemistry at the University of London and her diploma in environmental chemistry at the University of Edinburgh. In addition, she received her postgraduate diploma in environmental science and technology at the International Institute for Hydraulic and Environmental Engineering in Delft. She worked in the University of Patras as Research Fellow and laboratory assistant. She is currently engaged in the chemistry and biology in the marine environment.

LOUIS ALEXANDRE ROMAÑA (1945) is a Research Scientist with the Institut Français de Recherche pour l'Exploitation de la Mer (IFREMER). He received his Masters in Physical Science from the University of Buenos Aires, Argentina and his Diploma of Advanced Studies in Biological Occeanography of the University of Marseille, France. At the latter institute he completed his postgraduate thesis. He is also a Professor for environmental training at the University of Aix-Marseille.

VITTORIO SILANO (1940) received his M.D. (Laurea) in Biochemistry from the University of Naples, Italy and a Ph.D. degree (Libera Docenza) in Biochemistry from the University of Rome. He currently is Director General for Food Safety and Nutrition, Ministry of Health, Rome. Previously he was Director General of the Pollution Prevention and Reclamation Service at the Ministry of the Environment.

MILTIADIS VASSILOPOULOS (1935) is permanent representative and environmental councillor of Greece to the EEC. He received a Diploma in Chemical Engineering from the Technical University of Athens, a Dr. Ph. Degree from the Technical University of Karlsruhe, Germany and a Business Administration degree of ASOEE. He previously worked at the Ministry of Environment and National Council for Physical Planning and Environment.

APPENDIX

Investment needs

Investment Requirements: Spain

	Estimated Costs (US$ Millions)
Policy Actions	
1. Definition of roles and responsibilities at all levels of Government for environment protection	0.1
2. Strengthening of legislation for the effective protection of natural resources including birds and game.	0.2
3. Strengthening of legislation for effective coastal zone management.	0.2
Institutional Actions	
1. Strengthening of MAPA (Ministry of Agriculture Fisheries and Food) mentioned in text to enable it to develop and effectively monitor and enforce national environmental policies and guidelines.	1.5
2. Strengthening local, provincial and regional governments to enable effective coordination of implementation of policies on the environment.	2.0
Technical Assistance (Studies)	
1. Coastal Zone Management (one region)	3.0
2. Watershed Management (one region)	4.0
3. Integrated Planning and Management	3.0
4. Wetland Conservation	1.0
Investment Programs	
1. Water Pollution Control Facilities	1,800
2. Toxic Waste Collection and Disposal	20
3. Port Reception Facilities for Ships Wastes	10
4. Coastal Zone Management (one region)	40
5. Watershed Management (in same region as 4 above).	40
6. Integrated Planning and Management (one Balearic Island).	40
7. Wetland Conservation	10
Total	1,975

Investment Requirements: France

	Estimated Cost (US$ Million)
Policy Actions	
1. Continuation of implementation of the EEC environmental standards and regulations	5.0
2. Continuation of implementation of the Oslo-Paris convention on strongly polluted waters	3.0
3. Gradual privatization of the sewerage system	–
Institutional Actions	
1. Establishment of system for prevention of accidental pollution	3.0
2. Strengthening, within the Ministry of Environment, of the "Direction de la Protection de la Nature"	1.0
3. Strengthening of national, regional and local institutions in charge of coastal management	1.0
4. Improvement of the Rhône basin management	1.0
5. Strengthening research institutions working on treatment measures for anoxic conditions in the sea bed and on microtoxic pollution	2.0
6. Improvement of data collection and analysis, in particular that related to toxic and hazardous wastes	0.5
7. Strengthening of the Ministry of Environment for coordination of the nuclear energy industry	0.5

Investment Requirements: France (Cont'd)

	Estimated Cost (US$ Million)
Technical Assistance	
1. Improvement of operating procedures for existing treatment plants	100.0
2. Study of the potential for privatizing sewerage system	1.0
3. Study of the appropriate role of the Ministry of Environment in the nuclear energy industry	0.3
Investment Projects	
1. Hydrocarbon reception facilities	1.7
2. Treatment plants for domestic and industrial discharges	500.0
3. Water supply in the cities	250.0
4. Protected areas	150.0
5. National Parks	100.0
6. Recreational areas	100.0
7. Forest planting/replanting and protection	350.0
TOTAL ESTIMATED COST	US$1,570.0

Investment Requirements: Italy

	Estimated Costs (US$ Million)
Policy Actions	
1. Development of a hazardous waste management and disposal policy.	0.2
2. Revision and strengthening of environmental legislation for: (i) operation and maintenance of wastewater collection and treatment systems; (ii) effective protection of natural resources including birds and game; (iii) effective coastal zone and watershed management, including protection of vegetation from damage caused by acid rain.	0.3
Institutional Actions	
1. Promotion of public/private sector cooperation for: (i) implementation and management of hazardous waste disposal; (ii) wastewater treatment and disposal.	0.5
2. Strengthening of the Ministry of Environment	0.5
3. Strengthening of regional, provincial and local governments for an effective: (i) planning and enforcement of coastal zone and watershed management, and (ii) enforcement of regulations for the protection of natural resources, birds and game.	1.5
Technical Assistance	
1. Not separately identified.	
Investment Programs	
1. River Basin Management Environment Protection Projects (Po Basin including the Lambro, Olona and and Seveso river basin).	5,000
2. Provincial Wastewater Programs	4,000
3. Environmental protection of the Lagoon of Venice.	2,200
4. Solid Waste Management	800
5. Soil Conservation Programs	5,000

Investment Requirements: Italy (Cont'd)

	Estimated Costs (US$ Million)
Investment Programs	
6. Wild Life Protection	17
7. Historical Monuments and Sites	100
8. Coastal Zone Management	100
Total	17,220

Investment Requirements: Greece

	Estimated Cost (US$ Million)
Policy Actions	
1. Upgrading fuel quality standards	0.1
2. Development of policy on industrial wastes (Regulations/requirements regarding effluent quality, pre-treatment requirements, etc.).	0.2
3. Revision of system and responsibility for setting wastewater tariffs.	-
4. Revision of entry fees to monuments and historic sites.	-
Institutional Actions	
1. Clarification of responsibility and establishment of central government role with respect to hazardous wastes.	0.3
2. Exploration of the feasibility and legal requirements for forming consortia between municipalities for solid waste disposal.	0.1
3. Stricter monitoring and control of monuments and historic sites.	0.5
4. Improvement of coordination between the National Tourism Organization and organizations responsible for natural resources management.	0.2
5. Improvement of management of existing protected areas.	0.5
6. Enforcement of air quality standards	5.0
Technical Assistance	
1. Training in O&M of waste water treatment plants	1.0
2. Carrying out an inventory of coastal zones requiring special attention.	0.3
3. Identification of specific investments from the "Heraklion Coastal Management" proposal.	0.5

Investment Requirements: Greece (Cont'd)

		Estimated Cost (US$ Million)
Technical Assistance (Cont'd)		
4.	Study of effect of EEC countries complete integration (1992) on coastal zone management.	0.2
5.	Formulation of environmental management plans for (i) Rhodes (pilot), (ii) the larger islands, and (iii) the smaller islands.	0.8
6.	Additional studies for Athens metro	3.0
Investment Projects		
1.	Fuel desulphurization plant	150.0 [a]
2.	Athens traffic control improvements	100.0
3.	Replacement of existing sub-standard storage facilities for pesticides.	5.0
4.	Implementation of first phase identified from the "Heraklion Coastal Management" proposal.	50.0
5.	Preventive and remedial measures against deforestation (fire fighting equipment, tree planting).	100.0
6.	Continuation of the waste water treatment plant building program.	600.0
7.	Implementation of planned port reception facilities	13.6
8.	Protection of monuments and historic sites.	50.0
	Total estimated cost	1,031.0

TOTAL ESTIMATED COST FOR SPAIN, FRANCE, ITALY AND GREECE   21,796.0

SUBJECT INDEX

accidents 42, 273
acid rain 341-342, 356
activated sludge 269
Adige (river) 4, 24, 331, 336, 343, 345, 349-350
Adriatic (sea) 2-3, 6, 23-24, 35-38, 40-41, 43, 49, 59-61, 63, 98, 329, 331, 333-334, 336-339, 343-344, 350
Aegean (sea) 3-5, 27, 29, 36-37, 40, 43, 45, 48, 63, 426, 434, 440-442, 449
aerosol 54, 68, 342
aggravated continuing scenario 26, 364-366
aggregated alternative scenario 364-366
agricultural waste 88
agriculture 11, 26, 30, 65, 67, 88, 191, 335, 341, 343, 353, 357, 425, 450-451
agro-food industry 26, 364, 457
air pollution 25-26, 29, 86, 190, 266, 355-356, 446-447, 460
Aitolikon (lagoon) 441
Ajaccio 16, 259, 268
Albegna (river) 330
Alboran (sea) 40
Albufera (lake) 186
alcohol distilleries 18, 270
aldrin 336
Alexandria 40
Alexandroupolis (gulf) 426
algae 2, 33, 37-38, 68, 92, 353-354, 357, 432
algal toxins 38, 47
Aliakmon (river) 28, 435-436, 443
Alicante 11-12, 181, 183, 196
aliphatic hydrocarbons 183-184

alkalization 56
alkanes 42, 50, 55
allergies 196
Almería (gulf) 2
alternative "integration" scenario 82
alternative reference scenario 82
ammonia 16, 19, 27, 61, 85, 257, 272, 426, 431, 433-434, 436-437
anchovy 3, 22, 184, 334, 336
Ancona 3, 22, 329-330, 336
anoxic 19, 23, 27, 38, 271, 274, 432-434, 437
aquaculture 30, 53, 66, 451
aquifer 26, 29, 189, 192, 367, 444
Argeles 266
Argentario (promontory) 3, 22, 333
<u>Aristeus antennatus</u> (see also shrimp) 12, 181-183
Arles 272
Arno (river) 24, 339, 345, 349-350
Arochlor 184
aromatic hydrocarbons 72, 184, 192, 259
arsenic 24, 54, 98, 334, 350
artificial reef 8, 95
As see arsenic
asbestos 192
ascidians 46
Athens 4-5, 27, 29, 31, 73, 425, 427-429, 431-432, 442-443, 445-448, 454-457, 459-460
Atlantic (ocean) 2, 35-37, 54, 98, 257
atrazine 344
Augusta (bay) 334

Axios (river) 28, 435-437, 442-443, 445

Bacoli 22, 336
bacteria 68, 184, 186-187
Badalona 185-186, 190
Bagnoli 332, 336
Balearic (islands) 51, 63, 180, 190, 192, 202
ballast water 90, 195
bank swallows 65
Banyuls 264
Barcelona 3-5, 12, 41, 50, 73-74, 181-182, 184-185, 187, 190-191, 195
Barcelona Convention 8, 94, 445, 454
Bastia 20, 266, 268
beach 12, 17, 23, 27, 44, 51, 53, 56, 64, 67-68, 71, 98, 180, 184-187, 191, 196-197, 266, 329, 338, 340, 362, 427, 434, 449
Beach Control Law 202
Belone belone (see also garfish) 334
bentazone 344
benthic community 46, 98
benzene hexachloride see BHC
benzopyrene 337
Berre 16, 257, 266-267
Berre (lagoon) 2, 17-19, 21, 257, 261-263, 267, 270-271, 274
Besós (river) 4-5, 12, 14, 181-183, 185, 187, 195
BHC (benzene hexachloride) 22, 45, 336, 430
BHC compounds 64
bilge water 7, 77, 90, 195-196
biochemical oxygen demand see BOD
biocide 16, 259, 438
biodegradation 62
biogas 88
biogenic hydrocarbons 42

biological resources 362
biotope 5, 8, 10, 33, 445
bird 5, 33, 49, 57, 65
bivalve 50, 334
black ibis 65
blacklist substances 10
Blue Plan 73, 364
Blue Plan scenario 81, 99
blue whiting 451
BOD (biochemical oxygen demand) 1, 10, 14, 18-19, 23-24, 26, 28, 33, 60-61, 188, 193-195, 268, 270, 272-273, 343, 345-350, 366, 436, 443-444
bonito 49
breeding areas 5, 98
Brenta (river) 331
Brindisi 332
Brussels Convention 276

cadmium 3, 11, 16, 22, 28, 39-41, 48, 54, 69, 72, 92, 98, 181-182, 188, 203, 257-260, 264-267, 274, 330, 333-334, 345-346, 427, 435
calcium 188
California (bight) 45
Cannes 264, 267
Cannes (bay) 16, 257-258, 262
carbon dioxide 72, 86
carbon monoxide 353, 446
Caretta caretta (see also loggerhead turtle) 51
Caronte (canal) 261-263
carp 23, 343, 345
Cartagena 12, 181-182, 190, 195
Carteau (inlet) 261
Castellammare di Stabia 347
Castelldefels 181, 185-186
Castellón 48, 181, 190, 195
Cavalaire 20, 266
Cd see cadmium
cement works 85
cephalopods 67, 451
Cesano (river) 329
cetaceans 361

Chamelea gallina (see also bivalve) 334
Chattonella (see also algae, chloroflagellate) 38
chemical industry 18, 270
chemical oxygen demand see COD
Chioggia 337
chlor-alkali complex 54
chlor-alkali plant 41, 93, 331
chlorides 188
chlorinated hydrocarbons 4, 45, 54, 98, 336, 430
chloroflagellate 38
chloromonads 68
chlorophyll 37, 39, 196
cholera 186
chromium 6, 24, 40, 63, 188, 331, 345, 350, 428-429, 435, 439-440
chrysomonads 68
clams 4, 50
clean technologies 31, 84, 91
CO see carbon monoxide
CO_2 see carbon dioxide
Co see cobalt
coal 199
Coastal Law 200
coastal protection 359
cobalt 266-267, 332, 435, 440
COD (chemical oxygen demand) 10, 21, 28, 60, 271, 366, 436
coliform 28, 68, 78-79, 184, 186-188, 197, 266, 337, 344, 443
coliform bacteria 78, 186, 266
coliphages 68
color 37, 70, 337-338
Commission des Aides 20, 277-278
compost 13, 87, 191
compost plant 13, 191
composting 191
conductivity 188
congenital disease 69
conservation 5, 33, 95, 199, 276, 356, 362, 452
copper 16-17, 21, 24, 27, 39-40, 48, 188, 257-262, 266-267, 329-334, 343, 345, 350, 428, 435, 440
Corfu 27, 427, 458
CORINE program 9, 75
Corsica (island) 18, 41, 257-260, 264-265, 267-268, 271
Cortiou 4, 16-17, 257-258, 260-261, 263-264
Cortiou (cove) 17, 259, 263
cost 21, 26, 31-32, 74, 80, 83, 85, 88, 90-91, 197, 202-203, 273, 275, 354, 357, 366-368, 447, 453-454, 457-460
cost-effectiveness 11
Cr see chromium
Crete (island) 43-44, 427, 444-445, 449-450, 457
Crissolo 344
criteria 47, 78, 91, 359, 434
crustaceans 48-49, 334, 357
Cu see copper
Cullera 12, 183
cultural heritage 356
cuttlefish 451
cyanides 188
cyclodienes 64

Dalmatian pelican 57
Darse (inlet) 261
DDD 183, 263, 265, 430
DDE 22, 183, 263, 265, 335-336, 430
DDT 5, 12, 17, 22, 27, 45, 50, 182-184, 263, 265, 335-336, 344, 429-430, 434, 438
DDT compounds 64
decomposition 28, 38, 98, 444
deforestation 189, 460
desertification 11, 13, 189
detergents 61, 188, 259, 269, 337-339
diatoms 70
dieldrin 22, 336, 430, 438
diesel 447
digestive organs 183

dinoflagellate 37, 39, 68, 70, 432
directives 25, 28, 31, 78, 360-361, 443, 453
dissolved oxygen 188, 271, 346, 432, 436
Dodecanese (islands) 427
dolphins 5
domestic sewage 6, 28, 47, 59-61, 63-64, 98, 268, 443
domestic waste 27, 192, 432, 435, 444
domestic, industrial and agricultural wastes 434
Drapetsona 447
Drapetsona (bay) 27, 428, 431
drilling 349
drilling platform 50
dumps 353, 368
Durance (river) 16, 19, 257, 271
dust 85, 352-353

ear infections 197
ear and eye infections 67, 197
Ebro (river) 4-6, 12-14, 36, 50, 60, 64, 182-184, 187, 195
EC see European Community
EC directives 25, 28, 31, 360-361, 443, 453
economic growth 81-83, 85
ecosystem 2, 8, 19, 38, 45, 57, 70, 77, 91, 95, 97, 338, 340, 346, 450
EDF (canal) 262
education 96
eels 274
effluent 39, 47, 77, 91-94, 260, 270, 273, 426, 431, 435-436, 449, 457
effluent standard 455
eggs 49, 51
EIB see European Investment Bank
electricity-generating plant 199, 431
electrolysis industries 93
Elefsis 447, 458-459

Elefsis (bay) 27, 39, 426-429, 431-433
El Grau de Castellón 3, 11, 180
emission 29, 59, 80, 84-86, 267, 352-353, 446-448
endangered species 51
endemic species 5, 33
endrin 430
Engraulis (see also anchovy) 22, 184, 334, 336
enterococci 68, 78
environmental impact assessment 359, 456
epidemic 68
erosion 11, 13-14, 23, 55-56, 72-73, 86, 99, 188-189, 195, 197, 204, 340, 342, 362, 444, 460
erosion control 84
Escherichia coli (see also bacteria) 68, 78
ETBA (Hellenic Industrial Development Bank S.A.) 450
Euboikos (gulf) 426, 439
Euphrates (river) 56
European Commission 65-66
European Community (EC) 2-3, 9, 11-12, 16, 65, 72-73, 76-77, 82, 90, 180, 184, 191, 199-200, 258, 329, 338, 349, 360, 427, 447, 455, 458
European Economic Community 75
European Fund for Regional Development 31, 457
European Investment Bank (EIB) 31, 65, 362, 456-457
European Regional Development Fund Environmental Program 9, 76, 86
eutrophic 38, 340
eutrophication 2, 19, 23, 33, 36-39, 46, 54, 68, 70-71, 83, 98, 271, 338-339, 343, 353, 436
expenditure 25, 80-81, 86, 200, 256, 277-278, 354, 356, 362-

364, 368
eye infection 14, 196

factory 191, 261, 330, 333, 428, 431, 435, 452
fatty tissues 23, 344
Fe see iron
fertilizer application 84, 437
fertilizers 30, 73, 189, 203, 351, 431, 450
Figols 187
FIO see Investment and Employment Fund
fire 72, 94, 189, 341, 444-445
fish 5, 19, 22-23, 28, 38, 48-50, 67, 71, 274, 333-334, 336, 338, 343-344, 357, 430, 451
fishery 8, 50, 53, 57, 70-71, 95, 451
Fiumicino 346
flatfish 38
flooding 99, 204
fly ash 448
Follonica 330, 332-333
Follonica (bay) 22, 332-333
food chain 10, 12, 49, 98, 203
forest fires 13-14, 23, 57, 72-73, 188-189, 197, 340-341
Fos 16, 21, 44, 257, 260, 266-267, 270-271, 273, 278
Fos (gulf) 3, 17-18, 257, 260-265, 267, 270
Fourth Environmental Action Program 9, 75
freight transport 57
fuel oil 441, 447
fungicides 351

garfish 334
gases 28, 86, 444
gasoline 190, 349, 447
gastroenteritis 68
gastrointestinal disease 78, 186
GDP see Gross Domestic Product
General Quality Index (ICG) 13, 187-188

genetic resources 57
Genoa 41, 58, 332, 349
Gibraltar (strait) 35, 37, 39
GNP see Gross National Product
gobiids 38
Gobius jozo (see also gobiids) 38
gonads 183-184
Gonyaulax polyedra (see also algae, dinoflagellate) 39
grants 21, 31, 277-278
greenhouse effect 72, 86
Gross Domestic Product (GDP) 80, 89
Gross National Product (GNP) 25, 73, 80, 362
groundwater 344, 347, 351, 361, 444
guideline 78, 447
Gymnodinium corii (see also algae) 38

habitat 5, 8, 11, 50, 57, 95, 99, 444
hake 5, 12, 67, 183-184, 451
halogenated hydrocarbons 438
hazardous waste 7, 10, 28, 77, 191, 272
health 25, 39, 67, 180, 354-355
health services 354
heavy metal 1-2, 10-12, 23, 27-28, 33, 39-40, 47-48, 54, 91, 97-98, 180-181, 188, 192, 195, 203, 257-258, 262, 264, 329-334, 343, 345-346, 427-428, 440
Hellenic Industrial Development Bank S.A. (ETBA) 450
heptachlor epoxide 430
Hérault (river) 3, 265, 267
herbicides 345, 351
Hg see mercury
Hg-T see total mercury
historic settlements 58
historic sites 25, 34, 57-58, 355-356, 445, 449

hospital waste 25
housing density 192
human health 11, 354
hydrocarbon 4, 7, 10, 12, 16-17,
   21, 33, 42-45, 50, 62-63, 75,
   195, 259-260, 262-263, 265,
   271, 273, 275, 349, 441, 454
hydrocarbon residues 45, 77
hypertrophic 23, 339

Ialyssos 427
Ibiza 183
ICG see General Quality Index
IMPs see Integrated Mediterranean Programs
incentives 31, 88, 91, 454-455
incidents 69
incineration 276
indicators 2, 33, 68, 186, 449
industrial effluent 27, 426,
   428, 432, 434-435
industrial emissions 199
industrial waste 24, 40, 59, 79,
   87, 192, 201, 351, 353, 366,
   368, 425, 444, 459
industrial and municipal wastes
   332
industrialization 364
insecticide 5, 45, 182, 263,
   335, 351, 429, 434
Integrated Mediterranean Programs (IMPs) 65-66, 77
International Union for the Conservation of Nature and Natural Resources (IUCN) 94
Interregional Conference 367
intestinal infections 197
investment 2, 15, 29, 31, 33,
   85-86, 97, 99, 196, 200-203,
   275, 278, 369, 442, 446-447,
   453-454, 457, 460
Investment and Employment Fund
   (FIO) 362-363, 369
Ionian (sea) 27, 29, 42, 44, 51,
   333-334, 339, 426-427, 442,
   449

iron 22, 27, 37, 332-333, 428,
   435, 440
iron and steel industry 84-85
Ischia (island) 21, 329
Iskenderun (bay) 50
IUCN see International Union for
   the Conservation of Nature and
   Natural Resources

Júcar (river) 12, 183, 201

Kavala (gulf) 434, 438
Keratsini (bay) 27, 428-429, 431
kidney cortex 69
Kleissova (lagoon) 441

landfill 28, 443-444
land use 7, 13, 34, 55, 268
landscape conservation 97
La Spezia 336
La Spezia (gulf) 41, 330
Lavéra 17, 21, 261, 263, 273
Law on protected natural areas
   190
leaching 331, 450
lead 11-12, 16-17, 21-22, 24,
   28, 48, 54, 63, 98, 180-182,
   188, 190, 203, 257-262, 264,
   266-267, 274, 329-334, 343,
   345, 350, 427, 435, 440, 447
leather industry 331
Les Saintes-Maries-de-la-Mer
   259, 266
Letoianni 338
Levantin (sea) 40
license 8, 29, 95, 444, 451
Lido di Ostia 21, 329
lignite 448
Ligurian (sea) 3, 36, 48, 55,
   63, 330, 332
limpet 333
lindane 16, 259-260, 262-263,
   438
Lions (gulf) 2, 38-39, 98, 260
lipid 183-184
liquid waste 29

liver 5, 12, 49-50, 183-184
livestock farming 342, 353
Livorno 41, 54, 331
Llobregat (river) 4-5, 12, 182, 185, 187
Load-on-Top 89
lobster 22, 333-334, 336
loggerhead turtle 50-51, 98
Los Angeles 446
Loudias (river) 28, 435-436
Lyon 19, 272

mackerel 3, 48
magnesium 188
Maiori 338
Málaga 41, 58, 195
mammals 50-51, 333
manganese 266-267, 332, 435, 440
manufacturing industry 59, 365
MAP see Mediterranean Action Plan
marine plants 5, 71
Marine Pollution Monitoring Pilot Project (MAPMOPP) 43
marine turtle 57
Mar Menor (lagoon) 185
Mar piccolo di Taranto 50, 339
MARPOL Convention 90, 273, 457
Marseille 3-4, 16-18, 20-21, 41, 54, 58, 257-258, 260-261, 264-265, 267, 270, 273, 277
Martigues 261
maximum permissible concentration 5, 336
maximum permissible level 344
Mediterranean, central (sea) 4, 40, 43-44
Mediterranean, northwestern (sea) 6, 40-41, 48-49, 59-61, 63
Mediterranean, southwestern (sea) 44
Mediterranean Action Plan (MAP) 73, 446
Mediterranean Pollution Monitoring and Research Program see

MED POL
Mediterranean Strategy and Action Plan (MEDSAP) 10, 66, 76-77, 458
MED POL (Mediterranean Pollution Monitoring and Research Program) 47-48, 73, 78, 94
MEDSAP see Mediterranean Strategy and Action Plan
mercaptans 444
mercury 2-3, 6, 11-12, 16-17, 21-22, 24, 28, 41, 48-49, 54, 63, 69, 92-94, 98, 180-181, 188, 203, 258-262, 264-265, 271, 329-334, 350, 427, 435
"Merli Law" 359-361
Merluccius merluccius (see also hake) 5, 12, 183-184
Messolonghi (lagoon) 441, 445
metal 6, 11, 24, 39, 59, 63, 98, 180, 262, 264, 330, 332, 350, 440
metalloids 54, 98
metallurgical industries 29, 329, 332-333, 449
Metauro (river) 349
methane 444
methyl mercury 69, 344
microorganisms 47, 54, 98, 187, 269, 432
migration 8, 66, 82, 95, 348
military arsenals 265
mineral oil 62, 337
mines 63
mining 11, 54, 181, 188, 194, 332-333, 448
mining waste 181, 203
Mn see manganese
moderate continuing scenario 26, 364, 366
moderate trend scenario 82, 85
Mojácar 2, 11, 180
molinate 344
molluscs 38, 48-49, 68, 182, 333-334, 357, 435
Monachus monachus (see also monk

seal) 5, 51
Monaco 54
Mongat 180
monitoring program 47, 73, 429
monk seal 5, 50-52, 57, 98
monuments 356, 446
morbidity 67-68, 196-197
mortality 19, 38, 46, 52, 98, 274
mullet 3, 5, 22, 27, 48-49, 181-184, 333-334, 336, 429-430, 439
Mullus (see also mullet) 3, 5, 22, 27, 48-49, 181-184, 333-334, 336, 429, 439
municipal waste 444, 448
municipal and industrial wastes 87
muscle 28, 50, 183-184, 439
Musone (river) 329
mussel 3-4, 12, 17, 19, 22, 28, 48-50, 181, 183, 264-265, 274, 333-334, 336-337, 434
Mytilus (see also mussel) 3-4, 12, 22, 48-50, 182-184, 265, 333-334, 336, 434

N see nitrogen
Naples 21-22, 25, 332, 335-336, 347, 352
Naples (bay) 2, 335, 339, 347
national parks 20, 190, 277
natural heritage 99
natural resources 96, 199-200, 276, 425
nature conservation 199, 359
nature reserves 7, 34, 53, 57, 359
Navigational Code 361
Nephrops norvegicus (see also lobster) 22, 333-334, 336
nervous system 69
New York (bight) 45
NH_4^+ see ammonia
NH_4^+-N see ammonia
Ni see nickel

nickel 22, 24, 39-40, 266-267, 332-333, 350, 435, 440
nitrate 16, 23, 27, 36, 39, 188, 257, 343, 345-346, 426, 431-433, 436-437, 440
nitrite 433-434, 436-437
nitrogen 6, 19, 24, 32, 37-38, 61, 84, 257, 273, 339, 343, 345-346, 350-351, 353, 433-434, 436-437
nitrogen compounds 37
nitrogen dioxide 446
nitrogen oxides 352-353, 446, 448
NO_2 see nitrogen dioxide
NO_2 hourly limit 447
NO_2^--N see nitrite
NO_3^--N see nitrate
noise levels 347
North (sea) 4, 43
NO_x see nitrogen oxides
nutrient 2, 19, 24, 27, 36-39, 46, 54, 59, 61, 71, 83, 97, 188, 257, 274, 339-340, 343, 345, 350, 353, 357, 426-427, 431-433, 436-437, 440

octopods 451
OECD see Organization for Economic Cooperation and Development
offshore drilling 349
oil 6, 41-42, 44, 62, 72, 74, 89-90, 448, 453, 458-459
oil components 72
oil pollution 6, 42, 62, 85, 89-90, 427, 454
oil refineries 27, 432
oil residues 42, 459
oil terminals 45, 62, 195, 349
oily compounds 43, 427
oligotrophic 2, 27, 37, 70, 426
Orb (river) 3, 265, 267
organic carbon 27, 349, 428, 436, 441
organic compounds 446

organic matter 1, 19, 27, 33, 59-60, 193, 268-269, 271, 273-274, 343-344, 435-436, 441
Organization for Economic Cooperation and Development (OECD) 81, 347
organochlorine 10, 16, 23, 50, 64, 183, 259, 344, 431, 434
orthophosphate 346, 437
Oslo-Paris Convention 3, 17, 265
Ostrea edulis (see also oysters) 4, 50
outfall 4-5, 7, 14, 16-17, 27-28, 35, 47, 92-94, 185-186, 193-194, 201-203, 257, 259-261, 263-264, 340, 347, 428-429, 431-432, 434-437
overfishing 8, 67, 71, 83, 95
overgrazing 55, 57, 444
oxygen 37, 39, 71, 98, 337-338, 345, 357, 432
oysters 4, 17, 19, 50, 264, 274

P see phosphorus
Pagassitikos (gulf) 425, 439
PAH (polycyclic aromatic hydrocarbons) 22, 42, 45, 50, 336, 438
Palaiopotamos (lagoon) 441
Palazuelos 13, 187
Palma 3, 12, 181-182, 196, 202
PAP see Priority Actions Program
paper industry 18, 270
Paphos 44
Parapenaeus longirostris (see also shrimp) 181
particulate matter 41, 330
particulates 272, 446
Patella caerulea (see also limpet) 333
Patraikos (gulf) 439
Pb see lead
PCB 4-5, 12, 16-17, 22, 27, 45, 50, 182-184, 259-260, 262-263, 265, 274, 335-336, 344, 346, 429-430, 434, 437-439

PCB residues 22, 336
pelagic fish 3, 50
Penal Code 456
penalties 276, 456
Peñíscola 12, 184
peregrine falcon 57
perylene 337
Pesaro 21, 329
pesticide 16, 23, 27, 30, 32, 50, 54, 59, 64, 183, 188-189, 203, 259, 335, 344, 429-430, 434, 437-438, 450, 453
Petit Rhône (river) 16, 259
petrochemical complexes 262
petrochemical and refinery industry 270
petrogenic hydrocarbons 4, 43, 45
petrol 190
petroleum hydrocarbon 1, 12, 33, 41-42, 44, 49-50, 54, 62-63, 97-98, 183-184, 259, 337, 441
pH 188, 337
phages 68
phenols 62, 188, 271, 337
phosphate 16, 19, 27, 36, 39, 188, 257, 272-273, 339, 426, 431, 433-434, 436
phosphoric acid 431
phosphorus 6, 19, 23-24, 32, 37-38, 61, 84, 257, 273, 339, 343, 345, 350-351, 353, 433, 437
physical planning 30, 96, 452
phytoplankton 93, 426, 432
Pina 13, 187
Piombino 330
Piombino (channel) 332-333
Piraeus 27, 428-431, 443, 457-459
plankton 48
planning 7, 20, 58, 73, 83, 99, 275, 358-359, 425
plastics wastes 191
platform blowouts 42
Playa de Magalluf 180

Po (river) 2, 4, 6, 21-24, 26, 36-39, 60, 64, 331, 335-336, 339, 343-344, 349-352, 363, 367
Po di Levante (river) 331, 335
PO_4^{3-} see phosphate
PO_4^{3-}-P see orthophosphate, phosphate
polychlorinated biphenyl see PCB
polyethylene 191
poor cod 451
population 5, 13, 15, 18, 25, 51-52, 56, 59, 62, 68-69, 81, 84, 89, 95-96, 98, 192, 256, 268, 346, 348, 352, 354, 364, 425, 427, 443, 446, 448
Port-de-Bouc 17, 261, 263
Portman 3, 11-12, 181-183, 194
Porto Corsini 331
Porto Santo Stefano 330
Porto-Vecchio 266
<u>Posidonia</u> (see also marine plants) 5, 66, 71, 257
power plant 7, 53, 64, 332-333, 448
power station 25, 35, 85, 449
Pozzuoli (gulf) 22, 336
preservation 13, 58, 256
primary treatment 88, 273
Priority Actions Program (PAP) 58, 73
protected areas 8, 53, 74, 94-95, 190, 445
protection 8, 11, 14, 20, 25, 30-31, 51-52, 57, 65, 74-75, 77, 80, 83, 86, 94-97, 190, 197-199, 275-276, 356, 359-363, 451-454
pruritus 68
public health 67-68, 87, 187, 452
Public Investment Fund 31
Puerto de Alcudia 12, 184
Punta Ala 330

quality standards 78, 199, 446

radioactive substances 59
radionuclides 7, 60, 64
RAMSAR Convention 65, 444, 452
Ravenna (port canal) 331
reception facilities 2, 4, 7, 15, 21, 32-33, 44, 85, 90-91, 99, 195-196, 204, 273, 425, 455, 457-459
reclamation 342, 359, 362, 367-368
recreation 436, 452
recycling 77, 87, 275, 442
red coral 66
red tide 2, 23, 33, 37-39, 68, 339, 431
reference trend scenario 82
refineries 18, 29, 41, 53, 62, 263, 267, 270, 273, 449
refining 41
refuse 20, 28, 276, 352, 443, 460
release 32, 38, 43, 85-86, 91, 93-94, 98, 349-350, 353, 427, 431, 433, 459
Rentis 447
reptiles 50-51
required standard 27, 184, 266, 338, 427
resource management 8, 425
Rhodes (island) 427, 445, 449
Rhône (river) 3, 6, 15-19, 36, 39, 60-61, 64, 256-257, 259-261, 263, 267, 271-272, 278
Ripoll (river) 187
risk 14, 67, 85, 187, 196, 203, 342, 366-367
Roquetas de Mar 2, 11, 180
run-off 6, 11, 30, 35, 59, 61, 63-65, 181, 188, 333, 340, 351, 432, 437, 442, 445, 450

Saint-Chamas (bay) 17, 261
Saint-Florent 266
salinity 36, 39, 187, 257, 271, 329, 432
salinization 56, 65, 73

salmon 23, 72, 343
Salmonella (see also bacteria) 337
salts 271
sanitation 31, 201
San Javier 185
San Juan Despí 187
Santa Gilla (lagoon) 49
Sarda sarda (see also bonito) 49
Sardina pilchardus (see also sardines) 48
sardines 3, 48
Sarno (river) 335, 347
Saronikos (gulf) 2, 4-5, 27, 38-39, 66, 98, 425-432, 434, 436, 439, 441, 457
scad 184
Scomber scombrus (see also mackerel) 48
sea dumping 359
secondary treatment 7, 90, 195
seepage 441
Seine (bay) 17, 265
septic tanks 444
Sète 2, 16, 21, 259, 267, 273
settling tanks 271
sewage 14, 38, 47, 88, 193, 202, 277, 334, 428, 430-431, 439
sewage sludge 77, 88, 352
sewage treatment 26, 46, 88, 201-202, 359, 366
sewer networks 31, 86, 456
sewerage 26, 352, 359, 362, 368
shellfish farming 19, 274
shipbuilding yard 431
shipyard 265
shipping industry 4, 44, 330
shrimp 12, 181-183, 334
Sicily (island) 51, 54, 334, 339
Sile (river) 335
silicate 16, 27, 39, 257, 426, 431-434, 440
silt 56, 187, 271
SiO_4^{4-}-Si see silicate
Sirte (gulf) 36, 44
skin infection 14, 67, 197

skin rash 68
slaughterhouse and municipal wastes 436
smelting plant 439
smog 29, 356, 446
smoke 446
SO_2 see sulfur dioxide
sodium 188
soil conservation 26, 197, 369
solid waste 13, 25, 29, 77, 191-192, 352-354, 366, 453, 460
Sorrento 338
spill 89, 455
spillage 459
sponges 38
spoonbill 65
squids 451
Squilla mantis (see also shrimp) 334
standard 31, 67, 75, 77-78, 80, 82, 266, 270, 273, 275, 337, 453
staphylococci 68
streptococci 28, 185-186, 197, 337, 443
stresses 27, 432
Strymon (river) 28, 442-443
sturgeons 361
subsidence 23, 340
subsidies 195, 200, 453, 457
sulfates 188
sulfur 447
sulfur dioxide 199, 266-267, 352-353, 447-448
surface water 24, 36, 39, 192, 344-345, 347, 350-352, 360, 363, 432, 440, 442, 444
surfactants 345
suspended matter 59, 63, 188, 271
suspended solids 1, 6, 10, 14, 18, 33, 63, 188, 193-195, 268-270, 272, 340
swordfish 333

tank cleaning 7

tankers 89-90
tar 4, 27, 43-44, 62, 97, 427
tar residues 44
Taranto 332
tax 447, 453
terminal 90, 262
terraces 56
tetrachloroethene 438
Thassos (island) 427, 441, 444-445
Thau (lagoon) 16, 19, 257-258, 264, 274
Thermaikos (bay) 426
Thermaikos (gulf) 2-3, 28, 425, 434-436, 438
thermal waste 85
Thessaloniki (bay) 2, 426, 435-437
They de la Gracieuse 261, 263
throat infections 196
Thunnus thynnus (see also tuna) 22, 48-49, 183, 336
Tiber (river) 6, 24, 60, 339-340, 345-346, 349-350
tin 54, 98
Tinos (island) 427
tipping 87
titanium 266
Torderá (river) 12, 183
Torre Annunziata 347
total mercury 41, 48-49, 69, 93, 274, 330, 332, 344
Toulon 17-18, 264-265, 267, 270, 277
tourism 13, 23, 25-26, 29, 51, 53, 58, 67, 70, 72, 180, 192, 196, 203, 268, 341-342, 355-356, 364-365, 425, 449
toxic substances 74, 270
trace elements 40, 440
trace metals 72, 332, 440
Trachinidae 38
Trachurus (see also scad) 184
traffic 29, 43, 85, 195, 263, 441, 446-447, 458, 460
training 91, 96, 368, 456

transparency 70, 196, 337-338
transport 79, 85, 330
treatment plant 2, 7-8, 14-16, 18, 20-21, 31, 33, 36, 59, 79-81, 86, 88-89, 99, 193, 200-203, 269-270, 275, 277-278, 367, 456-457, 459
treatment stations 88
trichloroethene 438
trickling filter 269
Trieste (gulf) 2-3, 21, 38, 48, 331, 334, 339
tritium 6, 60, 64
tuna 3, 22, 48-49, 183, 333, 336
turtles 51, 361
typhoid fever 25
Tyrrhenian (sea) 3-6, 21-22, 40, 42, 45, 48-49, 60, 330-331, 333-334, 336, 339, 350

UNEP see United Nations Environment Programme
UNESCO see United Nations Educational, Scientific and Cultural Organization
United Nations Educational, Scientific and Cultural Organization (UNESCO) 446
United Nations Environment Programme (UNEP) 34, 73, 84, 349
urban planning 199, 359
urbanization 7, 51, 55, 79, 98-99, 340, 365
U.S. Environmental Protection Agency (EPA) 78, 437-438
U.S. EPA see U.S. Environmental Protection Agency

Vaine (lagoon) 17, 261-263
Valencia 48, 181, 186, 190-191, 196, 201
Venice 41, 336-337, 349, 357
Venice (gulf) 22, 335
Venice (lagoon) 21, 25-26, 331, 337, 341, 353-354, 368
Venus gallina (see also clams)

4, 50
Vibrio cholerae (see also bacteria, vibrios) 186-187
vibrios 186
Villefranche (bay) 16, 257, 262
Vinalopó (river) 14, 195, 201
virus 12, 184-187

waste 15, 17, 24-25, 42, 59, 66, 84, 87-88, 91-93, 191, 203-204, 264, 275, 331, 350-354, 425, 428, 436, 444-445, 458-460
waste collection 87, 191
waste disposal 11, 13, 25-26, 87, 91, 359, 363, 368
waste management 87
waste treatment 87
waste water 2, 4, 6, 13-14, 17, 20, 26, 28, 30, 37, 44, 61, 63-64, 92-93, 98-99, 185, 193-194, 201, 264, 269, 277, 343, 345-347, 352, 359, 366, 439, 442, 445, 450, 453
wastewater treatment 18, 25-26, 200, 352, 457
waterlogging 56
water pollution 11, 20-21, 29, 188, 196, 270, 275, 354-355, 357, 360, 449
water resources 442
water supply 31, 35, 53, 202, 345, 363, 442, 453, 457
weathering 22, 331-333
WHO Task Group 69
wildlife 453
wine-liquor industries 18, 270
World Bank 31, 456
World Health Organization 69, 447
worse trend scenario 82, 84

Xanthi 427

zinc 6, 11, 16-17, 21-22, 24, 27, 40, 48, 63, 180-181, 188, 257-260, 262, 266-267, 329, 331-334, 350, 427-428, 435, 440
Zn see zinc

ENVIRONMENT & ASSESSMENT

1. J. Rotmans: *IMAGE.* An Integrated Model to Assess the Greenhouse Effect. 1990 ISBN 0-7923-0957-X
2. H. Briassoulis and J. van der Straaten (eds.): *Tourism and the Environment.* Regional, Economic and Policy Issues. 1992 ISBN 0-7923-1986-9
3. A. Elzinga (ed.): *Changing Trends in Antarctic Research.* 1993 ISBN 0-7923-2267-3
4. M. Sadiq and J. C. McCain: *The Gulf War Aftermath.* An Environmental Tragedy. 1993 ISBN 0-7923-2278-9
5. F. B. de Walle, M. Nikolopoulou-Tamvakli and W. J. Heinen (eds.): *Environmental Condition of the Mediterranean Sea*: European Community Countries. 1993 ISBN 0-7923-2468-4

KLUWER ACADEMIC PUBLISHERS – DORDRECHT / BOSTON / LONDON